Discrete Mathematics Using a Computer

T0202514

John O'Donnell, Cordelia Hall
and Rex Page

Discrete Mathematics
Using a Computer

Second Edition

Springer

John O'Donnell, PhD
Cordelia Hall, PhD
Computing Science Department, University of Glasgow, Glasgow G12 8QQ, UK

Rex Page, PhD
School of Computer Science, University of Oklahoma, Norman, Oklahoma, USA

British Library Cataloguing in Publication Data
A catalogue record for this book is available from the British Library

Library of Congress Control Number: 2005935334

ISBN-10: 1-84628-241-1
ISBN-13: 978-1-84628-241-6

Printed on acid-free paper

© Springer-Verlag London Limited 2006

Apart from any fair dealing for the purposes of research or private study, or criticism or review, as permitted under the Copyright, Designs and Patents Act 1988, this publication may only be reproduced, stored or transmitted, in any form or by any means, with the prior permission in writing of the publishers, or in the case of reprographic reproduction in accordance with the terms of licences issued by the Copyright Licensing Agency. Enquiries concerning reproduction outside those terms should be sent to the publishers.

The use of registered names, trademarks, etc. in this publication does not imply, even in the absence of a specific statement, that such names are exempt from the relevant laws and regulations and therefore free for general use.

The publisher makes no representation, express or implied, with regard to the accuracy of the information contained in this book and cannot accept any legal responsibility or liability for any errors or omissions that may be made.

Printed in the United States of America   (HAM)

9 8 7 6 5 4 3 2 1

Springer Science+Business Media
springer.com

This book is dedicated to our parents.

Preface to the Second Edition

Computer science abounds with applications of discrete mathematics, yet students of computer science often study discrete mathematics in the context of purely mathematical applications. They have to figure out for themselves how to apply the ideas of discrete mathematics to computing problems. It is not easy. Most students fail to experience broad success in this enterprise, which is not surprising, since many of the most important advances in science and engineering have been, precisely, applications of mathematics to specific science and engineering problems.

To be sure, most discrete math textbooks incorporate some aspects applying discrete math to computing, but it usually takes the form of asking students to write programs to compute the number of three-ball combinations there are in a set of ten balls or, at best, to implement a graph algorithm. Few texts ask students to use mathematical logic to analyze properties of digital circuits or computer programs or to apply the set theoretic model of functions to understand higher-order operations. A major aim of this text is to integrate, tightly, the study of discrete mathematics with the study of central problems of computer science.

Concepts in discrete mathematics are illustrated through the solution of problems that arise in software development, hardware design, and other fundamental domains of computer science. The text introduces discrete math concepts and immediately applies them to computing problems. Applications of mathematical logic in design and analysis of hardware and software is an especially strong theme. The goal in this part of the material is to prepare students for a world that places a high value on the correct operation of computing systems in safety-critical, security-sensitive, and embedded systems and recognizes that formal methods based in mathematical logic are the primary tools for ensuring that computing systems function properly in such environments.

The emphasis, here, is on preparation. In commercial applications, mechanized logic engines are essential to the enterprise of applying logic to the design and implementation of computing hardware and software. This text introduces students to mechanized logic in the form of propositional proof checking, and, through numerous paper-and-pencil exercises in applying logic to mathematical verification of hardware and software artifacts, gives students experience with the fundamental notions used by engineers who apply mechanized logic engines to the design of commercial computing systems. We believe these skills will be of increasing value in computer and software engineering, and our experience suggests that such skills contribute positively, even in the short run, to the ability of students to successfully design and implement software.

The text is organized in four parts: reasoning with equations, formal logic, set theory, and applications. The principle of induction is introduced early, for reasoning with equations, and applied to problems throughout the text. Reasoning with equations covers examples in several domains, including natural numbers of course, but also including sequences and sets. The logic portion of the text discusses two frameworks for formal reasoning: the natural deduction format of Gentzen and another syntax-based reasoning system based in Boolean algebra. Propositional logic is introduced first, then predicate logic, both in a natural deduction and Boolean algebra setting. Set theory discusses the usual basics, and illustrates many of the concepts by applying induction to define the integers. The set theoretic definitions of relations and functions are discussed, along with the usual properties that categorize them and allow them to be combined and manipulated. The applications portion of the text covers two extended examples, one concerning the design of a circuit for n-bit, ripple-carry addition, the other on the implementation of AVL tree operations. These augment the many, smaller examples that occur throughout the text and, together, help students understand how discrete mathematics contributes to the solution of difficult and important problems in computing.

A website for the text contains a collection of tools for experimenting with most of the concepts introduced. Included among these is a proof-checking system for propositional calculus. Students can use this system to make sure their proofs are correct and, more importantly, to experience the notion that proofs can be entirely formal and, therefore, useful in verifying the correctness of software and digital circuits. Other tools allow experimentation with set operations, Boolean formulas, and the notions of predicate calculus. These tools are expressed in Haskell, and the various operations for experimentation, including proofs, are expressed using Haskell syntax. In addition, Haskell is used to express the software and hardware designs that illustrate practical uses of logic and other aspects of discrete mathematics in computer science.

We feel that Haskell is an ideal notational choice for these examples because of its close affinity with customary algebraic notation. The compactness of software and hardware artifacts expressed in Haskell is another important advantage. Haskell serves both as a formal, mathematical notation, and as a practical and powerful programming language. This helps to strengthen the tight connection between mathematics and applications. Thus Haskell is used in the text on an equal footing with other mathematical notations. Students see Haskell in its role as a programming language, as well as a hardware description language, and the emphasis in this book is on reasoning about programs and circuits, not just programming.

We hope that students will find the experience of learning about logic, sets, mathematical induction, and other concepts of discrete mathematics and its applications to computing problems interesting and enjoyable, and that they will be able to use these ideas in subsequent studies and professional work in computer science.

Software Tools for Discrete Mathematics

A central part of this book is the use of the computer to help learn the discrete mathematics. The software (which is free; see below) provides many facilities that aid the student in learning the material:

- Logic and set theory have many operators that are used to build mathematical expressions. The software allows the user to type in such expressions interactively and experiment with them.

- Predicate logic expressions with quantifiers can be expanded into propositional logic expressions, as long as the universe is finite and reasonably small. This makes the meaning of the quantifiers more concrete and helps the development of intuition.

- Students frequently misuse expressions in logic and set theory; a typical error that arises frequently is to write an expression that treats $A \subseteq B$ as a set rather than a Boolean value. The software tools will immediately flag such mistakes as type errors. Teaching experience shows that many students will have long-lasting misconceptions about basic notations without immediate feedback.

- A formal proof checker for natural deduction is provided. This allows students to find errors in their proofs before handing in exercises, and it also provides a quick and effective way for the instructor to check the validity of large numbers of proofs. Furthermore, the automated proof checker underscores the nature of formal proof; vague or ill-formed proofs are not acceptable.

- Using a proof checker gives a deeper appreciation of the relationship between discrete mathematics and computer science. The experience of debugging a proof is much like debugging a computer program; the proof checker is itself a computer program (which the students can read if they wish to); proof checking software makes formal proof feasible for larger scale problems.

- The techniques of recursion and induction are applied directly and formally to function definitions that the student can execute.

The version of Haskell used in the book is Haskell98. This is a standard pure functional language with excellent support. Several implementations are freely available and they are supported on most major computers and operating systems. Students can install the software on their own machines, and universities can, of course, install it on laboratory computers.

The Software Tools for Discrete Mathematics package is a library of definitions that are loaded into Haskell. This package is available on the book web page (see Appendix B).

Haskell is an ideal language for teaching discrete mathematics. It offers a powerful and concise expression language; many problems that would require writing a complete program of 10 to 100 lines of code in a language such as Pascal, C++, or Java can be written as a simple expression in Haskell, which is only a few lines long. This makes it possible to use Haskell interactively to experiment with the mathematical expressions of propositional logic, predicate logic, set theory, and relations. Such lightweight interactive exploration is infeasible in traditional imperative or object-oriented languages. Haskell is also well suited for complex applications, such as the proof checker used in Chapters 6 and 7, and the hardware description language used in Chapter 13.

It is assumed that the reader of the book has no knowledge in advance about Haskell or functional programming; everything that is needed is covered here. Because it is self-contained, this book can be used in any curriculum, regardless of what programming languages happen to be in use.

To the Student

It's best to read this book *actively* with pencil and paper at hand. As you read, try out the examples yourself. It is especially important to try some of the exercises, and solutions to many of them appear in Appendix C. Don't just read the exercise and then the solution—the benefit comes from trying to solve an exercise yourself, even if you don't get it right. When you find your own solution, or if you get stuck, then compare your solution with the one in the book.

The web page for this book has additional information that will be useful as you study discrete mathematics:

```
http://www.dcs.gla.ac.uk/
    ~jtod/discrete-mathematics/
```

Many of the exercises require the use of a computer with Haskell installed. The software is free, and it's straightforward to download it and install on your own machine. See the book web page for information on obtaining the software.

A good way to improve your understanding of the material is to read about it at a more advanced level and also to learn about its application to real problems. The Bibliography near the end of the book lists many good sources of information, and each chapter ends with some suggestions for further reading.

We wish you success with your studies in mathematics and computer science!

To the Instructor

This book is primarily intended for students of computer science, and applications of the mathematics to computing are stressed. No specific topics in computing are prerequisites, but some familiarity with elementary computer programming is assumed. The level is appropriate for courses in the first or second year of study. The contents of this book can be covered in a course of one semester.

The *Instructor's Guide* gives suggestions for organising the course, solutions to the exercises, additional problems with solutions and other teaching resources. It is available online:

```
http://www.dcs.gla.ac.uk/~jtod
        /discrete-mathematics/instructors-guide/
```

Because the four parts of the text are largely independent of one another, topics may be introduced in the order that best suits the needs of particular instructors and students. The only serious restriction on ordering is that Part I (reasoning with equations and induction), Part II (logic), and Part III (Sets) must be covered before Part IV (applications). Reasoning with equations, logic, and set theory may be covered in any order. Chapter 1 describes Haskell, which is used as a mathematical notation at many points in the text. Readers may need to refer to Chapter 1 as they read other portions of the text, but it is probably better to discuss that material on as as-needed basis instead of spending a block of time on it in the beginning. The following graph shows the dependencies in more detail.

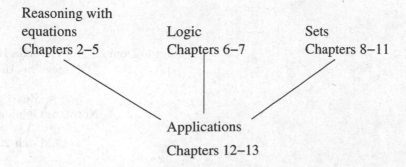

Reasoning with
equations
Chapters 2–5

Logic
Chapters 6–7

Sets
Chapters 8–11

Applications
Chapters 12–13

A website accessible to instructors includes lesson plans, slides for lectures, homework problems, and exam questions for a course based on the text. Altogether, the website contains over 100 homework problems (with solutions), about 350 lecture slides, and more than 300 exam questions (with solutions). These materials are accessible on the web:

```
http://www.dcs.gla.ac.uk/~jtod
      /discrete-mathematics/instructors-guide/
```

Notation

Standard mathematical notation is used in this book when discussing mathematics: $A \subseteq B$. A typewriter font is used for notations that are intended to be input to a computer: a 'subset' b. For example, a general discussion in English might say that a theorem is true; that theorem might make a statement about the proposition True, and a Haskell program would use the constant True. The end of a proof is marked by a square box □.

Acknowledgements

We would like to thank the following colleagues for their helpful feedback and encouragement during the process of writing this book: Tony Davie, Bill Findlay, Joy Goodman, Mark Harman, Greg Michaelson, Genesio Gomes da Cruz Neto, Thomas Rauber, Richard Reid, Gudula Rünger, and Noel Winstanley. We would also like to thank the students at the University of Glasgow and the University of Michigan, who gave both of us experience teaching with preliminary versions of this material, and our editors, Karen Barker, Rosie Kemp, and Catherine Brett, for their help in producing this book. Finally, we would like to thank the students and instructors who made use of the first edition of this text, especially those who took the time to let us know what they liked and disliked about it. We have benefitted from their comments and have tried to apply their ideas in this revision. All remaining errors are ours alone.

John O'Donnell and Cordelia Hall
Glasgow, Scotland

Rex Page
Norman, Oklahoma

March 2006

Contents

II   Logic   107

Part I

Programming and Reasoning with Equations

Chapter 1

Introduction to Haskell

The topic of this book is discrete mathematics with an emphasis on its connections with computers:

- *The computer can help you to learn and understand mathematics.* As various mathematical objects are defined, software tools will enable you to perform calculations with those objects. Exploring and experimenting with mathematical ideas gives a practical intuition.

- *The mathematics has widespread applications in computing.* We will focus on the topics of discrete mathematics that are most important in modern computing and will look at many examples.

- *Software tools make it possible to use the mathematics more effectively.* Mathematical structures are frequently large and complex, and computers are often necessary to bring out their full potential.

In order to achieve these goals, it won't be enough to provide the occasional pseudo-code program. We need to work with real programs throughout the book. The programming language that will be used is Haskell, which is a modern standard functional programming language.

It is *not* assumed that you know anything about Haskell or about functional programming. Everything you need to know about it is covered in this book. You will need only a small part of the language for this book, and that part is introduced in this chapter. If you would like to learn more about Haskell or functional programming, Section 1.11 recommends a number of sources for further reading.

Why use a functional language, and why Haskell in particular? Because:

- Haskell allows you to compute directly with the fundamental objects of discrete mathematics.

- It is a powerful language, allowing programs that would be long and complicated in other languages to be expressed simply and concisely.

3

- You can reason mathematically about Haskell programs in the same way you do in elementary algebra.

- The language provides a strong type system that allows the compiler to catch a large fraction of errors; it is rare for a Haskell program to crash.

- Haskell is an excellent language for *rapid prototyping* (i.e., implementing a program quickly and with minimal effort in order to experiment with it).

- There is a stable, standard, and well-documented definition of Haskell.

- A variety of implementations are available that are free and run on most computers and operating systems.

- Haskell can be used interactively, like a calculator; you don't need a heavyweight compiler.

1.1   Obtaining and Running Haskell

Appendix A gives a pointer to the web page for this book. On that page you can find pointers to the most up-to-date implementations of Haskell as well as the software tools used in this book. There is an active Haskell development community, and new tools are constantly emerging, so you should check the book web page for current information. The book web page also contains additional documentation for the software.

We will use the computer interactively, like a desk calculator. Haskell itself provides a powerful set of built-in operations, but others that you will need are defined in the *Software Tools for Discrete Mathematics* file `stdm.hs`, which you can download from the book web page.

To give an idea of what it's like to use the computer with this book, here is a typical interactive session, using the `stdm` file and an interactive Haskell implementation, such as ghci, Hugs, or nhc.

First we start the Haskell system, and it will give an introductory screen followed by a prompt '`>`'. Haskell is now acting like an interactive calculator; you enter an expression after the `>` prompt, and it will evaluate the expression, print it, and give another prompt:

```
... the introductory message from Haskell system ...
Type :? for help
> 1 + 2
3
> 3*4
12
>
```

You have just typed in an *expression*, and it was evaluated. An expression is a combination of operators and values that are defined by the programming language. For example, 1 + 2 * x is an expression in Haskell. We write 1+2 ⇒ 3, meaning that when the expression 1+2 is evaluated, the result is 3.

Now you should load the software tools file, which defines the programs used in the examples and exercises in this book:

```
> :load stdm
```

Notice the colon in the :load stdm. If the first character after a prompt is : then the rest of the line is a command to the interpreter, rather than an expression. One good command to remember is :?, which prints a help screen listing all the other commands you can enter.

At this point, you are ready to start writing and testing definitions. These should be saved in a file, so that you don't have to enter them in again and again. To do this, create a file with a name like mydefs.hs. The extension .hs stands for *Haskell script*. Store the following two lines in your file:

```
y = x+1
x = 2*3
```

The syntax of Haskell is designed to look similar to normal mathematical notation, and to be easy to read. Programs should be properly indented, and Haskell will use the indentation of the code in your script file in order to determine its structure. For example, if have an equation that is too long to fit on one line, you need to indent the subsequent lines. Here is one way to do it:

```
x = a + b + c
      + d  + e
y = 2 * x
```

Notice that you don't need to write any extra punctuation, semicolons, braces, begin/end, etc. The indentation makes it clear (both to the Haskell system and to human readers) that the second line is a continuation of an equation. Rather than learning the precise details of indentation in Haskell, we recommend that you just follow the indentation style used in this book. One advantage of Haskell's approach is that the programmer specifies the structure of the program just once (through the indentation), rather than twice (through braces and semicolons, and again through indentation).

A double minus sign indicates that the rest of the line is a comment. You can also indicate that a portion of text is a comment by enclosing it in {- and -}. This convention can be used to turn a part of a line into a comment, or a group of several lines.

```
x = 2 + 2   -- the result should be 4
```

Now enter the command :load mydefs, which tells the Haskell system to read your file. The effect of this is to define all the names that appear on the left-hand sides of the equations in the file. From now on, the expressions you enter interactively can make use of the variables x and y:

```
:load mydefs
> :load mydefs
Reading file "mydefs.hs":
mydefs.hs
> x
6
> y
7
> x*y
42
>
```

You can always edit the file and save it, and then reload the file into Haskell. This will replace all the old definitions with the ones in the edited file. For example, if you modify the file so that it says x = 4*5 and reload it, then x $\Rightarrow$ 20 and y $\Rightarrow$ 21.

When you would like to leave Haskell, you can enter the quit command :quit, and with some operating systems you can also quit by entering control D.

1.2   Expressions

You can do a lot just by writing simple expressions in Haskell—far more than just basic arithmetic (although that can be interesting, too). In the following sections, we will show some of the most useful kinds of expression, organised according to the *type* of value.

As you will see later, types are of fundamental importance, both in Haskell and in discrete mathematics. For now, however, you can just think of a type as being something like Integer, Float, Char, etc. The essential point is that an operator is defined for a specific type. For example, the + operator specifies addition, which makes sense only for number types, while the length operation gives the length of a list, but makes no sense for types that are not lists.

1.2.1   Integer and Int

Integer constants are written as a sequence of digits. For example, 2, 0, 12345, and -72 are all integer constants.

Like most programming languages, Haskell provides operators for addition (+), subtraction (-), and multiplication (*). There is also an exponentiation operator ^; for example, 2^3 means '2 raised to the power 3', or 2^3.

For the time being, don't use / for division; instead, write x `div` y in order to divide x by y. This is an integer division, and any remainder is thrown away. For example, 5 `div` 2 ⇒ 2, 17 `div` 3 ⇒5 and -100 `div` 20 ⇒ -5. You can get the remainder or modulus with the `mod` operator: 8 `mod` 3 ⇒ 2.

Notice how `div` and `mod` are *operators* (an operator is written in between its two arguments), but these operators are names made up of letters. Haskell has many named operators like this, and they are always enclosed in back-quote characters: `opname`.

There are functions that find the larger and smaller of two numbers: max 3 8 ⇒ 8, and min 3 8 ⇒ 3. There is a further discussion about operators like `div` and functions like max in Section 1.4.3.

Computers store numbers in *words*, and the word size of modern processors is 64 bits long. Whatever the word size on your computer, however, there is a limit to how large an integer it will hold. Fortunately, Haskell does not limit you to numbers that fit into a word. It provides two distinct integer types:

- Int is the type of integers that fit into a word on the computer (naturally this will vary from one computer to another);

- Integer is the type of mathematical integers.

Since there are two different integer types, we need a way of saying which type we want an expression to have. The :: operator (read it as *has type*) is used in Haskell to specify what type an expression has. Thus 2::Int says '2 has type Int', and its representation in the computer is different from 2::Integer.

Here is an example that will illustrate the difference between Int and Integer. First, evaluate 2^2, which means 2 squared and gives 4:

```
> 2^2
4
```

Now, 2^20 presents no problem to most computers, as the word size will be well above 20:

```
> 2^20
1048576
```

However, most computers have a word size much less than 200 bits, and the expression 2^200 will not give the right answer if it's evaluated with the default Int type:

```
> 2^200
0
```

Therefore we say explicitly that we want the unlimited Integer type to be used instead; now we are guaranteed to get the right answer:

```
> (2^200)::Integer
1606938044258990275541962092341162602522202993782792835301376
```

One might wonder just how large an `Integer` number can be. When it performs arithmetic on these numbers, Haskell allocates enough memory dynamically in order to store the result. There is of course a limit to the size of number that can be stored, but a modern machine with a large memory can easily accommodate numbers that contain millions of digits.

Besides actually needing large numbers, there is a theoretical benefit from using `Integer`: the arithmetic operations on this type satisfy the algebraic laws. For example, we know from algebra that $(x + y) - y = x$, but this is *not* always true in a computer if the arithmetic is performed with the `Int` type, or in a language that offers only fixed-word arithmetic. It might happen that x and y fit in a machine word, but the intermediate result $x + y$ does not. In contrast, if the arithmetic is performed on the `Integer` type then the computer program will definitely satisfy the mathematical law.

1.2.2   Rational and Floating Point Numbers

Single-precision floating point numbers have type `Float`, and double-precision numbers have type `Double`. Besides the operators `+`, `-` and `*`, you can divide floating point numbers with the `/` operator. The floating point exponentiation operation is `**`. For example, `2**0.5` $\Rightarrow$ `1.41421`.

There are also a number of functions that can be applied to floating point numbers. A *function application* in Haskell requires no parentheses; you just write the name of the function, followed by a space, followed by the argument. For example, the square root function is `sqrt`, and a typical application is `sqrt 9` $\Rightarrow$ `3.0`.

Floating point representations are approximations, and they are *not* guaranteed to satisfy the algebraic laws as `Integer` numbers do. Try evaluating the following expressions on your computer:

```
0.11 - 0.10
2.11 - 2.10
```

If the arithmetic were performed exactly, these would both give the same result, but they do not on some computers. This is a property of floating point representation, not Haskell. It is possible to round the numbers so that they look the same when printed, but the internal representations will still be different. It is important to remember that when you use mathematics to reason about real numbers, the results may not apply exactly to a program that uses floating point. It is particularly important to be careful when comparing floating point numbers: the right way to compare them is to determine whether the absolute value of their difference falls within an acceptable error tolerance.

Haskell supports *exact* arithmetic on rational numbers, allowing you to work with fractions as well as with their decimal equivalents (approximations). `Ratio Integer` is the type of rational numbers; these are numbers in the form of a fraction, where the numerator and denominator are represented with the `Integer` type. A rational constant is written in the form *numerator%denominator*. You can divide two integers using the / operator, and the result is an exact rational number. Haskell automatically reduces fractions; for example:

```
2/3                              ⇒  0.66667 ::  Double
2/3 ::  Ratio Integer            ⇒  2%3 ::  Ratio Integer
2%3 + 1%6                        ⇒  5%6 ::  Ratio Int
(1/3 + 1/4) ::  Ratio Integer    ⇒  7%12
```

1.2.3 Booleans

The `Bool` type is used to represent the result of comparisons, and similar operations, where the result must be either `True` or `False`. These are the only two values with type `Bool`.

The following operators can be used to compare two numbers, and they produce a result of type `Bool`:

```
==  -- equality
/=  -- not equal
<   -- less than
<=  -- less than or equal
>   -- greater than
>=  -- greater than or equal
```

For example, 9>3 ⇒True, and 5<=5 ⇒True.

There are also some operators that take arguments of type `Bool`, again returning a `Bool` result:

```
&&  -- Boolean and
||  -- Boolean or
not -- Boolean not
```

The expression x&&y evaluates to `True` if *both* x and y are True; x||y evaluates to `True` if *either* of the arguments is `True`. Finally, not x is `True` if x is `False`, and vice versa.

Exercise 1. Evaluate these expressions, and check with the computer:

```
True && False
True || False
not False
3 <= 5 && 5 <= 10
```

```
3 <= 20 && 20 <= 10
False == True
1 == 1
1 /= 2
1 /= 1
```

1.2.4 Characters

A character has type `Char`, and a constant is written as the character surrounded by single-quote characters; for example, `'a'` and `'*'`. Recall that the back-quote character is used for operators, *not* for characters. Thus `'?'` is a character and `` `div` `` is an operator.

The `:type` command will tell you the type of an expression. As you experiment with the language by evaluating expressions, it is also a good idea to use `:type` to be sure that they have the same type you think they should. Often you will see a type with `=>` in it; the meaning of this is explained in Section 1.10. This book (but *not* Haskell) also uses $\Longrightarrow$ to indicate evaluation of an expression to a specified value.

The comparison operators can be used on characters. Two useful built-in functions are `toUpper` and `toLower`, which take a character and convert it from one case to another. For example:

```
'c' < 'Z'
toUpper 'w' => 'W'
toLower 'Q' => 'q'
```

There is a special character called *newline* which causes a line break when it is printed. This character is written `'\n'`.

1.2.5 Strings

A `String` is a sequence of zero or more characters. A string constant is written inside double-quote marks. For example, `"tree"::String`.

There is a difference between `'a'::Char` and `"a"::String`. The first is the character a, while the second is a string that *contains* the character a. A string can have any length, unlike a character.

The `length` function can be used to determine how many characters are in a string. For example, `length "dog"` $\Rightarrow$ 3, and `length "a"` $\Rightarrow$ 1. You cannot apply `length` to a character.

Two strings can be *concatenated* together with the operator `++` to form a larger string: `"abc" ++ "defg"` $\Rightarrow$ `"abcdefg"`. A common example is to place a newline character at the end of a line: `"Here is a line" ++ "\n"`.

The `length` function and `++` operator are actually more general than described here; this will be discussed later.

1.3 Basic Data Structures: Tuples and Lists

The most commonly used data structures in Haskell are tuples and lists. Both of them are used to gather several data values together into one data structure.

1.3.1 Tuples

Tuples package up several values into one. They are written as a sequence of data values, separated by commas, inside parentheses. For example, (2,"dog") is a tuple with two components; the first is 2 and the second is "dog". Its type is written (Int,"dog"). In general, if x::A and y::B, then (x,y)::(A,B). For example:

```
("dog","cat") :: (String,String)
(True, 5) :: (Bool,Int)
('a',"b") :: (Char,String)
("bat",(3.14,False)) :: (String,(Double,Bool))
```

The usual way to extract the components from a tuple is pattern matching, which will be covered later. For tuples with 2 components, the function fst returns the first component and snd returns the second. These functions have the following types:

```
fst :: (a,b) -> a
snd :: (a,b) -> b
```

This says that fst is a function; its argument has a tuple type of the form (a,b), and the result it returns has type a.

A tuple can have any number of components. If it contains n components, it is called an n-tuple, but 2-tuples are often just called pairs. You can even have a 0-tuple, which contains nothing, written (); this is often used as a dummy value. (But you cannot have a 1-tuple, as that would conflict with the use of parentheses to control precedence of operators.)

Tuples have two important characteristics. First, a tuple has a fixed number of components. If you have a 3-tuple (a,b,c), and you add an extra data value to obtain (a,b,c,d), the new tuple has a different type from the old one. Second, there is no restriction on the type of any component: it is common for a tuple to contain data values with different types.

1.3.2 Lists

Lists are the most commonly used data structure in functional languages. A list is written as a sequence of elements, separated by commas, surrounded by square brackets: for example, [1,2,3] is a list of integers.

A list may contain any number of elements, but all of the elements *must* have the same type. The type of a list is written [A], where A is the element type. For example:

```
[13,9,-2,100] :: [Int]
["cat","dog"] :: [String]
[[1,2],[3,7,1],[],[900]] :: [[Int]]
```

Expressions can appear in a list expression; all the elements of a list are evaluated. If your script file contains an equation defining x to be 10, then [13,2+2,5*x] evaluates to the list [13,4,50].

A String is actually just a list of characters. The constant "abc" is exactly equivalent to writing ['a','b','c'].

You can easily specify a list that is a sequence of numbers. For example, the Haskell notation [1..10] means the list [1,2,3,4,5,6,7,8,9,10], and the .. is filled in with the missing numbers. Normally the numbers are incremented by 1, but if you give two numbers before the .. any increment can be used. For example, [1,3..10] counts up by 2, and its value is [1,3,5,7,9]. Sequences of characters can be created in the same way. Here are some examples:

```
['a'..'z'] => "abcdefghijklmnopqrstuvwxyz" :: String
['0'..'9'] => "0123456789" :: String
[0..9]     => [0,1,2,3,4,5,6,7,8,9] :: [Int]
[10,9..0]  => [10,9,8,7,6,5,4,3,2,1,0] :: [Int]
```

Haskell has a large number of features to make lists easy to use. Some of these are presented in Section 1.8.

1.3.3 List Notation and (:)

A new element can be added to the front of a list using the operator (:), which is usually pronounced 'cons' (because it constructs a list). This function takes a value and a list and inserts the value at the front of the list:

```
(:) :: a -> [a] -> [a]
```

```
1:[2,3] => [1,2,3]
1:[] => [1]
```

Every list is built up with the (:) operator, starting from the empty list []. Thus 1:[] is a list containing one element, the number 1. This list can be written just as [1]; in fact, the usual notation for lists is merely a nicer syntax for an expression that builds the list. This notational convention allows the following equations to be written:

```
[1,2,3,4] = 1 : (2 : (3 : (4 : [])))
"abc" = 'a' : ('b' : ('c' : []))
```

The parentheses can be omitted, because the (:) operator associates to the right. This enables you to write the equations more simply:

```
[1,2,3,4] = 1 : 2 : 3 : 4 : []
"abc" = 'a' : 'b' : 'c' : []
```

1.3.4 List Comprehensions

The list structure has been so useful and influential in functional languages that Haskell offers a variety of features to make them convenient to use. The *list comprehension* is a simple but powerful syntax that lets you define many lists directly, without needing to write a program to build them. List comprehensions are based on the standard set comprehension notation in mathematics. For example, the set comprehension $\{x^2 \mid x \in S\}$ describes the set of squares of elements of another set S.

The basic form of list comprehension is a list in which an expression appears first, followed by a vertical bar (read 'such that'), followed by a *generator*:

```
[expression | generator]
```

The generator specifies a sequence of values that a variable takes on; this is written in the form `var <- list`, and it means that the variable `var` will take on each of the values in `list`, one by one. For each of those values, the expression to the left of the bar is evaluated, and the result goes into the list. For example, the following list comprehension says that x should take on the values 1, 2, and 3; for each of these the value of 10*x goes into the result list, which is [10,20,30]:

```
[10*x | x <- [1,2,3]]
   => [10,20,30]
```

Often a sequence with the `..` notation is used to define the list from which the variable draws its values. For example:

```
[x^2 | x <- [1..5]]
   => [1,4,9,16,25]
[y 'mod' 3 | y <- [1..6]]
   => [1,2,0,1,2,0]
[toLower c | c <- "Too Many CAPITALs"]
   => "too many capitals"
```

If the list to the right of the `<-` is actually a list of tuples, then the variable can be replaced by a tuple of variables. In the following example, the tuple (a,b) takes on the values (1,2), (10,20), and (6,6). For each of these list elements, the expression a*b multiplies the components of the tuple:

```
[a*b | (a,b) <- [(1,2), (10,20), (6,6)]]
   => [2,200,36]
```

List comprehensions can have more than one generator. For each value of the first generator taken by the first variable, the second variable gets all of the values in its generator. This is similar to a nested loop, where the first generator is the outer loop and the second generator is the inner one. For example,

```
[(x,y) | x <- [1,2,3], y <- ['a','b']]
  => [(1,'a'),(1,'b'),(2,'a'),(2,'b'),(3,'a'),(3,'b')]
```

In the examples so far, all of the elements of the generator list cause a corresponding value to appear in the result. Sometimes we want to throw out some of the elements of the generator. This is performed by using a *filter*. A filter is an expression of type `Bool`, which (normally) uses the generator variable. After the variable takes on a value of the generator, the filter is evaluated, and if it is `False` then the value is thrown out. For example, the following list comprehension makes a list of all the integers from 0 to 100 that are multiples of both 2 and 7:

```
[x | x <- [0..100], x 'mod' 2 == 0 && x 'mod' 7 == 0]
  => [0,14,28,42,56,70,84,98]
```

The next comprehension produces a list of all of the factors of 12. It works by considering all combinations of x and y between 1 and 12; the value of x is retained if the product is exactly 12. If you replace 12 with any other number, it will make a list of the factors of that number:

```
[x | x <- [1..12], y <- [1..12], 12 == x*y]
  => [1,2,3,4,6,12]
```

List comprehensions provide an easy way to implement database queries. For example, `db` is a database represented as a list of tuples. Each tuple contains a person's name, their street address, age, and annual income:

```
db = [("Ann Smith",  "29 Byres Road",  30, 48000),
      ("Alan Jones", "36 High Street", 25, 17000),
       ...
     ]
```

Now we can use a list comprehension to make a mailing label for every employee under 30 who is making at least £15000 per year:

```
[name ++ "\n" ++ addr ++ "\n"
  | (name,addr,age,sal) <- db, age<30, sal>=15000]
```

Exercise 2. Work out the values of the following list comprehensions; then check your results by evaluating them with the computer:

```
[x | x <- [1,2,3], False]

[not (x && y) | x <- [False, True],
                y <- [False, True]]

[x || y | x <- [False, True],
```

```
                    y <- [False, True],
                    x /= y]

        [[x,y,z] | x <- [1..50], y <- [1..50], z <- [1..50],
                   x ** 2 + y ** 2 == z ** 2]
```

1.4 Functions

Functions are, unsurprisingly, central to functional programming languages like
Haskell. We have already seen a few brief examples. Now we take a closer look
at functions: how to use them, their types, and how to define them.

1.4.1 Function Application

An expression using a function to perform a computation is called a *function
application*, and we say that a function is *applied* to its arguments. For example,
we might apply the function sqrt to the argument 9, obtaining the result 3.0.

The notation for a function application consists of the function name, followed by a space, followed by the function argument(s). If there are several
arguments, they should be separated by spaces. Unlike some other languages,
you do not put parentheses around the arguments of a function.

```
sqrt 9.0
3.4 + sqrt 25 * 100
2 * sqrt (pi * 5 * 5) + 10
```

1.4.2 Function Types

Just as data values have types, so do functions. Function types are important
because they say what type of argument a function requires and what type of
result it will return.

A function that takes an argument of type a and returns a result of type
b has a *function type* which is written $a \rightarrow b$ (read a arrow b). To say that f
has this type, we write $f :: a \rightarrow b$. In Haskell, the $\rightarrow$ symbol is written as ->.
For example, some of the Haskell functions that we have already seen have the
following types:

```
sqrt :: Double -> Double
max :: Integer -> Integer -> Integer
not :: Bool -> Bool
toUpper :: Char -> Char
```

(You can use max with the type given above, but Haskell actually gives it
a more general type, allowing max to be used with all the integer types. This
involves type classes, discussed in Section 1.10.)

1.4.3   Operators and Functions

An *operator* is a function in Haskell, but an operator must take exactly two arguments, and it is written between the arguments. For example, the + operator takes two numbers and adds them, and is written between the numbers: 2+3. Since an operator is a function, it has a function type. When you use the operator by itself, for example to state its type, it must be enclosed in parentheses:

```
(+)  :: Integer -> Integer -> Integer
(&&) :: Bool -> Bool -> Bool
```

An operator is just a function, so you can do everything with an operator that you can do with a function, but it must be enclosed in parentheses if it does not appear between its arguments. For example, an alternative way to write 2+3 is (+) 2 3, which says that the (+) function is applied to two arguments, 2 and 3. Later in this book you will see a number of examples where it is useful to do this.

Just as you can treat an operator as a function by putting parentheses around it, you can treat a function as an operator by putting single-back-quote characters around it. For example, the function max can be applied to two arguments in a function application: max 4 7. If you prefer, you can use it as an operator, like this: 4 `max` 7.

1.4.4   Function Definitions

You can define new functions by giving the *type declaration* followed by the *defining equation*. The type declaration has the form:

$$function\_name :: argType_1 \rightarrow argType_2 \rightarrow \ldots \rightarrow argType_n \rightarrow resultType$$

The arrows are written as $\rightarrow$ in mathematical notation, and they are written as -> in Haskell programs. The defining equation has the form:

$$function\_name\ arg_1\ arg_2\ \ldots\ arg_n = expression\ that\ can\ use\ the\ arguments$$

The function definition should be inserted in a Haskell script file. When you load the file, you will then be able to use your functions. For example, suppose we want a function square that takes an Integer and squares it. Here is the function definition:

```
square :: Integer -> Integer
square x = x*x
```

When a function is applied to an argument, the value of the argument is available to the expression on the right-hand side of the function's defining equation. For example, if we evaluate square 5, then x is temporarily defined to be 5, and the right-hand side x*x is evaluated, producing the result 25. The scope of the argument (that is, the part of the program where it is 'visible' and can be used) is just the right-hand side of the defining equation. If you just evaluate the expression x at the top level, Haskell will give an error message because x is not defined:

```
> x
ERROR: Undefined variable "x"
```

1.4.5 Pattern Matching

As we have seen, the left-hand side of the defining equation has the form of an application. If the argument in that application is a name (for example, x in the defining equation square x = x*x), then when an application is evaluated, the name will take on the argument value, and the right-hand side of the function is evaluated.

There is another option: if the argument on the left-hand side is a constant, then the right-hand side will be used only if the function is applied to that value. This makes it possible for a function definition to consist of several defining equations. For example, here is a definition of f with three defining equations, each with a constant argument:

```
f :: Integer -> String
f 1 = "one"
f 2 = "two"
f 3 = "three"
```

When the application f 2 is evaluated, the computer begins by checking the first defining equation, and it discovers that the application f 2 does not match the left-hand side of the first line, f 1. Therefore the next equation is tried; this one does match, so the corresponding right-hand side is evaluated and returned. If f is applied to an argument like 4, which does not match any of the defining equations, an error message is printed and the program is terminated.

The following function tests its value, returning True if the value is a 3 and False otherwise. It has two defining equations. If the argument pattern 3 in the first defining equation matches the argument, then that equation is used. If the argument is not 3, then the second equation is checked; the argument x will match *any* argument, so the corresponding value False is returned.

```
is_three :: Int -> Bool
is_three 3 = True
is_three x = False
```

For example, given the application is_three (1+1), the argument is evaluated to 2; since this does not match 3, the second defining equation is used.

A function may have more than one argument. If this is the case, then each argument in a given defining equation is checked, from left to right. For example, here is the `nor` function, which returns `True` if and only if both of the arguments are `False`:

```
nor :: Bool -> Bool -> Bool
nor False False = True
nor a     b     = False
```

Consider the evaluation of the application `nor False True`. The first defining equation is checked from left to right at runtime. The first pattern matches, so the second argument on that line is also checked. It does not match, so the second defining equation is checked; its left-hand side matches both arguments.

Pattern matching is commonly used when the argument to a function is a tuple. For example, suppose we want a function `fst` that takes a pair and returns its first element, and a similar function `snd` to return the second element. These functions are easily defined with pattern matching:

```
fst :: (a,b) -> a
fst (x,y) = x

snd :: (a,b) -> b
snd (x,y) = y
```

When `fst (5,6)` is evaluated, the argument value `(5,6)` is matched against the argument pattern `(x,y)` in the defining equation; this causes `x` to be defined as 5 and `y` as 6. Then the right-hand side `x` is returned, giving the result 5. The `fst` and `snd` functions are very useful, so they are already defined in the standard Haskell library.

An argument pattern may also be a list. Since a list is either `[]` or else it is constructed with the cons operator `:`, the patterns take the forms `[]` and `x:xs`. For example, the following function determines whether a list is empty:

```
isEmpty :: [a] -> Bool
isEmpty [] = True
isEmpty (x:xs) = False
```

The parentheses around the pattern `x:xs` in the left-hand side are required because otherwise the compiler would parse it incorrectly.

When `isEmpty []` is evaluated, the argument matches the pattern in the first defining equation, so `True` is returned. However, when the application `isEmpty (1:2:3:[])` is evaluated, the argument fails to match `[]` and the second equation is tried. Here the match succeeds, with `x` defined as 1 and `xs` defined as `2:3:[]`, and `False` is returned. It makes no difference if the argument is written with the simpler syntax `[1,2,3]`; this is merely an abbreviation for `1:2:3:[]`, so the evaluation is identical.

The following functions, which are defined in the standard Haskell libraries, can be used to return the first element of a list (the head), or everything *except* the first element (the tail):

```
head :: [a] -> a
head (x:xs) = x

tail :: [a] -> [a]
tail (x:xs) = xs
```

Exercise 3. Write a function that takes a character and returns True if the character is 'a' and False otherwise.

Exercise 4. Write a function that takes a string and returns True if the string is "hello" and False otherwise. This can be done by specifying each element of the string in the list pattern (e.g. 'h':'i':[]).

Exercise 5. Write a function that takes a string and removes a leading space if it exists.

1.4.6   Higher Order Functions

Functions in Haskell are 'first class objects'; that is, you can store them in data structures, pass them as arguments to functions, and create new ones. A function is called *first order* if its arguments and results are ordinary data values, and it is called *higher order* if it takes another function as an argument, or if it returns a function as its result. Higher order functions make possible a variety of powerful programming techniques.

The twice function takes another function f as its first argument, and it applies f two times to its second argument x:

```
twice :: (a->a) -> a -> a
twice f x = f (f x)
```

We can work out an application using equational reasoning. For example, twice sqrt 81 is evaluated as follows:

$$
\begin{aligned}
twice\ sqrt\ 81 \\
= sqrt\ (sqrt\ 81) \\
= sqrt\ 9 \\
= 3
\end{aligned}
$$

Let's examine the type of twice in detail. Assuming the second argument has type a, then the argument function has to accept an argument of type a and also return a value of the same type (because the result of the inner application becomes the argument to the outer application). Hence the argument function must have type a->a.

In Haskell, functions receive their arguments one at a time. That is, if a function takes two arguments but you apply it to just one argument, the evaluation gives a new function which is ready to take the second argument and finish the computation. For example, the following function takes two arguments and returns their product:

```
prod :: Integer -> Integer -> Integer
prod x y = x*y
```

A *full application* is an expression giving `prod` all of its arguments: for example, prod 4 5 ⇒ 20. However, the *partial application* prod 4 supplies just one argument, and the result of this is a new function that takes a number and multiplies it by 4. For example, suppose the following equations are defined in the script file:

```
g = prod 4
p = g 6
q = twice g 3
```

Then p ⇒ 24, and q ⇒ 48.

1.5   Conditional Expressions

A conditional expression uses a `Bool` value to make a choice. Its general form is:

if Boolean_expression then exp1 else exp2.

The Boolean expression is first evaluated; if it is `True` then the entire conditional expression has the value of *exp1*; if it is `False` the whole expression has the value of *exp2*.

A conditional expression must always have both a `then` expression and an `else` expression. Both of these expressions must have the same type, which is the type of the entire conditional expression. For example,

```
if 2<3 then "bird" else "fish"
```

has type `String` and its value is `"bird"`. However, the following expressions are incorrect:

```
if 2<3 then 10            -- no else expression
if 2+2 then 1 else 2      -- must be Bool after if
if True then "bird" else 7  -- different types
```

The `abs` function returns the absolute value of its argument, and it uses a conditional expression because the result depends on the sign of the argument:

```
abs :: Integer -> Integer
abs x = if x<0 then -x else x
```

1.6 Local Variables: `let` Expressions

There are many times when we need to use computed values more than once.
Instead of repeating the expression several times, it is better to give it a local
name that can be reused. This can be done with a `let` expression. The general
form is:

$$\begin{array}{c} \text{let } equation \\ equation \\ \vdots \\ equation \\ \text{in } expression \end{array}$$

This entire construct is just one big expression, and it can be used anywhere
an expression would be valid. When it is evaluated, the local equations give
temporary values to the variables in their left-hand sides; the final expression
after `in` is the value of the entire `let` expression. Each of the local definitions
can refer to the others.

For example, consider the following function definition which defines the
two real solutions (x_1, x_2) of the quadratic formula $a \times x^2 + b \times x + c = 0$:

```
quadratic :: Double -> Double -> Double -> (Double,Double)
quadratic a b c
   = let d = sqrt (b^2 - 4*a*c)
         x1 = (-b + d) / (2*a)
         x2 = (-b - d) / (2*a)
     in (x1,x2)
```

The `let` expression gives the three variables `d`, `x1`, and `x2` values that can
be used locally. For example, the value of `d` is used in the equations for `x1` and
`x2`. Outside of this `let` expression, the names `d`, `x1`, and `x2` are not defined by
these equations (although they might be defined by equations in some enclosing
`let` expression).

Let expressions are *expressions*! For example, this is not a block of statements that gets executed; it is an expression that uses a local definition:

```
2 + let x = sqrt 9 in (x+1)*(x-1)
  => 10.0
```

1.7 Type Variables

Recall the functions `fst` and `snd`, which return the first and second component
of a pair. There is actually an infinite number of different types these functions
could be applied to, including:

```
(Integer,Float)
([Char], [Integer])
(Double, (Int,Char))
```

It would not be good to give `fst` a restrictive type, like `(Integer,Float)`
`-> Integer`; this would work for the first example above, but not the others.
We really want to say that `fst` takes a pair whose components may have *any*
type, and its result has the same type as the first component. This is done by
using *type variables*:

```
fst :: (a,b) -> a
snd :: (a,b) -> b
```

Type variables must begin with a lower case letter, and it is a common
convention to use `a`, `b`, and so on.

A function with a type variable in its type signature is said to be *polymorphic*, because its type can take many forms. Many important Haskell functions
are polymorphic, enabling them to be used in a wide variety of circumstances.
Polymorphism is a very important invention because it makes it easier to reuse
programs.

1.8   Common Functions on Lists

Haskell provides a number of operations on lists. A few of the most important
ones are presented in this section. We will have more to say about these
functions later in the book, where we will use them in a variety of practical
applications, show how they are implemented, and prove theorems stating some
of their mathematical properties.

The `length` Function

The `length` function returns the number of elements in a list:

```
length :: [a] -> Int

length [2,8,1] => 3
length [] => 0
length "hello" => 5
length [1..n] => n
length [1..] => <infinite loop>
```

The `!!` (index) Operator

The `(!!)` operator lets you access a list element by its index. Indices start
from 0.

```
(!!) :: [a] -> Int -> a

[1,2,3] !! 0 => 1
"abcde" !! 2 => 'c'
```

The take Function

This function takes the first n elements from a list:

```
take :: Int -> [a] -> [a]

take 2 [1,2,3] => [1,2]
take 0 [1,2,3] => []
take 4 [1,2,3] => [1,2,3]
```

The drop Function

The drop function drops the first n elements from a list; it removes exactly the same elements that take would return:

```
drop :: Int -> [a] -> [a]

drop 2 [1,2,3] => [3]
drop 0 [1,2,3] => [1,2,3]
drop 4 [1,2,3] => []
```

The ++ (append) Operator

Two lists can be joined together using the append (also called concatenation) (++) operation. All of the elements in the resulting list must have the same type, so the two lists must also have the same type.

```
(++) :: [a] -> [a] -> [a]

[1,2] ++ [3,4,5] => [1,2,3,4,5]
[] ++ "abc" => "abc"
```

The map Function

Often, we want to apply a function to each element in a list. For example, we may have a string of lower case letters and want them to be upper case letters. To do this, we use map, which takes a function of one argument and a list. It applies the function to each element of the list.

```
map :: (a -> b) -> [a] -> [b]

map toUpper "the cat and dog" => "THE CAT AND DOG"
map (* 10) [1,2,3] => [10,20,30]
```

The map function is often used in functional programming to solve problems where you would use a **for** loop in an imperative language. When you need to perform some computation on all the elements of a list xs, define a function f to do the computation, and write map f xs.

Exercise 6. Suppose a program has read in a list of numbers of type Int. Each number is intended to represent a Boolean value, where 0 means False, 1 means True, and any other number constitutes invalid input. Write a function convert :: [Int] -> [Bool] that converts a list of numbers to the corresponding Booleans.

Exercise 7. Write a function member0 :: String -> Bool that takes a list of Char values (i.e., a String), and returns True if at least one of the characters is '0' and False otherwise. Hint: use map and the function or, which takes a list of Boolean values and returns True if at least one of them is True; otherwise it returns False. Thus or [False, True, False] => True, but or [False, False, False] => True.

The zip Function

The function zip pairs up the elements of two lists.

```
zip :: [a] -> [b] -> [(a,b)]

zip [1,2,3] "abc" => [(1,'a'),(2,'b'),(3,'c')]
zip [1,2,3] "ab" => [(1,'a'),(2,'b')]
zip [1,2] "abc" =>  [(1,'a'),(2,'b')]
```

If the lists are not of the same length, then the longer one's extra elements do not appear in the result.

The zipWith Function

The function zipWith is a function like map except that it takes two lists. Like zip, the longer list's extra elements are ignored.

```
zipWith :: (a -> b -> c) -> [a] -> [b] -> [c]

zipWith (+) [2,4..10] [1,3..10] => [3,7,11,15,19]
zipWith (*) [1,2,3] [1,2,3] => [1,4,9]
```

The foldr and foldl Functions

As noted above, you can think of map as an iteration over a list, producing a list of results. Another kind of iteration produces a single result, by combining all the elements of the list with some operator like (+). For example, you could compute the sum of a list by iterating over it, using (+) to combine the elements. This kind of iteration is called a *fold*. Since you could iterate over the list either direction, there are two versions: `foldr` is the "fold from the right" function, and `foldl` is the "fold from the left" version.

The fold functions take three arguments. The first argument is the operator to be used to combine the elements of the list. The second argument is a default or starting value, which is needed so that the fold can produce a result even if the list is empty. For example, we use this default value to ensure that the sum of an empty list is 0. The third argument is the list over which the iteration takes place.

The meaning of the functions can be illustrated by a couple of examples, where we use the (+) operator, a default value of a, and a list [p,q,r,s].

```
foldl (+) a [p,q,r,s]
  = (((a+p) + q) + r) + s

foldr (+) a [p,q,r,s]
  = p + (q + (r + (s+a)))
```

Note that in both cases, the result is a single value, not a list, and it is computed by a sequence of calculations using the (+) operator. There are two key differences. First, the parentheses are grouped to the right in `foldr` but they are grouped to the left in `foldl`. Second, the default value a comes in at the left for `foldl` and at the right for `foldr`.

Since the definitions of `foldl` and `foldr` require recursion, the subject of Chapter 3, we will not give those here. As you can imagine, however, the folds are implemented by iterations over the list, with a "running value" or "accumulator" that gives the current intermediate result.

Let's look at an accumulator more closely. Suppose that we apply (+) to the elements of the list [1,2,3] in this way:

```
(+) 1
  ((+) 2
    ((+) 3 0))
```

On the third line, there is an expression that forms the second argument of the function application that appears on the second line. That application in turn forms the second argument of the application appearing on the first line. In each case, the second argument of the (+) function is the result of a previous computation, handed on to the next application. This second argument is an accumulator.

The order in which the list elements are used is from right to left. First the 3 is added to 0, then the result of that addition is added to 2, the second element of the list. Finally 1, the first element of the list, is added to the result of the previous addition.

The `foldr` function does something very similar. It takes a function of two arguments, and the second argument is an accumulator. Initially, the accumulator has to have a value, so `foldr` receives an initial value as well. Its last argument is a list of values, and it returns the accumulated value. If we were to use `foldr` to implement the example above, we would write:

```
foldr (+) 0 [1,2,3]
```

The type of `foldr` is complex. Its function argument takes two values and returns the accumulator, so the type of this argument is `a -> b -> b`. The first value of the accumulator is the initial value, so its type is `b`. The type of the list argument is `[a]`, and the result type is `b`, the final accumulator value.

```
foldr :: (a -> b -> b) -> b -> [a] -> b
```

The accumulator makes it possible to do a surprising variety of things with `foldr`. For example, the `sum` function sums the numbers in its list argument. We can implement `sum` with `foldr` like this:

```
foldr (+) 0 [1,2,3,4,5]
=> (+) 1 ((+) 2 ((+) 3 ((+) 4 ((+) 5 0))))
=> 15
```

We can also implement (++) with `foldr`:

```
foldr (:) [3,4,5] [1,2]
=> (:) 1 ((:) 2 [3,4,5])
=> [1,2,3,4,5]
```

Other functions such as `and`, which returns True if its list argument contains only True values, and `or` (which is similar) can also be implemented using `foldr`:

```
foldr (&&) True [True, False, True] => False
```

```
foldr (||) False [True, False, True] => True
```

The `foldl` function also uses an accumulator, but it processes the list elements from left to right. We look again at applying (+) to the elements of the list [1,2,3]:

```
        (+)                     3
            ((+)           2)
                ((+) 0 1)
```

As you can see, the first argument is the accumulator, and the *first* element of the list, not the last, is given to the accumulator initially.

In many cases, this makes no difference, and `foldl` returns the same result as `foldr`. For example, the following two expressions produce the same result because the multiplication operation is associative, and 1 is an identity for multiplication:

```
foldl (*) 1 [1,2,-3]
foldr (*) 1 [1,2,-3]
```

However, these expressions do not give the same result because subtraction is not associative:

```
foldl (-) 0 [1,2,3,4]
```

```
foldr (-) 0 [1,2,3,4]
```

It can be challenging to write applications of `foldr` and `foldl`. The best way to do this effectively is to do an example by hand until you get used to writing applications that work. This process is called 'expanding' an application. Here are some examples of expanding applications of `foldr` and `foldl`:

```
foldr (++) [] [[1],[2],[3]]
=> (++) [1] ((++) [2] ((++) [3] []))
=> (++) [1] ((++) [2] [3])
=> (++) [1] [2,3]
=> [1,2,3]

foldr (&&) False [True,False]
=> (&&) True ((&&) False False)
=> (&&) True False
=> False

foldl (-) 0 [1,2,3]
=> (-) ((-) ((-) 0 1) 2) 3
=> (-) ((-) -1 2) 3
=> (-) -3 3
=> -6

foldl max 0 [1,2,3]
=> max (max (max 0 1) 2) 3
=> max (max 1 2) 3
=> max 2 3
=> 3
```

Exercise 8. Expand the following application:

```
foldr max 0 [1,5,3]
```

The . (composition) Operator

Now let's take a look for just a moment at some more notation. The *composition* operator (.) allows us to create a pipeline of function applications, each of which is waiting for an argument. For example,

```
(toUpper.toLower) 'A' => 'A'
```

```
(toLower.toUpper) 'Z' => 'z'
```

In each case, the first (rightmost) function receives its argument, then gives its result to the function on the left. That function returns the final result of the application.

This notation also allows us to form function applications like this:

```
((:).toUpper) 'a' "bc" => "Abc"
```

Again, the first function receives the 'a' and returns 'A'. Then the *cons* function receives the 'A' and the string, and creates a new string.

With this in mind, new possibilities for using `foldr` become available. We can use it as an alternative to `map`, which we discussed earlier, as well as `elem`, a function that returns True if its first argument occurs within its second (list) argument:

```
map toUpper "abc"
  = foldr ((:).toUpper) [] "abc"
    => "ABC"
```

```
elem 3 [1,2,3,4,5]
  = foldr ((||).(== 3)) False [1,2,3,4,5]
    => True
```

1.9   Data Type Definitions

Tuples and lists are very useful, but there comes a point when you would like to define the shape of your data so that it fits the problem being solved. Haskell is particularly good at doing this because it supports *algebraic data types*, a flexible form of user-defined data structure. Furthermore, pattern matching, which we have already used for tuples and lists, can be used with the algebraic data types. Together, these allow you to define and use made-to-order structures.

Suppose that you wanted to specify an enumerated type in Haskell. For example, you would like a type that enumerates colours:

```
data Colour = Red | Orange | Yellow
            | Green | Blue | Violet
```

This declaration states that the `Colour` type contains each of the values `Red`, `Orange`, and so on. If these appeared in a list, as in [`Red, Yellow, Green`], the type of the list would be [`Colour`]. Each of these values is a *constructor*, and constructors are the only Haskell values that start with an upper case letter.

The `Bool` type that we have been using throughout is not actually a built-in type; it is defined as follows in the standard library:

```
data Bool = False | True
```

Now suppose that you would like values that contain fields, because information of some kind must be associated with each of the values. We might define an `Animal` type in which each value indicates an animal species and some additional information about a particular animal, such as its breed. The type would contain entries for dogs and cats, and a string for each of these which could be used to store their breed, but our entry for rats would not need a string for the breed because we are not particularly interested in rat breeding.

```
data Animal = Cat String | Dog String | Rat
```

Here are some values with type `Animals`:

```
Cat "Siamese"
Cat "Tabby"
Dog "Spaniel" ·
Dog "big and hungry"
Rat
```

As you can see, there is a field next to the `Cat` constructor, which can be accessed by a function defined on the type, using pattern matching.

Sometimes a more flexible approach is needed. Instead of storing a string with each cat or dog, we might want to store some arbitrary information. This can be accommodated by letting `Animal` take two *type variables*, `a` and `b`, which can stand for any data type:

```
data Animal a b
  = Cat a | Dog b | Rat
```

Type variables must start with lowercase letters. It is a common convention, but not required, to use `a`, `b`, and so on for type variables. Type variables allow `Animal` to be used with arbitrary annotations, for example:

```
data BreedOfCat = Siamese | Persian | Moggie

Cat Siamese  :: Animal BreedOfCat b
Cat Persian  :: Animal BreedOfCat b
Cat "moggie" :: Animal String b
Dog 15       :: Animal a Integer
Rat          :: Animal a b
```

Data types with type variables can be very useful. Suppose that you are writing a function that may succeed in computing its result, which has type a, but may also fail to get a result. A common example is a search function, which is looking for a value in a table, and which might fail. Instead of letting the program crash, it is better to return a value that says either 'I succeeded, and the result is x' or else 'I failed to get a result'. The Maybe type is just what we need:

```
data Maybe a = Nothing | Just a
```

Suppose we are writing a function to look up someone's telephone number, using a database represented as a list of pairs. The first component of each pair is a person's name, and the second component is their telephone number (an integer). It isn't a good idea to have this function simply return an integer; a search might be made for a person not in the database. The solution is to use Maybe:

```
phone_lookup :: [(String,Integer)] -> String -> Maybe Integer
```

Now the lookup function can be defined so that it always returns something sensible.

You may have tried some examples by this time and noticed that Haskell will not print values of your new type. This is because it uses a function called show to convert a data value to a string that can be printed, and you have not constructed a version of show that it can use.

Fortunately, you do not have to define show yourself. Haskell will do it automatically for you when you end a data type definition with the words deriving Show, as in:

```
data Colour = Red | Orange | Yellow
            | Green | Blue | Violet
          deriving Show
```

Pattern matching can take place over values of any types, including algebraic data types defined by the user. For example, let's define a function that takes the result of a phone number search and produces a comprehensible string to be printed.

```
phone_message :: Maybe Integer -> String
phone_message Nothing = "Telephone number not found"
phone_message (Just x) = "The number is " ++ show x
```

Exercise 9. Write a function that takes two lists of type [Maybe Int] and examines the pair of list heads before looking at the rest of the lists. It returns a list in which the Ints of each pair have been added if both are of the form Just n, preserving any Just n value otherwise. For example,

```
addJust [Just 2, Nothing, Just 3]
        [Nothing, Nothing, Just 5]
  => [Just 2, Nothing, Just 8]
```

Exercise 10. Define a data type that represents six different metals and automatically creates versions of (==) and show.

Exercise 11. Suppose that you have some coins that have been sorted into piles, each of which contains only one kind of coin. Define a data type that can be used to represent the piles of coins.

Exercise 12. A *universal* type is one in which any type can be represented. Each different type is identified by its own constructor, which serves as a distinguishing tag. For example, here is a universal type that represents three different types of number:

```
data Number = INT Int | INTEGER Integer | FLOAT Float
            deriving (Eq, Show)
```

Define a universal type that contains Booleans, characters, and integers (Ints).

Exercise 13. Define a type that contains tuples of up to four elements.

Exercise 14. The quadratic equation $a \cdot x^2 + b \cdot x + c = 0$ has two roots, given by the formula

$$x = \frac{-b \pm \sqrt{b^2 - 4 \cdot a \cdot c}}{2 \cdot a},$$

as long as the discriminant (the expression under the square root sign) is non-negative. If the discriminant is negative the roots are complex. Define a function that finds the real solutions of the quadratic equation, and reports failure if they don't exist.

1.10 Type Classes and Overloading

There are some operations that can be used on several different types, but not on all types. For example, the (+) function can be applied to integers or floating point numbers (as well as several other kinds of number). This means that there are two completely different implementations of this function, one for integers and another for floating point. And that raises a question: what is the type of (+)? It would be too restrictive to specify

```
(+) :: Integer -> Integer -> Integer
```

because then it would be illegal to write 3.14+2.7. On the other hand, it would be too general to specify

```
(+) :: a -> a -> a
```

because this says that *any* type can be added, allowing nonsensical expressions like True+False and "cat"+"mouse". What we need to say is that (+) can be used to add two values of any type, provided that type is numeric. The (+) function actually has the following type:

```
(+) :: Num a => a -> a -> a
```

Num is the name of a *type class*, a set of types sharing a property. The members of the set Num include Int, Integer, Float, Double, and many others, and their essential property is that arithmetic makes sense on these types. However, types like Bool and String, where addition is meaningless, are *not* members of the class Num. In the type of (+), the notation Num a => is called a *class constraint* or a *context*, and it means that (+) can only be applied to arguments with types that belong to the Num set.

Haskell allows you to define new type classes and to specify that a type is a member of a class. This is a powerful corner of the language, but we will not go into it in detail; see the references in Section 1.11 for a complete explanation.

To read this book, you don't need to be able to define new type classes or instances, but error messages sometimes mention type class constraints, so it is helpful to know what they mean.

There are a few commonly used type classes that are ubiquitous in Haskell; the most important are Num, Show, and Eq. Show is the class of types that can be converted in a meaningful way into a character string. Most ordinary data values are in Show, but functions are not. Eq is the class of types that can be compared for equality.

When you are faced with a type error in a program, it is good to realise that values (with their class constraints) can migrate from a long way away. For example, if we have the definition

```
fun a b c = if a then b == c else False
```

then the type of fun has to reflect the fact that b and c have to be in the Eq class. This means that the fun function also has a type constrained by the Eq class, so the definition should be written as:

```
fun :: Eq b => a -> b -> b -> Bool
fun a b c = if a then b == c else False
```

If you forget to include the class constraint Eq b => in the type signature, the Haskell compiler will give an error message. The context on the type of fun declares that whatever calls fun and supplies it with arguments must ensure that there is a meaningful way to compare them.

You will inevitably come up against functions that have contexts in their types. The common sense rule is: if your function definition uses an overloaded operator (one with a type that has a context), then *its* type must contain that context as well. If your function has more than one such operator and the operator types have different contexts, then each new context must appear in the type of the function.

For example, suppose you want to write a function that checks whether a value appears in a list and returns a corresponding message. Here's a definition:

```
detect :: (Eq a, Show a) => a -> [a] -> String
detect v list =
  if elem v list
    then "List contains " ++ show v
    else "List does not contain" ++ show v
```

The function `elem` has a type that contains the context `Eq a` because it uses the overloaded operator `==`. The type of `detect` now has to have this context as well, because it uses `elem` and it also needs the `Show a` context, because it uses the overloaded function `show`.

1.11 Suggestions for Further Reading

A good source of information on the web about Haskell, and functional programming in general, is the Haskell Home Page: `www.haskell.org`. This contains pointers to a variety of relevant books and papers, as well as the language reference manuals.

Several good textbooks on Haskell are available. A book covering the basics of Haskell, *Two Dozen Short Lessons in Haskell*, by Rex Page, can be downloaded from the web. Books covering advanced programming techniques include *Introduction to Functional Programming using Haskell*, by Richard Bird [4]; *An Introduction to Functional Programming Systems Using Haskell*, by Tony Davie [8]; and *Haskell: The Craft of Functional Programming*, by Simon Thompson [32]. *The Haskell School of Expression: Learning Functional Programming through Multimedia*, by Paul Hudak [18], shows how to use Haskell through a series of applications to graphics, animations, and music.

The use of equations rather than assignments gives functional programming a very different style from imperative programming. *Purely Functional Data Structures* [24], by Okasaki, explores this topic in depth, and is an excellent intermediate level text on functional programming.

1.12 Review Exercises

Exercise 15. Define a function

```
showMaybe :: Show a => Maybe a -> String
```

that takes a `Maybe` value and prints it.

Exercise 16. A Bit is an integer that is either 0 or 1. A Word is a list of bits that represents a binary number. Here are some binary values that can be represented by Words:

```
[1,0]   => 2
[1,0,0,1] => 9
[1,1,1] => 7
```

We can define functions that are the Bit equivalent of or and **and** as follows:

```
bitOr :: Int -> Int -> Int
bitOr 0 0 = 0
bitOr x y = 1

bitAnd :: Int -> Int -> Int
bitAnd 1 1 = 1
bitAnd x y = 0
```

Now it is possible to take the 'bitwise' *and* of two words as follows:

```
bitwiseAnd [1,0,0] [1,0,1]
 => [bitAnd 1 1, bitAnd 0 0, bitAnd 0 1]
 => [1,0,0]

bitwiseAnd [0,0,0] [1,1,0]
 => [0,0,0]
```

Write a function `bitwiseAnd` that takes two Words and creates a third Word that is the bitwise *and* of the two Words.

Exercise 17. Each of the following expressions has a type error. Change the expression so that the type error no longer occurs.

```
[1, False]              '2' ++ 'a'
[(3,True), (False,9)]   2 == False
'a' > "b"               [[1],[2],[[3]]]
```

Exercise 18. What caused the type error in this definition and application?

```
f :: Num a => (a,a) -> a
f (x,y) = x + y

f (True,4)
```

Exercise 19. Why does this definition produce an error when used?

```
f :: Maybe a -> [a]
f Nothing = []

f (Just 3)
```

Exercise 20. Write a list comprehension that takes a list of `Maybe` values and returns a list of the `Just` constructor arguments. For example,

```
[Just 3, Nothing, Just 4] => [3,4]
```

Exercise 21. Using a list comprehension, write a function that takes a list of Int values and an Int value **n** and returns those elements in the list that are greater than **n**.

Exercise 22. Write a function

```
f :: [Int] -> Int -> [Int]
```

that takes a list of Int values and an Int and returns a list of indexes at which that Int appears.

Exercise 23. Write a list comprehension that produces a list giving all of the positive integers that are *not* squares in the range 1 to 20.

Exercise 24. Write a function that uses `foldr` to count the number of times a letter occurs in a string.

Exercise 25. Write a function using `foldr` that takes a list and removes each instance of a given letter.

Exercise 26. Using `foldl`, write a function

```
rev :: [a] -> [a]
```

that reverses its list argument.

Exercise 27. Using `foldl`, write a function

```
maybeLast :: [a] -> Maybe a
```

that takes a list and returns the last element in it if there is one, otherwise it returns `Nothing`.

Chapter 2

Equational Reasoning

There are many different formal methods used in computer science, and several of them will be introduced throughout this book. This chapter discusses one of the most powerful and yet simplest of the formal methods: *equational reasoning*.

You don't need to learn any fancy new mathematics to use equational reasoning; it is just like simple school algebra. Despite (or perhaps because of) this simplicity, equational reasoning is an effective tool for solving extremely complicated problems. Much of the most advanced current research on formal methods is based on equational reasoning, and we will use it often through the rest of this book.

2.1 Equations and Substitutions

An equation $x = y$ says that x and y have the same value; any time you see one of them you can replace it by the other. Such a replacement is called a *substitution*. For example, suppose we are given the following definitions, which are written as equations:

```
x = 8
y = 4
```

These equations are going to be cited as justifications for reasoning steps, so it helps to have a standard way to name equations. The notation { x } means "the equation that defines x". Now, suppose the problem is to evaluate $2 * x + x/y$. We do this by writing a chain of expressions, each one equal to the previous, beginning with the original problem and ending with the final result. Each step is justified by a reason given in braces { ... }.

```
2*x + x/y
  = 2*8 + 8/y      { x }
  = 2*8 + 8/4      { y }
```

```
= 16 + 2              { arithmetic }
= 18                  { arithmetic }
```

Sometimes the justification is an explicit reference to an equation; thus the justification { x } means "the equation defining x allows this step to be taken". In other cases, where we don't want to give a formal justification because it is too trivial and tedious, an informal reason is given, for example { arithmetic }.

The format for writing a chain of equational reasoning is not rigidly specified. The reasoning above was written with each step on a separate line, with the justifications aligned to the right. This format is commonly used; it is compact and readable for many cases.

Sometimes the expressions or the justifications in a chain of reasoning become very long, and the format used above becomes unreadable. In such cases, a common approach is to place the = sign on a separate line, followed by the justification. Here is the same chain of reasoning written in the alternative notation:

```
2*x + x/y
  = { x }
2*8 + 8/y
  = { y }
2*8 + 8/4
  = { multiplication, division }
16 + 2
  = { addition }
18
```

This style takes twice as many lines, and it may seem harder to read, but experience has shown that it scales up better to very large cases. As we said before, equational reasoning is a practical tool used for large-scale proofs about realistic programs, and the expressions sometimes become quite big. The second style is becoming standard for research papers. In doing your own proofs, you may use whichever style you prefer, although it's important to be able to read both styles.

The example above is just elementary mathematics, but exactly the same technique can be used to reason about Haskell programs, because *equations in Haskell are true mathematical equations*—they are not assignment statements.

2.2   Equational Reasoning as Hand-execution

An important skill for all programmers is *hand execution*: taking a program and its data, and working out by hand what the result should be. Hand execution tells you what you think the values being computed should be, so you can locate errors by comparing the hand execution with the results of an actual execution by the computer.

For imperative programming languages, hand execution involves simulating the actions of the computer as it obeys the commands in the program. For a pure functional language, like Haskell, hand execution is done by equational reasoning instead. Thus it is not just a tool for theoreticians trying to prove theorems about software—equational reasoning is essential for all functional programmers.

For example, suppose we have the following script file:

```
f :: Integer -> Integer -> Integer
f x y = (2+x) * g y

g :: Integer -> Integer
g z = 8-z
```

Now the expression f 3 4 can be evaluated (or "hand-executed") by equational reasoning.

```
f 3 4
   = (2+3) * g 4        { f }
   = (2+3) * (8-4)      { g }
   = 20                 { arithmetic }
```

There is an important point to notice about that example: sometimes you need to introduce parentheses to prevent different terms from getting mixed up. In the second line, we had two subexpressions being multiplied together (the first subexpression, 2+3, is enclosed in parentheses, whereas the second needs no parentheses as function application has the highest precedence). Now, suppose that somebody just blindly replaces g 4 by 8-4, resulting in (2+3) * 8-4. Now the 8-4 is really an expression on its own, yet the precedence conventions of mathematics would cause this to be interpreted as ((2+3)*8) - 4. Notice that we prevented this error by introducing explicit parentheses, (8-4), in the third line.

There is nothing really difficult going on here; you just have to be careful to introduce parentheses whenever there is a danger of confusion. An alternative approach would simply be to use fully parenthesized expressions, without relying on precedence rules at all. If you find that helpful, by all means do it! However, as you gain more experience you will probably prefer to omit most of the redundant parentheses, for the sake of improved readability.

Another issue you need to be careful with is variable names. Again, there is nothing difficult about names, but careless reasoning can lead to mistakes. The problem is that a set of equations may use the same name for different purposes. For example, consider the following equations:

```
f x = x+1
g x = x*5
```

Now the name x is used locally as the parameter name for both f and g, but there should be no confusion between them. When the function is applied to an argument, the argument is substituted for the parameter:

```
f 8 * g 9
  = (8+1) * (9*5)   { f, g }
  = 405             { arithmetic }
```

In the second line, we used x twice, once with the value 8 and once with the value 9.

The best way to think of this is to consider every function as having its own local parameters. The parameters of f and g are completely different, they just happen both to be called x. The situation is just like real life: there may be (and actually are) several people named John Smith. The fact that the names are the same does not mean that the people are the same!

You can define names within a *scope* in Haskell, just as with other programming languages. A scope can be introduced implicitly: for example, the f and g defined above shouldn't be confused with unrelated functions defined in another chapter. A let expression or a **where** clause introduces a scope explicitly. Consider the following nested let expression:

```
let a = 1
    b = 2
    f x = x + b  -- b=2
in
  let b = 5
      c = 6
      g x = [x,a,b,c]  -- a=1, b=5, c=6
```

The scope of a is the entire piece of code—it is available for use everywhere. However, the b defined in b = 2 is hidden inside the inner let because the b = 5 is visible instead. The c is in scope only inside the inner let expression; you could not use it in the right hand side of f.

Many function definitions contain several equations. When you use one of these equations to justify a step of equational reasoning, it's necessary to specify which equation is being used, not just the name of the function. For example, suppose we have the following definition:

```
f :: Bool -> Int
f True = 5
f False = 6
```

If you simply give a justification { f }, it may not be clear to the reader which equation you are using. (At least, with large and complicated definitions that can happen, although it should be clear here!) There are two styles for writing justifications using such definitions. One way to do it is to imagine all the equations in the definition as being numbered 1, 2, 3, and so on. Then the justification { f.2 } means "equation 2 of the definition of f". Another way to do it is to specify the pattern used on the left hand side of the equation, for example { f.False }. Either style is perfectly acceptable, and we will use both from time to time in this book.

2.2.1 Conditionals

A conditional expression satisfies the following equations:

```
if True  then e2 else e3  =  e2      { if True }
if False then e2 else e3  =  e3      { if False }
```

These equations are used for reasoning about programs that contain if expressions. For example, suppose f is defined as follows:

```
f :: Double -> Double
f x =
  if x >= 0
    then sqrt x
    else 0
```

In order to calculate the value of an application of the function, we need to use one of the if-equations.

```
f (-3)
  = if (-3) >= 0 then sqrt (-3) else 0   { f }
  = if False then sqrt (-3) else 0       { arithmetic }
  = 0                                    { if False }
```

This example raises an interesting issue. When you are using equational reasoning, there will always be an enormous number of valid steps that could be taken. Here is another piece of equational reasoning:

```
f (-3)
  = if (-3) >= 0 then sqrt (-3) else 0   { f }
  = if (-3) >= 0 then ERROR else 0       { sqrt }
  = if False then ERROR else 0           { arithmetic }
  = 0                                    { if False }
```

The ERROR denotes a value representing a failed computation (because the square root of a negative number is undefined). In most imperative languages, the calculation of an ERROR will abort the program. Haskell will also throw

an exception if it encounters an ERROR. However, Haskell also is smart about
its equational reasoning: when evaluating if e1 then e2 else e3, it won't
attempt to evaluate either e2 or e3 until it knows which one is needed. Therefore Haskell begins by evaluating the conditional; then it evaluates whichever
expression is required. It is perfectly acceptable to write equations that sometimes define error conditions, as long as these error values aren't actually used
by the program.

2.3   Equational Reasoning with Lists

So far, the examples we have given of equational reasoning have been quite
simple. To give a more realistic impression of how equational reasoning is
actually used, we will give the proofs of a few simple but important theorems.
First, however, we will state a few theorems without proof; those theorems will
be proved in the chapter on induction, but for now we will assume their validity
and use them as justifications in some simple proofs of further theorems.

The following theorem states the obvious result that the length of a concatenation is the sum of the lengths of the pieces that were concatenated. This
theorem can be proved using list induction (see Chapter 4). We can use the
theorem directly to justify equational reasoning steps, and an example is given
below.

Theorem 1 (length (++)). Let xs, ys :: [a] be arbitrary lists. Then
length (xs ++ ys) = length xs + length ys.

The next theorem says that if you map a function over a list, the list of
results is the same size as the list of arguments.

Theorem 2 (length map). Let xs :: [a] be an arbitrary list, and f :: a
-> b an arbitrary function. Then length (map f xs) = length xs.

We also need the following result, which says that if you map a function over
a concatenation of two lists, the result could also be obtained by concatenating
separate maps over the original lists. This theorem is frequently used in parallel
functional programming, because it shows how to decompose a large problem
into two smaller ones than can be solved independently.

Theorem 3 (map (++)). Let xs, ys :: [a] be arbitrary lists, and let f
:: a -> b be an arbitrary function. Then map f (xs ++ ys) = map f xs
++ map f ys.

To illustrate how a theorem can be proved using direct equational reasoning,
we will give a proof for this theorem:

Theorem 4. For arbitrary lists xs, ys :: [a], and arbitrary f :: a ->
b, the following equation holds: length (map f (xs ++ ys)) = length xs
+ length ys.

Proof. We prove the theorem by equational reasoning, starting with the left hand side of the equation and transforming it into the right hand side.

```
length (map f (xs ++ ys))
  = length (map f xs ++ map f ys)          { map (++) }
  = length (map f xs) + length (map f ys)  { length (++) }
  = length xs + length ys                  { length map }
```

There are usually many ways to prove a theorem; here is an alternative version:

```
length (map f (xs ++ ys))
  = length (xs ++ ys)            { length map }
  = length xs + length ys       { length (++) }
```

Either proof is fine, and you don't need to give more than one! It is a matter of taste and opinion as to which proof is better. Sometimes a shorter proof seems preferable to a longer one, but issues of clarity and elegance may also affect your choice. □

2.4   The Role of the Language

By this time, you may be wondering why you haven't been using equational reasoning for years, as it is so simple and powerful.

The reason is that equational reasoning requires *equations*, which say that two things are *equal*. Imperative programming languages simply don't contain any equations. Consider the following code, which might appear in a C or Java program:

```
n = n+1
```

That might look like an equation, but it isn't! It is an *assignment*. The meaning of the statement is a command: "fetch the value of n, then add 1 to it, then store it in the memory location called n, discarding the previous value". This is not a claim that n has the same value as n+1. Because of this, many imperative programming languages use a distinctive notation for assignments. For example, Algol and its descendants use the := operator:

```
n := n+1
```

What is different about pure functional programming languages like Haskell is that x = y *really is an equation*, stating that x and y have the same value. Furthermore, Haskell has a property called *referential transparency*. This means that if you have an equation x = y, you can replace any occurrence of x by y, and you can replace any occurrence of y by x. (This is, of course, subject to various restrictions—you have to be careful about names and scopes, and grouping with parentheses, etc.)

The imperative assignment statement n := n+1 increases the value of n, whereas the Haskell equation n = n+1 defines n to be the solution to the mathematical equation. This equation has no solution, so the result is that n is undefined. If the program ever uses an undefined value like n, the computer will either·give an error message or go into an infinite loop.

Assignments have to be understood in the context of time passing as a program executes. To understand the n := n+1 assignment, you need to talk about the old value of the variable, and its new value. In contrast, equations are timeless, as there is no notion of changing the value of a variable.

Indeed, there is something peculiar about the word "variable". In mathematics and also in functional programming, a variable does not vary at all! To vary means to change over time, and this simply does not happen. A mathematical equation like $x = 2^3$ means that x has the value 8, and it will still have that value tomorrow. If the next chapter says $x = 2 + 2$, then this is a *completely different variable* being defined, which just happens to have the same name; it doesn't mean that 8 shrank over the course of a few pages. A variable in mathematics means "a name that stands for some value", while a variable in an imperative programming language means "a memory location whose value is modified by certain instructions".

As one would expect, there is a price to be paid for the enormous benefits of equational reasoning. Pure functional languages like Haskell, as well as mathematics itself, are demanding in that they require you to think through a problem deeply in order to express its solution with equations. It's always possible to hack an imperative program by sticking an assignment in it somewhere in order to patch up a problem, but you cannot do that in a functional program. On the other hand, if our goal is to build correct and reliable software—and this should be our goal!—then the discipline of careful thought will be repaid in higher quality software.

2.5   Rigor and Formality in Proofs

A computer program written in a programming language like Haskell is either syntactically correct, or it is incorrect. Any violation of the syntax of the language—no matter how trivial it may seem—renders the program invalid. The reason we are so picky about the rules of the language is that software (compilers, etc.) will process programs, and software is unable to cope with the ambiguities introduced by sloppy input.

Indeed, there were attempts in the 1970s to make compilers that could figure out what the programmer meant when syntax errors were present. The compiler would go ahead and translate what the programmer obviously intended into machine language, saving the programmer from the bother of getting the program right. This worked in many cases, and yet the whole approach has fallen out of favor. The difficulty is that occasionally the compiler would get confused and the "obvious" meaning of the program was not at all what the programmer meant. The program would then go ahead and execute, producing bugs that *could not possibly* be found by comparing the output with the program text. And the more intelligent and accurate the compiler becomes, the worse the final result: when translation errors are rare, the programmer is less likely to find the ones that do occur.

Are mathematical proofs as precisely defined as computer programs? It is tempting to say that they are even more precise and picky, since they are claiming to establish fundamental mathematical truths. The reality, however, is more subtle, for two reasons: proofs may be intended for humans or for computers to read, or both; and proofs may be small (where all details can be included) or enormous (where most details must be omitted).

Consider the following equational reasoning, which was given earlier in the chapter:

```
    ...
  = (8+1) * (9*5)    { f, g }
  = 405              { arithmetic }
```

Although this is highly detailed, a number of details are left to the reader. The step taken here requires an addition and two multiplications, yet it was taken in one step rather than three, and the vague justification { `arithmetic` } was given. This proof should be convincing to a human, but it is not completely formal. We never even proved that 8+1=9.

Mathematicians make a distinction between *rigorous proofs* and *formal proofs*.

A rigorous proof is thought through clearly and carefully, and does not contain sloppy shortcuts, but it may omit trivial details that the readers should be able to work out for themselves. The point about a rigorous proof is that it includes the essential details, and skips the inessential ones. Normally, professional mathematicians will agree about the steps that can be omitted. Nonetheless, it has happened on occasion that a mathematical proof has been published and generally accepted to be valid, only for an error to be discovered years later.

A formal proof consists of solid reasoning based on a clearly specified set of axioms. Since no details are omitted and no sloppiness is allowed, computer software can be used to check the validity of a formal proof.

The degree of rigor or formality that is needed in a proof depends on the purpose for which it is intended. It is unrealistic to say that all proofs should be formal, and inadequate to say that rigor is always adequate.

Sometimes it is reasonable to give an equational reasoning proof in a *semi-formal* style, where details are omitted in order to shorten the reasoning. It is common to perform several substitutions in one step (as in the example above). Sometimes a generic justification like "arithmetic" or "calculation" is appropriate.

Strictly speaking, justifications should always be given in equational reasoning. In some cases, however, the justifications may be omitted entirely.

Sometimes this is done where the justifications should be obvious and don't really make the proof more convincing. Sometimes the justifications are omitted because they would require a lot of additional machinery (definitions, theorems, proofs) and the author feels that this would detract from the main point.

Many more examples of equational reasoning appear in this book. An excellent textbook with a focus on functional programming with equational reasoning is by Bird [4]. Equational reasoning is central to modern programming language theory, and you can see it in action in many papers appearing in *Journal of Functional Programming* and other leading publications in theoretical computer science.

Chapter 3

Recursion

Recursion is a *self referential* style of definition commonly used in both mathematics and computer science. It is a fundamental programming tool, particularly important for manipulating data structures.

This book looks at recursion from several points of view. This chapter introduces the idea by showing how to write recursive functions in computer programs. Chapter 4 introduces induction, the mathematical technique for proving properties about recursive definitions, allowing you to use mathematics to prove the correctness of computer software. Chapter 5 extends induction to handle recursion over trees. Chapter 9 applies the same ideas to mathematics, where recursion is used to define sets inductively. You will find examples of recursion and induction throughout many branches of computer science. In addition to the examples given through these three chapters, we will study a larger case study in Chapter 13, where recursion and induction are applied to the problem of digital circuit design.

The idea in recursion is to break a problem down into two cases: if the problem is 'easy' a direct solution is specified, but if it is 'hard' we proceed through several steps: first the problem is redefined in terms of an easier problem; then we set aside the current problem temporarily and go solve the easier problem; and finally we use the solution to the easier problem to calculate the final result. The idea of splitting a hard problem into easier ones is called the *divide and conquer* strategy, and it is frequently useful in algorithm design.

The factorial function provides a good illustration of the process. Often the factorial function is defined using the clear but informal '...' notation:

$$n! \; = \; 1 \times 2 \times \cdots \times n$$

This definition is fine, but the '$\cdots$' notation relies on the reader's understanding to see what it means. An informal definition like this isn't well suited for proving theorems, and it isn't an executable program in most programming languages[1]. A more precise recursive definition of factorial consists of the

---

[1] Haskell is a very high level programming language, and it actually allows this style:

following pair of equations:

$$0\,! \;=\; 1$$
$$(n+1)\,! \;=\; (n+1) \times n\,!$$

In Haskell, this would be written as:

```
factorial :: Int -> Int
factorial 0 = 1
factorial (n+1) = (n+1) * factorial n
```

The first equation specifies the easy case, where the argument is 0 and the answer 1 can be supplied directly. The second equation handles the hard case, where the argument is of the form $n+1$. First the function chooses a slightly easier problem to solve, which is of size n; then it solves for $n!$ by evaluating `factorial n`, and it multiplies the value of $n!$ by $n+1$ to get the final result.

Recursive definitions consist of a collection of equations that state properties of the function being defined. There are a number of algebraic properties of the factorial function, and one of them is used as the second equation of the recursive definition. In fact, the definition doesn't consist of a set of commands to be obeyed; it consists of a set of true equations describing the salient properties of the function being defined.

Programming languages that allow this style of definition are often called *declarative languages*, because a program consists of a set of declarations of properties. The programmer must find a suitable set of properties to declare, usually via recursive equations, and the programming language implementation then finds a way to solve the equations. The opposite of a declarative language is an *imperative* language, where you give a sequence of commands that, when obeyed, will result in the computation of the result.

3.1   Recursion Over Lists

For recursive functions on lists, the 'easy' case is the empty list [] and the 'hard' case is a non-empty list, which can be written in the form (x:xs). A recursive function over lists has the following general form:

```
f ::  [a] ->  type of result
f [] =  result for empty list
f (x:xs) =  result defined using (f xs) and x
```

A simple example is the `length` function, which counts the elements in a list; for example, `length [1,2,3]` is 3.

---

in Haskell you can define `factorial n  = product [1..n]`. Most programming languages, however, do not allow this, and even Haskell treats the [1..n] notation as a high level abbreviation that is executed internally using recursion.

```
length :: [a] -> Int
length []    = 0
length (x:xs) = 1 + length xs
```

An empty list contains no elements, so `length []` returns 0. When the list contains at least one element, the function calculates the length of the rest of the list by evaluating `length xs` and then adds one to get the length of the full list. We can work out the evaluation of `length [1,2,3]` using equational reasoning. This process is similar to solving an algebra problem, and the Haskell program performs essentially the same calculation when it executes. Simplifying an expression by equational reasoning is the right way to 'hand-execute' a functional program. In working through this example, recall that `[1,2,3]` is a shorthand notation for `1:(2:(3:[]))`.

```
  length [1,2,3]
= 1 + length [2,3]
= 1 + (1 + length [3])
= 1 + (1 + (1 + length []))
= 1 + (1 + (1 + 0))
= 3
```

It is better to think of recursion as a systematic calculation, as above, than to try to imagine low-level subroutine operations inside the computer. Textbooks on imperative programming languages sometimes explain recursion by resorting to the underlying machine language implementation. In a functional language, recursion should be viewed at a high level, as an equational technique, and the low-level details should be left to the compiler.

The function `sum` provides a similar example of recursion. This function adds up the elements of its list; for example, `sum [1,2,3]` returns $1+2+3 = 6$. The type of `sum` reflects the fact that the elements of a list must be numbers if you want to add them up. The type context 'Num a =>' says that a can stand for any type provided that the numeric operations are defined. Thus `sum` could be applied to a list of 32-bit integers, or a list of unbounded integers, or a list of floating point numbers, or a list of complex numbers, and so on.

The definition has the same form as the previous example. We define the sum of an empty list to be 0; this is required to make the recursion work, and it makes sense anyway. It is common to define the base case of functions so that recursions work properly. This also usually gives good algebraic properties to the function. (This is the reason that 0! is defined to be 1.) For the recursive case, we add the head of the list x to the sum of the rest of the elements, computed by the recursive call `sum xs`.

```
sum :: Num a => [a] -> a
sum [] = 0
sum (x:xs) = x + sum xs
```

The value of sum [1,2,3] can be calculated using equational reasoning:

```
sum [1,2,3]
= 1 + sum [2,3]
= 1 + (2 + sum [3])
= 1 + (2 + (3 + sum []))
= 1 + (2 + (3 + 0))
= 6
```

So far, the functions we have written receive a list and return a number. Now we consider functions that return lists. A typical example is the (++) function (its name is often pronounced either as 'append' or as 'plus plus'). This function takes two lists and appends them together into one bigger list. For example, [1,2,3] ++ [9,8,7,6] returns [1,2,3,9,8,7,6].

Notice that this function has *two* list arguments. The definition uses recursion over the first argument. It's easy to figure out the value of [] ++ [a,b,c]; the first list contributes nothing, so the result is simply [a,b,c]. This observation provides a base case, which is essential to make recursion work: we can define [] ++ ys = ys. For the recursive case we have to consider an expression of the form (x:xs) ++ ys. The first element of the result must be the value x. Then comes a list consisting of all the elements of xs, followed by all the elements of ys; that list is simply xs++ys. This suggests the following definition:

```
(++) :: [a] -> [a] -> [a]
[] ++ ys = ys
(x:xs) ++ ys = x : (xs ++ ys)
```

Working out an example is a good way to check your understanding of the definition:

```
[1,2,3] ++ [9,8,7,6]
= 1 : ([2,3] ++ [9,8,7,6])
= 1 : (2 : ([3] ++ [9,8,7,6]))
= 1 : (2 : (3 : ([] ++ [9,8,7,6])))
= 1 : (2 : (3 : [9,8,7,6]))
= 1 : (2 : [3,9,8,7,6])
= 1 : [2,3,9,8,7,6]
= [1,2,3,9,8,7,6]
```

Once we know the structure of the definition—a recursion over xs—it is straightforward to work out the equations. The trickiest aspect of writing the definition of (++) is deciding over which list to perform the recursion. Notice that the definition just treats ys as an ordinary value, and never checks whether it is empty or non-empty. But you can't always assume that if there are several list arguments, the recursion will go over the first one. Some functions perform recursion over the second argument but not the first, and some functions perform a recursion simultaneously over *several* list arguments.

The `zip` function is an example of a function that takes two list arguments and performs a recursion simultaneously over both of them. This function takes two lists, and returns a list of pairs of elements. Within a pair, the first value comes from the first list, and the second value comes from the second list. For example, `zip [1,2,3,4] ['A', '*', 'q', 'x']` returns `[(1,'A'), (2,'*'), (3,'q'), (4,'x')]`. There is a special point to watch out for: the two argument lists might have different lengths. In this case, the result will have the same length as the shorter argument. For example, `zip [1,2,3,4] ['A', '*', 'q']` returns just `[(1,'A'), (2,'*'), (3,'q')]` because there isn't anything to pair up with the 4.

The definition of `zip` must do a recursion over *both* of the argument lists, because the two lists have to stay synchronised with each other. There are two base cases, because it's possible for either of the argument lists to be empty. There is no need to write a third base case `zip [] [] = []`, since the first base case will handle that situation as well.

```
zip :: [a] -> [b] -> [(a,b)]
zip [] ys = []
zip xs [] = []
zip (x:xs) (y:ys) = (x,y) : zip xs ys
```

Here is a calculation of the example above. The recursion terminates when the second list becomes empty; the second base case equation defines the result of `zip [4] []` to be `[]` even though the first argument is non-empty.

```
  zip [1,2,3,4] ['A', '*', 'q']
= (1,'A') : zip [2,3,4] ['*', 'q']
= (1,'A') : ((2,'*') : zip [3,4] ['q'])
= (1,'A') : ((2,'*') : ((3,'q') : zip [4] []))
= (1,'A') : ((2,'*') : ((3,'q') : []))
= (1,'A') : ((2,'*') : [(3,'q')])
= (1,'A') : [(2,'*'), (3,'q')]
= [(1,'A'), (2,'*'), (3,'q')]
```

The `concat` function takes a list of lists and flattens it into a list of elements. For example, `concat [[1], [2,3], [4,5,6]]` returns one list consisting of all the elements in the argument lists; thus the result is `[1,2,3,4,5,6]`.

```
concat :: [[a]] -> [a]
concat [] = []
concat (xs:xss) = xs ++ concat xss
```

In working out the example calculation, we just simplify all the applications of `++` directly. Of course each of those applications entails another recursion, which is similar to the examples of `(++)` given above.

```
concat [[1], [2,3], [4,5,6]]
= [1] ++ concat [[2,3], [4,5,6]]
= [1] ++ ([2,3] ++ concat [[4,5,6]])
= [1] ++ ([2,3] ++ [4,5,6])
= [1] ++ [2,3,4,5,6]
= [1,2,3,4,5,6]
```

The base case for a function that builds a list must return a list, and this
is often simply []. The recursive case builds a list by attaching a value onto
the result returned by the recursive call.

In defining a recursive function f, it is important for the recursion to work
on a list that is shorter than the original argument to f. For example, in the
application sum [1,2,3], the recursion calculates sum [2,3], whose argument
is one element shorter than the original argument [1,2,3]. If a function were
defined incorrectly, with a recursion that is bigger than the original problem,
then it could just go into an infinite loop.

All of the examples we have seen so far perform a recursion on a list that is
one element shorter than the argument. However, provided that the recursion
is solving a smaller problem than the original one, the recursive case can be
anything—it doesn't necessarily have to work on the tail of the original list.
Often a good approach is to try to cut the problem size in half, rather than
reducing it by one. This organisation often leads to highly efficient algorithms,
so it appears frequently in books on the design and analysis of algorithms. We
will look at the quicksort function, a good example of this technique.

Quicksort is a fast recursive sorting algorithm. The base case is simple:
quicksort [] = []. For a non-empty list of the form (x:xs), we will first
pick one element called the *splitter*. For convenience that will be x, the first
element in the list. We will then take all the elements of the rest of the list,
xs, which are *less than or equal to* the splitter. We will call this list the *small
elements* and define it as a list comprehension [y | y <- xs, y<splitter].
In a similar way we define the list of *large elements*, which are greater than
the splitter, as [y | y <- xs, y>=splitter]. Now the complete sorted list
consists first of the small elements (in sorted order), followed by the splitter,
followed by the large elements (in sorted order).

```
quicksort :: Ord a => [a] -> [a]
quicksort [] = []
quicksort (splitter:xs) =
  quicksort [y | y <- xs, y<splitter]
  ++ [splitter]
  ++ quicksort [y | y <- xs, y>=splitter]
```

It is interesting to compare this definition with a conventional one in an
imperative language; see any standard book on algorithms.

Exercise 1. Write a recursive function copy :: [a] -> [a] that copies its
   list argument. For example, copy [2] ⇒[2].

Exercise 2. Write a function `inverse` that takes a list of pairs and swaps the pair elements. For example,

```
inverse [(1,2),(3,4)] ==> [(2,1),(4,3)]
```

Exercise 3. Write a function

```
merge :: Ord a => [a] -> [a] -> [a]
```

which takes two sorted lists and returns a sorted list containing the elements of each.

Exercise 4. Write (`!!`), a function that takes a natural number n and a list and selects the nth element of the list. List elements are indexed from 0, not 1, and since the type of the incoming number does not prevent it from being out of range, the result should be a `Maybe` type. For example,

```
[1,2,3]!!0 ==> Just 1
[1,2,3]!!2 ==> Just 3
[1,2,3]!!5 ==> Nothing
```

Exercise 5. Write a function `lookup` that takes a value and a list of pairs, and returns the second element of the pair that has the value as its first element. Use a Maybe type to indicate whether the lookup succeeded. For example,

```
lookup 5 [(1,2),(5,3)] ==> Just 3
lookup 6 [(1,2),(5,3)] ==> Nothing
```

Exercise 6. Write a function that counts the number of times an element appears in a list.

Exercise 7. Write a function that takes a value e and a list of values xs and removes all occurrences of e from xs.

Exercise 8. Write a function

```
f :: [a] -> [a]
```

that removes alternating elements of its list argument, starting with the first one. For examples, f [1,2,3,4,5,6,7] returns [2,4,6].

Exercise 9. Write a function `extract :: [Maybe a] -> [a]` that takes a list of `Maybe` values and returns the elements they contain. For example, `extract [Just 3, Nothing, Just 7] = [3, 7]`.

Exercise 10. Write a function

```
f :: String -> String -> Maybe Int
```

that takes two strings. If the second string appears within the first, it returns the index identifying where it starts. Indexes start from 0. For example,

```
f "abcde" "bc" ==> Just 1
f "abcde" "fg" ==> Nothing
```

3.2   Higher Order Recursive Functions

Many of the recursive definitions from the previous section are quite similar to
each other. An elegant idea is to write a function that expresses the general
computation pattern once and for all. This general function could then be
reused to define a large number of more specific functions, without needing
to write out the complete recursive definitions. In order to allow the general
function to know exactly what computation to perform, we need to supply it
with an extra argument which is itself a function. Such a general function is
called a *higher order function* or a *combinator*.

Several of the functions defined in the previous section take a list argument,
and return a list result, where each element of the result is computed from the
corresponding element of the input. We can write a general function, called
map, which expresses all particular functions of that form. The first argument
to map is another function, of type a->b, which takes a single value of type
a and returns a single result of type b. The purpose of map is to apply that
auxiliary function to all the elements of the list argument.

```
map :: (a->b) -> [a] -> [b]
map f []     = []
map f (x:xs) = f x : map f xs
```

Here is an example of the use of map. Suppose the function we define is
to multiply every list element by 5. We could write a special function to do
just that, using recursion. A simpler method is to specify the 'multiply by 5'
function (*5) as an argument to map. Here is an example:

```
map (*5) [1,2,3]
= 1*5 : map (*5) [2,3]
= 1*5 : (2*5 : map (*5) [3])
= 1*5 : (2*5 : (3*5 : map (*5) []))
= 1*5 : (2*5 : (3*5 : []))
= 5 : (10 : (15 : []))
= [5,10,15]
```

The map function takes an auxiliary function requiring one argument, and
one list argument. Sometimes we have an auxiliary function that takes *two*
arguments, and we want to apply it to all the corresponding elements of two
lists. If the argument lists are of different sizes, the result will have the length
of the shorter one, This is performed by the zipWith function, which performs
a recursion simultaneously over both list arguments:

```
zipWith :: (a->b->c) -> [a] -> [b] -> [c]
zipWith f [] ys = []
zipWith f xs [] = []
zipWith f (x:xs) (y:ys) = f x y : zipWith f xs ys
```

Some of the recursive functions from the previous section, such as `length` and `sum`, have a structure that is slightly different from that of `map`. These functions take a list but return a singleton result which is calculated by combining elements from the list. A general function for this is `foldr`, which is defined as follows:

```
foldr :: (a->b->b) -> b -> [a] -> b
foldr f z [] = z
foldr f z (x:xs) = f x (foldr f z xs)
```

Here is an example where `foldr` is used to add up the elements of a list. The singleton argument `z` can be thought of as an initial value for the sum, so we specify 0 for its value.

```
foldr (+) 0 [1,2,3]
= 1 + foldr (+) 0 [2,3]
= 1 + (2 + foldr (+) 0 [3])
= 1 + (2 + (3 + foldr (+) 0 []))
= 1 + (2 + (3 + 0))
= 6
```

The function's name suggests that a list is folded into a singleton value. The recursion produces a sequence of applications of the `f` function, starting from the right end of the list (hence the name, which stands for *fold from the right*). The singleton argument `z` serves to initialise the accumulator, and it also provides a result in case the entire list argument is empty.

Now we can define all the functions that follow this general pattern with a single equation; there is no need to keep writing out separate recursive definitions. Here is a collection of useful functions, all definable with `foldr`:

```
sum xs        = foldr (+) 0 xs
product xs    = foldr (*) 1 xs
and xs        = foldr (&&) True xs
or xs         = foldr (||) False xs
factorial n   = foldr (*) 1 [1..n]
```

We now have two definitions of sum, a recursive one and one using `foldr`. These two definitions ought to produce the same result, and it is interesting to consider in more detail how this comes about. The recursive definition, which we saw in the previous section, is

```
sum [] = 0
sum (x:xs) = x + sum xs
```

The definition using `foldr` is

```
sum xs        = foldr (+) 0 xs
```

First, consider `sum []`. The recursive definition says directly that this has the value 0. We can calculate what the `foldr` definition says:

```
sum []
= foldr (+) 0 []
= 0
```

The two definitions give the same result for the base case, as they should. Now consider the recursive case, where the argument has the structure `(x:xs)`. The recursive definition returns `x + sum xs`. The result produced by the definition with `foldr` is calculated:

```
sum (x:xs)
= foldr (+) 0 (x:xs)
= x + foldr (+) 0 xs
= x + sum xs
```

We have just proved that the two definitions of `sum` always produce the same result, because *every* list must either be `[]` or have the form `(x:xs)`.

Often there is a choice between using a recursive definition or using one of the existing higher order functions like `map`, `foldr`, etc. For example, consider the problem of writing a function `firsts`, which should take a list of pairs and returns a list of the first elements. For example,

```
firsts [(1,3),(2,4)] ==> [1,2]
```

Here is a recursive definition:

```
firsts :: [(a,b)] -> [a]
firsts [] = []
firsts ((a,b):ps) = a : firsts ps
```

And here is an alternative definition using `map`:

```
firsts :: [(a,b)] -> [a]
firsts xs = map fst xs
```

The definition using `map` is generally considered better style. It is shorter and more readable, but its real advantage is modularity: the definition of `map` expresses a specific form of recursion, and the definition of `firsts` with `map` makes it explicitly clear that this form of recursion is being used.

Exercise 11. Write `foldrWith`, a function that behaves like `foldr` except that it takes a function of three arguments and two lists.

Exercise 12. Using `foldr`, write a function `mappend` such that

```
mappend f xs = concat (map f xs)
```

Exercise 13. Write `removeDuplicates`, a function that takes a list and removes all of its duplicate elements.

Exercise 14. Write a recursive function that takes a value and a list of values and returns True if the value is in the list and False otherwise.

3.3 Peano Arithmetic

We turn now to the implementation of arithmetic operations over a simple data structure representing the natural numbers. The reason for working with this representation is that it provides a good introduction to recursion over general algebraic data types.

```
data Peano = Zero | Succ Peano deriving Show
```

For example,

```
1 = Succ Zero
3 = Succ (Succ (Succ Zero))
```

As you can see, the Peano data type is recursive. In this case, the recursion builds up a series of constructor applications, somewhat like the list data type:

```
data List a = Empty | Cons a (List a)

[1]     = Cons 1 Empty
[1,2,3] = Cons 1 (Cons 2 (Cons 3 Empty))
```

The simplest Peano function is decrement, which removes a Succ constructor if possible, returning Zero otherwise:

```
decrement :: Peano -> Peano
decrement Zero     = Zero
decrement (Succ a) = a
```

The definition of add is

```
add :: Peano -> Peano -> Peano
add Zero      b = b
add (Succ a) b = Succ (add a b)
```

This definition looks a lot like that of (++)!

However, the definition of subtraction is more complex. If the second argument is Zero, then sub returns the first argument. Negative numbers cannot be represented in this scheme, so they are approximated by Zero. Otherwise sub must decrement both numbers, as follows:

```
sub :: Peano -> Peano -> Peano
sub a         Zero     = a
sub Zero      b        = Zero
sub (Succ a) (Succ b) = sub a b
```

We can do more with recursion and Peano numbers. Here are two predicates, equals and lt:

```
equals :: Peano -> Peano -> Bool
equals Zero     Zero        = True
equals Zero     b           = False
equals a        Zero        = False
equals (Succ a) (Succ b)    = equals a b

lt :: Peano -> Peano -> Bool
lt a        Zero        = False
lt Zero     (Succ b)    = True
lt (Succ a) (Succ b)    = lt a b
```

3.4   Data Recursion

So far we have used recursion as a technique for defining functions. Recursive functions are useful in nearly all programming languages, and they are especially important for programming with data structures like trees and graphs. Another important programming technique uses recursion to define data structures; this is called *data recursion*.

The idea is to define *circular* data structures. Here is one way to define an infinitely long list where every element is 1:

```
f :: a -> [a]
f x = x : f x
ones = f 1
```

Each time the recursive function f is applied, it uses (:) to construct a new list element containing x. As a result, there is no bound on how much computer memory will be required to represent ones; this depends on how much of the computation is actually demanded by the rest of the program.

However, it's possible to represent ones very compactly with a circular list, defined with recursion in the data rather than in a function:

```
twos = 2 : twos
```

Figure 3.1 shows the internal representations of ones and twos. Mathematically, the *value* of both is a list that is infinitely long. However, the circular definition requires far less memory.

A data structure consisting of nodes connected by links is called a *graph*. Graphs can be constructed by defining each node with an equation in a let expression; this means that each node can be referred to by any other node (including even itself). For example, the following Haskell definition creates a circular data structure, where the internal names a, b, and c are used to set up the links:

```
object = let a = 1:b
```

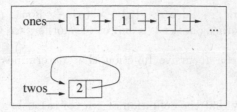

Figure 3.1: Circular and Non-circular Infinite Lists

```
        b = 2:c
        c = [3] ++ a
   in a
```

Data recursion does not work in all programming languages. If you were to try defining `twos` in C, for example, the program would go into an infinite loop. Haskell, however, is careful to delay the evaluation of an expression until it is actually needed. This technique is called *lazy evaluation*, and it makes possible a number of useful programming techniques including data recursion. Lazy evaluation also plays a role in making it easier to use mathematics to reason formally about Haskell programs.

3.5 Suggestions for Further Reading

Most programming languages have a variety of control constructs for looping, but recursion is general enough to encompass all of them. Textbooks on Haskell give many examples, and Abelson and Sussman [2] use recursion to provide a variety of control abstractions in the programming language Scheme.

Systems with self-reference, including especially recursion, are one of the major themes of Hofstadter's book *Gödel, Escher, Bach* [17].

Recursion is central to the theory of computation; one of the main theoretical subjects in computer science is called *recursive function theory*. This is one of the standard topics in computer science, and many textbooks are available.

3.6 Review Exercises

Exercise 15. Write a function that takes two lists, and returns a list of values that appear in both lists. The function should have type `intersection :: Eq a => [a] -> [a] -> [a]`. (This is one way to implement the intersection operation on sets; see Chapter 8.)

Exercise 16. Write a function that takes two lists, and returns `True` if all the elements of the first list also occur in the other. The function should have type isSubset ::   Eq a => [a] -> [a] -> Bool. (This is one way to determine whether one set is a subset of another; see Chapter 8.)

Exercise 17. Write a recursive function that determines whether a list is sorted.

Exercise 18. Show that the definition of `factorial` using `foldr` always produces the same result as the recursive definition given in the previous section.

Exercise 19. Using recursion, define `last`, a function that takes a list and returns a `Maybe` type that is `Nothing` if the list is empty.

Exercise 20. Using recursion, write two functions that expect a string containing a number that contains a decimal point (for example, `23.455`). The first function returns the whole part of the number (i.e., the part to the left of the decimal point). The second function returns the fractional part (the part to the right of the decimal point).

Chapter 4

Induction

A common type of problem is to prove that an object x has some property P. The mathematical notation for this is $P(x)$, where P stands for predicate (or property). For example, if x is 6 and $P(x)$ is the predicate 'x is an even number', then we could express the statement '6 is an even number' with the shorthand mathematical statement $P(6)$.

Many computer science applications require us to prove that *all* the elements of a set S have a certain property P. The statement 'for any element x in the set S the predicate P holds for x' can be written as

$$\forall x \in S . P(x).$$

Statements like this can be used to assert properties of data: for example, we could state that every item in a database has a certain property. They can also be used to describe the behaviour of computer programs. For example, $P(n)$ might be a statement that a loop computes the correct result if it terminates after n iterations, and the statement $\forall n \in N . P(n)$ says that the loop is correct if it terminates after *any* number of iterations. N denotes the set of natural numbers, so $n \in N$ just says that n is a natural number, and the complete expression $\forall n \in N . P(n)$ says that for any natural number n, the predicate $P(n)$ holds.

One approach to proving an assertion about all the elements of a set is to write out a separate direct proof for each element of the set. That would be all right if the set were small. For example, to prove that all the elements of the set $\{4, 6\}$ are even, you could just prove that 4 is even and also that 6 is even. However, this direct approach quickly becomes tedious for large sets: to prove that all the elements of $\{2, 4, 6, \ldots, 1000\}$ are even, you would need 500 separate proofs! Even worse, the brute-force method doesn't work at all—even in principle—for infinite sets, because proofs are always required to be finitely long.

Induction is a powerful method for proving that every element of a set has a certain property. Induction is a valuable technique, because it is not tedious even if the set is large, and it works even for countably infinite sets. Induction is used frequently in computer science applications. This chapter shows you how inductive proofs work and gives many examples, including both mathematical and computing applications. We will use two forms of induction: *mathematical induction* for proving properties about natural numbers, and *structural induction* for proving properties about lists.

4.1   The Principle of Mathematical Induction

Induction is used to prove that a property $P(x)$ holds for every element of a set S. In this section, we will restrict the set S to be the set N of natural numbers; later we will generalise induction to handle more general sets. Instead of proving $P(x)$ separately for every x, induction relies on a systematic procedure: you just have to prove a few facts about the property P, and then a theorem called the Principle of Mathematical Induction allows you to conclude $\forall x \in N . P(x)$.

The basic idea is to define a systematic procedure for proving that P holds for an arbitrary element. To do this, we must prove two statements:

1. The *base case* $P(0)$ says that the property P holds for the base element 0.

2. The *inductive case* $P(i) \to P(i+1)$ says that if P holds for an arbitrary element i of the set, then it must also hold for the successor element $i+1$. The symbol $\to$ is read as *implies*.

Because every element of the set of natural numbers can be reached by starting with 0 and repeatedly adding 1, you can establish that P holds for any particular element using a finite sequence of steps. Given that $P(0)$ holds (the base case) and that $P(0) \to P(1)$ (an instance of the inductive case), we can conclude that $P(1)$ also holds. This is a simple example of *logical inference*, which will be studied in more detail in the chapters on logic. The same procedure can be used sytematically to prove $P(i)$ for *any* element of the natural numbers. For example, $P(4)$ can be proved using the following steps (where the $\wedge$ symbol denotes *and*):

Conclusion	Justification		
$P(0)$	the base case		
$P(1)$	$P(0) \to P(1)$	$\wedge$	$P(0)$
$P(2)$	$P(1) \to P(2)$	$\wedge$	$P(1)$
$P(3)$	$P(2) \to P(3)$	$\wedge$	$P(2)$
$P(4)$	$P(3) \to P(4)$	$\wedge$	$P(3)$

Given any element k of N, you can prove $P(k)$ using this strategy, and the proof will be $k+1$ lines long. We have not yet used the Principle of Mathematical Induction; we have just used ordinary logical reasoning. However, proving $P(k)$ for an arbitrary k is not the same as proving $\forall k \in N . P(k)$, because there are an infinite number of values of k, so the proof would be infinitely long. The size of a proof must always be finite.

What the Principle of Mathematical Induction allows us to conclude is that P holds for *all* the elements of N, even if there is an infinite number of them. Thus it introduces something new—it isn't just a macro that expands out into a long proof.

Theorem 5 (Principle of Mathematical Induction). Let $P(x)$ be a predicate that is either true or false for every element of the set N of natural numbers. If $P(0)$ is true and $P(n) \to P(n+1)$ holds for all $n \geq 0$, then $\forall x \in N \, . \, P(x)$ is true.

This theorem can be proved using axiomatic set theory, which is beyond the scope of this book. For most applications in computer science, it is sufficient to have an intuitive understanding of what the theorem says and to have a working understanding of how to use it in proving other theorems.

The proof of the base case will usually turn out to be a straightforward calculation.

The expression $P(n) \to P(n+1)$ means that 'if $P(n)$ is true then so is $P(n+1)$'. We can establish this by temporarily assuming $P(n)$, and then— in the context of this assumption—proving $P(n+1)$. The assumption $P(n)$ is called the *induction hypothesis*. It is important to understand that *assuming the induction hypothesis does not actually mean that it is true!* All that matters is that *if* the assumption $P(n)$ enables us to prove $P(n+1)$, *then* the implication $P(n) \to P(n+1)$ holds. We will study logical inference in more detail in Chapter 6.

4.2 Examples of Induction on Natural Numbers

There is a traditional story about Gauss, one of the greatest mathematicians in history. In school one day the teacher told the class to work out the sum $1 + 2 + \cdots + 100$. After a short time thinking, Gauss gave the correct answer— 5050—long before it would have been possible to work out all the additions. He had noticed that the sum can be arranged into pairs of numbers, like this:

$$(1 + 100) + (2 + 99) + (3 + 98) + \cdots + (50 + 51)$$

The total of each pair is 101, and there are 50 of the pairs, so the result is $50 \times 101 = 5050$.

Methods like this can often be used to save time in computing, so it is worthwhile to find a solution to the general case of this problem, which is the sum

$$\sum_{i=1}^{n} i \; = \; 1 + 2 + \cdots + n.$$

Figure 4.1: Geometric Interpretation of Sum

If n is even, then we get $\frac{n}{2}$ pairs that all total to $n{+}1$, so the result is $\frac{n}{2}\times(n{+}1) = \frac{n\times(n+1)}{2}$. If n is odd, we can start from 0 instead of 1; for example,

$$\sum_{i=0}^{7} i = (0+7)+(1+6)+(2+5)+(3+4).$$

In this case there are $\frac{n+1}{2}$ pairs each of which totals to n, so the result is again $\frac{n\times(n+1)}{2}$. For any natural number n, we end up with the result

$$\sum_{i=0}^{n} i = \frac{n\times(n+1)}{2}.$$

This formula is useful because it reduces the computation time required. For example, if n is 1000 then it would take 999 additions to work out the summation by brute force, but the formula always requires just one addition, one multiplication, and one division.

Figure 4.1 shows another way to understand the formula. The rectangle is covered half by dots and half by stars. The number of stars is $1+2+3+4$, and the area of the rectangle is $4\times(4+1)$, so the total number of stars is $\frac{4\times(4+1)}{2}$.

So far we have only guessed the general formula for $\sum_{i=0}^{n} i$, and we have considered two ways of understanding it. The next step is to *prove* that this formula always gives the right answer, and induction provides a way to do it.

Theorem 6.
$$\forall n \in N \;.\; \sum_{i=0}^{n} i = \frac{n\times(n+1)}{2}$$

Proof. Define the property P as follows:

$$P(n) = \left(\sum_{i=0}^{n} i = \frac{n\times(n+1)}{2} \right)$$

Thus the aim is to prove that

$$\forall n \in N \;.\; P(n)$$

and we proceed by induction on n.

Base case. We need to prove $P(0)$.

$$\sum_{i=0}^{0} i = 0$$

$$= \frac{0 \times (0+1)}{2}$$

Thus we have established the property $P(0)$.

Induction case. The aim is to prove that $P(n) \to P(n+1)$, and we will prove this implication by assuming $P(n)$ (this is called the *induction hypothesis*) and then using that assumption to prove $P(n+1)$. We start by writing out in full detail the assumption and the aim of the proof. Assume for an arbitrary natural number n that

$$\sum_{i=0}^{n} i = \frac{n \times (n+1)}{2}.$$

The aim is to show (for this particular value of n) that

$$\sum_{i=0}^{n+1} i = \frac{(n+1) \times (n+2)}{2}.$$

We do this by starting with the left-hand side of the equation and using algebra to transform it into the right-hand side. The first step uses the assumption given above. This is called the *induction hypothesis.*

$$\sum_{i=0}^{n+1} i = \sum_{i=0}^{n} i + (n+1)$$

$$= \frac{n \times (n+1)}{2} + (n+1)$$

$$= \frac{n \times (n+1)}{2} + \frac{2 \times (n+1)}{2}$$

$$= \frac{n \times (n+1) + 2 \times (n+1)}{2}$$

$$= \frac{(n+1) \times (n+2)}{2}$$

Now we have established that

$$\left(\sum_{i=0}^{n} i = \frac{n \times (n+1)}{2} \right) \to \left(\sum_{i=0}^{n+1} i = \frac{(n+1) \times (n+2)}{2} \right).$$

That is,

$$P(n) \to P(n+1).$$

Use the principle of induction. To summarise, we have proved the base case $P(0)$, and we have proved the induction case $P(n) \to P(n+1)$. Therefore the principle of induction allows us to conclude that the theorem $\forall n \in N.P(n)$ is true. □

Exercise 1. Let a be an arbitrary real number. Prove, for all natural numbers m and n, that $a^{m \times n} = (a^m)^n$.

Exercise 2. Prove that the sum of the first n odd positive numbers is n^2.

Exercise 3. Prove that $\sum_{i=1}^{n} a^i = (a^{n+1} - 1)/(a - 1)$, where a is a real number and $a \neq 1$.

4.3   Induction and Recursion

Induction is a common method for proving properties of recursively defined functions. The *factorial* function, which is defined recursively, provides a good example of an induction proof.

```
factorial :: Natural → Natural
factorial 0 = 1
factorial (n + 1) = (n + 1) * factorial n
```

A common problem in computer science is to prove that a program computes the correct answer. Such a theorem requires an abstract mathematical specification of the problem, and it has to show that for all inputs, the program produces the same result that is defined by the specification. The value of $n!$ (factorial of n) is defined as the product of all the natural numbers from 1 through n; the standard notation for this is $\prod_{i=1}^{n} i$. The following theorem says that the *factorial* function as defined above actually computes the value of $n!$.

Theorem 7. For all natural numbers n, *factorial* $n = \prod_{i=1}^{n} i$.

Proof. Induction on n. The base case is

$$factorial\ 0$$
$$= 1 \qquad\qquad\qquad\qquad\qquad\qquad\qquad \{\ factorial.1\ \}$$
$$= \textstyle\prod_1^0 i \qquad\qquad\qquad\qquad\qquad\qquad \{\ \text{def. of } \prod\ \}$$

For the induction case, it is necessary to prove that

$$\left(factorial\ n = \prod_{i=1}^{n} i \right) \to \left(factorial\ (n+1) = \prod_{i=1}^{n+1} i \right).$$

This is done by assuming the left side as the inductive hypothesis: for a particular n, the hypothesis is *factorial* $n = \prod_{i=1}^{n} i$. Given this assumption, we must prove that *factorial* $(n+1) = \prod_{i=1}^{n+1} i$.

$factorial\ (n+1)$
$= (n+1) \times factorial\ n$ { *factorial*.2 }
$= (n+1) \times \prod_{i=1}^{n} i$ { hypothesis }
$= \prod_{i=1}^{n+1} i$ { def. $\prod$ }

□

Exercise 4. (This problem is from [12], where you can find many more.) The nth Fibonacci number is defined as follows:

```
fib :: Integer -> Integer
fib 0 = 0
fib 1 = 1
fib (n+2) = fib n + fib (n+1)
```

The first few numbers in this famous sequence are $0, 1, 1, 2, 3, 5, \ldots$. Prove the following:

$$\sum_{i=1}^{n} fib\ i = fib\ (n+2) - 1$$

4.4 Induction on Peano Naturals

The Peano representation of natural numbers is a rich source of examples for induction. Actually, it's hard to do anything at all in Peano arithmetic without induction. For example, how do we even know that a natural number is equal to itself? The following theorem says so, and its proof requires induction.

Theorem 8 (Self equality). $\forall x :: Nat.\ equals\ x\ x = True$

Proof. Base case:

equals Zero Zero
$= True$ { *equals*.1 }

For the inductive case, assume that *equals* $x\ x = True$. We must prove that *equals* $(Succ\ x)\ (Succ\ x) = True$.

equals $(Succ\ x)\ (Succ\ x)$
$= equals\ x\ x$ { *equals*.2 }
$= True$ { hypothesis }

□

This proof illustrates a subtle issue. When we are using the languages of English and mathematics to talk *about* natural numbers, we can assume that anything is equal to itself. We don't normally prove theorems like $x = x$. However, that is not what we have just proved: the theorem above says that x is the same as itself according to *equals*, which is defined *inside* the Peano system. Therefore the proof also needs to work inside the Peano system; hence the induction.

We'll look at a few more typical Peano arithmetic theorems, both to see how the Peano natural numbers work and to get more practice with induction. The following theorem says, in effect, that $(x + y) - x = y$. In elementary algebra, we would prove this by calculating $(x + y) - x = (x - x) + y = 0 + y = y$. The point here, however, is that the addition and subtraction are being performed by the recursive *add* and *sub* functions, and we need to prove the theorem in terms of these functions.

Theorem 9. *sub* (*add x y*) $x = y$

Proof. Induction over x. The base case is

    sub (*add Zero y*) *Zero*
      = *sub y Zero*   { *add*.1 }
      = y { *sub*.1 }

For the inductive case, assume *sub* (*add x y*) $x = y$; the aim is to prove *sub* (*add* (*Succ x*) *y*) (*Succ x*) $= y$.

    sub (*add* (*Succ x*) *y*) (*Succ x*)
      = *sub* (*Succ* (*add x y*)) (*Succ x*)   { *add*.2 }
      = *sub* (*add x y*) x { *sub*.3 }
      = y { hypothesis }

$\square$

The proof above happens to go through directly and easily, but many simple theorems do not. For example, it is considerably harder to prove $(x+y)-y = x$ than to prove $(x + y) - x = y$. Even worse, there is no end to such theorems. Instead of continuing to choose theorems based on their ease of proof, it is better to proceed systematically by developing the standard properties of natural numbers, such as associativity and commutativity. The attractive feature of the Peano definitions is that all these laws are theorems; the only definitions we need are the basic ones given already. It is straightforward to prove that addition is associative, so we begin with that property.

Theorem 10 (add is associative). *add x* (*add y z*) $=$ *add* (*add x y*) *z*

Proof. Induction over x. The Base case is

$add\ Zero\ (add\ y\ z)$
 $=\ add\ y\ z$ { $add.1$ }
 $=\ add\ (add\ Zero\ y)\ z$ { $add.1$ }

Inductive case. Assume $add\ x\ (add\ y\ z)\ =\ add\ (add\ x\ y)\ z$. Then

$add\ (Succ\ x)\ (add\ y\ z)$
 $=\ Succ\ (add\ x\ (add\ y\ z))$ { $add.2$ }
 $=\ Succ\ (add\ (add\ x\ y)\ z)$ { hypothesis }
 $=\ add\ (Succ\ (add\ x\ y))\ z$ { $add.2$ }
 $=\ add\ (add\ (Succ\ x)\ y)\ z$ { $add.2$ }

□

Next, it would be good to prove that addition is commutative: $x + y = y + x$. To prove this, however, a sequence of simpler theorems is needed—each providing yet another example of induction.

First we need to be able to simplify additions where the *second* argument is Zero. We know already that $add\ Zero\ x\ =\ x$; in fact, this is one of the Peano axioms. It is *not* an axiom that $add\ x\ Zero\ =\ x$; that is a theorem requiring proof.

Theorem 11. $add\ x\ Zero\ =\ x$

Proof. Induction over x. The base case is

$add\ Zero\ Zero$
 $=\ Zero$ { $add.1$ }

For the inductive case, we assume $add\ x\ Zero\ =\ x$. Then

$add\ (Succ\ x)\ Zero$
 $=\ Succ\ (add\ x\ Zero)$ { $add.2$ }
 $=\ Succ\ x$ { hypothesis }

□

The next theorem allows us to move a *Succ* from one argument of an addition to the other. It says, in effect, that $(x + 1) + y = x + (y + 1)$. That may not sound very dramatic, but many proofs require the ability to take a little off one argument and add it onto the other.

Theorem 12. $add\ (Succ\ x)\ y\ =\ add\ x\ (Succ\ y)$

Proof. Induction over x. Base case:

$add\ (Succ\ Zero)\ y$
 $=\ Succ\ (add\ Zero\ y)$ { $add.2$ }
 $=\ Succ\ y$ { $add.1$ }
 $=\ add\ Zero\ (Succ\ y)$ { $add.1$ }

Inductive case:

$add\ (Succ\ (Succ\ x))\ y$
$= Succ\ (add\ (Succ\ x)\ y)$ { $add.2$ }
$= Succ\ (add\ x\ (Succ\ y))$ { hypothesis }
$= add\ (Succ\ x)\ (Succ\ y)$ { $add.2$ }

□

Now we can prove that Peano addition is commutative.

Theorem 13 (add is commutative). $add\ x\ y\ =\ add\ y\ x$

Proof. Induction over x. Base case:

$add\ Zero\ y$
$= y$ { $add.1$ }
$= add\ y\ Zero$ { Theorem 11 }

Inductive case: assume that $add\ x\ y\ =\ add\ y\ x$. Then

$add\ (Succ\ x)\ y$
$= Succ\ (add\ x\ y)$ { $add.2$ }
$= Succ\ (add\ y\ x)$ { hypothesis }
$= add\ (Succ\ y)\ x$ { $add.2$ }
$= add\ y\ (Succ\ x)$ { Theorem 12 }

□

4.5   Induction on Lists

Lists are one of the most commonly used data structures in computing, and there is a large family of functions to manipulate them. These functions are typically defined recursively, with a base case for empty lists [] and a recursive case for non-empty lists, i.e. lists of the form $(x\ :\ xs)$. List induction is the most common method for proving properties of such functions.

Before going on, we discuss some practical techniques that help in coping with theorems. If you aren't familiar with them, mathematical statements sometimes look confusing, and it is easy to develop a bad habit of skipping over the mathematics when reading a book. Here is some advice on better approaches; we will try to illustrate the advice in a concrete way in this section, but these methods will pay off throughout your work in computer science, not just in the section you're reading right now.

- When you are faced with a new theorem, try to understand what it means before trying to prove it. Restate the main idea in English.

- Think about what applications the theorem might have. If it says that two expressions are the same, can you think of situations where there might be a practical advantage in replacing the left-hand side by the right-hand side, or vice versa? (A common situation is that one side of the equation is more natural to write, and the other side is more efficient.)

- Try out the theorem on some small examples. Theorems are often stated as equations; make up some suitable input data and evaluate both sides of the equation to see that they are the same.

- Check what happens in boundary cases. If the equation says something about lists, what happens if the list is empty? What happens if the list is infinite?

- Does the theorem seem related to other ones? Small theorems about functions—the kind that are usually proved by induction—tend to fit together in families. Noticing these relationships helps in understanding, remembering, and applying the results.

The principle of list induction states a technique for proving properties about lists. It is similar to the principle of mathematical induction; the main difference is that the base case is the empty list (rather than 0) and the induction case uses a list with one additional element (x:xs) rather than $n + 1$. List induction is a special case of a more general technique called *strutural induction*. Induction over lists is used to prove that a proposition $P(xs)$ holds for every list xs.

Theorem 14 (Principle of list induction). Suppose $P(xs)$ is a predicate on lists of type $[a]$, for some type a. Suppose that $P([])$ is true (this is the base case). Further, suppose that if $P(xs)$ holds for arbitrary $xs :: [a]$, then $P(x : xs)$ also holds for arbitrary $x :: a$. Then $P(xs)$ holds for every list xs that has finite length.

Thus the base case is to prove that the predicate holds for the empty list, and the inductive case is to prove that *if* P holds for a list xs, then it must *also* hold for any list of the form $x : xs$. When the base and inductive case are established, then the principle of induction allows us to conclude that the predicate holds for all finite lists.

Notice that the principle of list induction cannot be used to prove theorems about infinite lists. This point is discussed in Section 4.8.

We will now work through a series of examples where induction is used to prove theorems about the properties of recursive functions over lists.

The *sum* function takes a list of numbers and adds them up. It defines the sum of an empty list to be 0, and the sum of a non-empty list is computed by adding the first number onto the sum of all the rest.

```
sum :: Num a => [a] -> a
```

```
sum [] = 0
sum (x:xs) = x + sum xs
```

The following theorem states a useful fact about the relationship between two functions: *sum* and ++. It says that if you have two lists, say *xs* and *ys*, then there are two different ways to compute the combined total of all the elements. You can either append the lists together with ++, and then apply *sum*, or you can apply *sum* independently to the two lists, and add the two resulting numbers.

Theorems of this sort are often useful for transforming programs to make them more efficient. For example, suppose that you have a very long list to sum up, and two computers are available. We could use parallelism to cut the execution time almost in half. The idea is to split up the long list into two shorter ones, which can be summed in parallel. One quick addition will then suffice to get the final result. This technique is an example of the *divide and conquer* strategy for improving the efficiency of algorithms. Obviously there is more to parallel computing and program optimisation than we have covered in this paragraph, but theorems like the one we are considering really do have practical applications.

Theorem 15. $sum\ (xs{+}{+}ys)\ =\ sum\ xs + sum\ ys$

The proof of this theorem is a typical induction over lists, and it provides a good model to follow for future problems. The justifications used in the proof steps are written in braces $\{\ldots\}$. Many of the justifications cite the definition of a function, along with the number of the equation in the definition; thus $\{\ (+\!+).1\ \}$ means 'this step is justified by the first equation in the definition of ++'. The most crucial step in an induction proof is the one where the induction hypothesis is used; the justification cited for that step is $\{$ hypothesis $\}$.

Proof. Induction over *xs*. The base case is

$$\begin{aligned}
sum\ ([\,]\ {+}{+}\ ys) &\\
= sum\ ys & \qquad \{\ (+\!+).1\ \}\\
= 0 + sum\ ys & \qquad \{\ 0 + x = x\ \}\\
= sum\ [\,] + sum\ ys & \qquad \{\ sum.1\ \}
\end{aligned}$$

The inductive case is

$$\begin{aligned}
sum\ ((x : xs){+}{+}ys) &\\
= sum\ (x : (xs{+}{+}ys)) & \qquad \{\ (+\!+).2\ \}\\
= x + sum\ (xs{+}{+}ys) & \qquad \{\ sum.2\ \}\\
= x + (sum\ xs + sum\ ys) & \qquad \{\ \text{hypothesis}\ \}\\
= (x + sum\ xs) + sum\ ys & \qquad \{\ +.\ \text{is associative}\ \}\\
= sum\ (x : xs) + sum\ ys & \qquad \{\ sum.2\ \}
\end{aligned}$$

$\square$

Many theorems describe a relationship between two functions; the previous one is about the combination of *sum* and (++), while this one combines *length* with (++). Its proof is left as an exercise.

Theorem 16. *length* $(xs+\!\!+ys)$ = *length xs* + *length ys*

We now consider several theorems that state crucial properties of the *map* function. These theorems are important in their own right, and they are commonly used in program transformation and compiler implementation. They are also frequently used to justify steps in proofs of more complex theorems.

The following theorem says that *map* is 'length preserving': when you apply *map* to a list, the result has the same length as the input list.

Theorem 17. *length* (*map f xs*) = *length xs*

Proof. Induction over *xs*. The base case is

$length\ (map\ f\ [])$
$= length\ []$ { *map*.1 }

For the inductive case, assume *length* (*map f xs*) = *length xs*. Then

$length\ (map\ f\ (x:xs))$
$= length\ (f\ x\ :\ map\ f\ xs)$ { *map*.2 }
$= 1 + length\ (map\ f\ xs)$ { *length*.2 }
$= 1 + length\ xs$ { hypothesis }
$= length\ (x:xs)$ { *length*.2 }

□

The next theorem is reminiscent of Theorem 15; it says you can get the same result by either of two methods: (1) mapping a function over two lists and then appending the results together, and (2) appending the input lists and then performing one longer map over the result. Its proof is yet another good example of induction, and is left as an exercise.

Theorem 18. *map f* $(xs+\!\!+ys)$ = *map f xs* ++ *map f ys*

One of the most important properties of map is expressed precisely by the following theorem. Suppose that you have two computations to perform on all the elements of a list. First you want to apply *g* to an element, getting an intermediate result to which you want to apply *f*. There are two methods for doing the computation. The first method uses two separate loops, one to perform *g* on every element and the second loop to perform *f* on the list of intermediate results. The second method is to use just one loop, and each iteration performs the *g* and *f* applications in sequence. This theorem is used commonly by optimising compilers, program transformations (both manual and automatic), and it's also vitally important in parallel programming. Again, we leave the proof as an exercise.

Theorem 19. $(map\ f\ .\ map\ g)\ xs\ =\ map\ (f.g)\ xs$

For a change of pace, we now consider an intriguing theorem. Once you understand what it says, however, it becomes perfectly intuitive. (The notation $(1+)$ denotes a function that adds 1 to a number.)

Theorem 20. $sum\ (map\ (1+)\ xs)\ =\ length\ xs\ +\ sum\ xs$

Proof. Induction over xs. The base case is

$$
\begin{aligned}
&sum\ (map\ (1+)\ [])\\
&= sum\ [] && \{\ map.1\ \}\\
&= 0 + sum\ [] && \{\ 0 + x = x\ \}\\
&= length\ []\ +\ sum\ [] && \{\ length.1\ \}
\end{aligned}
$$

For the inductive case, assume $sum\ (map\ (1+)\ xs)\ =\ length\ xs\ +\ sum\ xs$. Then

$$
\begin{aligned}
&sum\ (map\ (1+)\ (x:xs))\\
&= sum\ ((1+x)\ :\ map\ (1+)\ xs) && \{\ map.2\ \}\\
&= (1+x)\ +\ sum\ (map\ (1+)\ xs) && \{\ sum.2\ \}\\
&= (1+x)\ +\ (length\ xs\ +\ sum\ xs) && \{\ hypothesis\ \}\\
&= (1 + length\ xs)\ +\ (x\ +\ sum\ xs) && \{\ (+).algebra\ \}\\
&= length\ (x:xs)\ +\ sum\ (x:xs) && \{\ length.2,\ sum.2\ \}
\end{aligned}
$$

$\square$

The *foldr* function is important because it expresses a basic looping pattern. There are many important properties of *foldr* and related functions. Here is one of them:

Theorem 21. $foldr\ (:)\ []\ xs\ =\ xs$

Some of the earlier theorems may be easy to understand at a glance, but that is unlikely to be true for this one! Recall that the *foldr* function takes apart a list and combines its elements using an operator. For example,

$$foldr\ (+)\ 0\ [1,2,3]\ =\ 1 + (2 + (3 + 0))$$

Now, what happens if we combine the elements of the list using the cons operator (:) instead of addition, and if we use [] as the initial value for the recursion instead of 0? The previous equation then becomes

$$
\begin{aligned}
foldr\ (:)\ []\ [1,2,3]\ &=\ 1:(2:(3:[]))\\
&=\ [1,2,3].
\end{aligned}
$$

We ended up with the same list we started out with, and the theorem says this will always happen, not just with the example $[1,2,3]$ used here.

Proof. Induction over xs. The base case is

$foldr\ (:)\ []\ []$
$= []$ { $foldr.1$ }

Now assume that $foldr\ (:)\ []\ xs\ =\ xs$; then the inductive case is

$foldr\ (:)\ []\ (x : xs)$
$= x\ :\ foldr\ (:)\ []\ xs$ { $foldr.2$ }
$= x : xs$ { hypothesis }

□

Suppose you have a list of lists, of the form $xss\ =\ [xs_0, xs_1, \ldots, xs_n]$. All of the lists xs_i must have the same type $[a]$, and the type of xss is $[[a]]$. We might want to apply a function $f :: a \to b$ to all the elements of all the lists, and build up a list of all the results. There are two different ways to organise this computation:

- Use the *concat* function to make a single flat list of type $[a]$ containing all the values, and then apply *map f* to produce the result with type $[b]$.

- Apply *map f* separately to each xs_i, by computing *map (map f) xss*, producing a list of type $[[b]]$. Then use *concat* to flatten them into a single list of type $[b]$.

The following theorem guarantees that both approaches produce the same result. This is significant because there are many practical situations where it is more convenient to write an algorithm using one approach, yet the other is more efficient.

Theorem 22. *map f (concat xss) = concat (map (map f) xss)*

Proof. Proof by induction over xss. The base case is

$map\ f\ (concat\ [])$
$= map\ f\ []$ { $concat.1$ }
$= []$ { $map.1$ }
$= concat\ []$ { $concat.1$ }
$= concat\ (map\ (map\ f)\ [])$ { $map.1$ }

Assume that $map\ f\ (concat\ xss)\ =\ concat\ (map\ (map\ f)\ xss)$. The inductive case is

$map\ f\ (concat\ (xs : xss))$
$= map\ f\ (xs \mathbin{+\!\!+} concat\ xss)$ { $concat.2$ }
$= map\ f\ xs \mathbin{+\!\!+} map\ f\ (concat\ xss)$ { Theorem 18 }
$= map\ f\ xs \mathbin{+\!\!+} concat\ (map\ (map\ f)\ xss)$ { hypothesis }
$= concat\ (map\ f\ xs\ :\ map\ (map\ f)\ xss)$ { $concat.2$ }
$= concat\ (map\ (map\ f)\ (xs : xss))$ { $map.2$ }

□

Sometimes you don't need to perform an induction, because a simpler proof technique is already available. Here is a typical example:

Theorem 23. *length* $(xs ++ (y : ys))$ $=$ 1 $+$ *length xs* $+$ *length ys*

This theorem could certainly be proved by induction (and that might be good practice for you!) but we already have a similar theorem which says that *length* $(xs ++ ys)$ $=$ *length xs* $+$ *length ys*. Instead of starting a new induction completely afresh, it's more elegant to carry out a few steps that enable us to apply the existing theorem. Just as reuse of software is a good idea, reuse of theorems is good style in theoretical computer science.

$$length \ (xs ++ (y : ys))$$
$$= length \ xs + length \ (y : ys)$$
$$= length \ xs \ + \ (1 + length \ ys)$$
$$= 1 + length \ xs + length \ ys$$

Exercise 5. Prove Theorem 16.

Exercise 6. Prove Theorem 18.

Exercise 7. Prove Theorem 19.

Exercise 8. Recall Theorem 20, which says

$$sum \ (map \ (1+) \ xs) \ = \ length \ xs \ + \ sum \ xs.$$

Explain in English what this theorem says. Using the definitions of the functions involved (*sum*, *length* and *map*), calculate the values of the left and right-hand sides of the equation using $xs = [1, 2, 3, 4]$.

Exercise 9. Invent a new theorem similar to Theorem 20, where $(1+)$ is replaced by $(k+)$. Test it on one or two small examples. Then prove your theorem.

4.6   Functional Equality

Many theorems used in computer science (including most of the ones in this chapter) say that two different algorithms are always guaranteed to produce the same result. The algorithms are defined as functions, and the theorem says that when you apply two different functions to the same argument, they give the same result.

It is simpler and more direct simply to say that the two functions are equal. However, this raises an interesting question: what does it mean to say $f = g$ when f and g are functions? (This issue will be revisited in Chapter 11.) There are at least two standard notions of functional equality that are completely different from each other, so it pays to be careful!

- *Intensional equality.* Two functions f and g are intensionally equal if their definitions are identical. This means, of course, that the functions are not equal if their types are different. If they are computer programs, testing for intensional equality involves comparing the source programs, character by character. The functions are intensionally equal if their definitions *look* the same.

- *Extensional equality.* Two functions f and g are extensionally equal if they have the same type $a \to b$ and $f\,x = g\,x$ for all well typed arguments $x :: a$. More precisely, $f = g$ if and only if

$$\forall x :: a \,.\, f\,x \;=\; g\,x.$$

The functions are extensionally equal if they deliver the same results when given the same inputs.

In computer science we are almost always interested in extensional equality. A typical situation is that we have an algorithm, expressed as a function f, and the aim is to replace it by a more efficient function g. This will not affect the correctness of the program as long as f and g are extensionally equal, but they are obviously not intensionally equal.

Some of the theorems given in the previous section can be stated in a simpler fashion using extensional equality. For example, recall Theorem 17, which says that map doesn't change the length of its argument:

$$length\,(map\,f\,xs) = length\,xs$$

A more direct way to state the same fact is to omit the irrelevant argument xs, and just say that these two functions are equal:

$$length \,.\, (map\,f) = length$$

To prove such a theorem of the form $f = g$, we need only prove that $\forall x :: a.\ f\,x = g\,x$, and this can be achieved by choosing an arbitrary $x :: a$, and proving the equation $f\,x = g\,x$.

Theorem 24. *foldr* $(:)\,[\,] \;=\; id$

Proof. The equation states that two functions are equal: the right-hand side, *id*, is a function, and the left-hand side is a partial application (*foldr* takes three arguments, but it has been applied to only two), so that is also a function. Therefore we use the definition of extensional equality of functions; thus we choose an arbitrary list xs, and we must prove that *foldr* $(:)\,[\,]\,xs \;=\; id\,xs \;=\; xs$. Now the right-hand side is just xs, by the definition of *id*, so the equation is proved by Theorem 21. $\qquad\square$

Theorem 25. *map* $f\,.\,concat \;=\; concat\,(map\,(map\,f))$.

Exercise 10. Prove Theorem 25.

Exercise 11. Prove that the ++ operator is associative.

Exercise 12. Prove *sum . map length = length . concat*.

4.7   Pitfalls and Common Mistakes

After a bit of practice, induction can come to seem almost too easy. You just
set up the base and inductive cases, crank the handle, and out comes a proof.

There are many kinds of bad inductive proofs. Their flaws are often due to
suspicious base cases, although there are a variety of dubious ways in which to
prove the induction cases too. Here is an interesting example.

4.7.1   A Horse of Another Colour

The following theorem is a famous classic.

Theorem 26. All horses are the same colour.

Proof. Define $P(n)$ to mean 'in any set containing n horses, all of them have
the same colour'. We proceed by induction over n.

Base case. Every horse has the same colour as itself, so $P(1)$ is true.

Inductive case. Assume $P(n)$, and consider a set containing $n + 1$ horses;
call them $h_1, h_2, \ldots, h_{n+1}$. We can define two subsets $A = h_1, \ldots, h_n$ and
$B = h_2, \ldots h_{n+1}$. Both sets A and B contain n horses, so all the horses in A
are the same colour (call it C_A), and all the horses in B are the same colour (call
it C_B). Pick one of the horses that is an element of both A and B. Clearly this
horse has the same colour as itself; call it C_h. Thus $C_A = C_h = C_B$. Therefore
all the horses $h_1, \ldots, h_{n+1}$ have the same colour.

Now we have proved $P(n) \rightarrow P(n + 1)$, so it follows by mathematical
induction that $\forall n \in \text{Nat} . P(n)$. Thus all horses are the same colour.   □

Exercise 13. What is the flaw in the proof given above? (Please try to work
   this out yourself, and then check the answer in the Appendix.)

4.8   Limitations of Induction

Induction can be used to prove that every element of a set satisfies a certain
property. The set may be finite or infinite. When the theorem states a property of an arbitrary natural number, or an arbitrary list, the inductive proof
establishes that an infinite number of values satisfy the theorem.

Nevertheless, there are some limits on what can be proved using mathematical induction. One such limit is that if the set is infinite, it must be countable

(that is, it must be possible to enumerate its elements, so that each one is associated with a unique natural number). Another limitation, which is particularly important for computing applications, is that ordinary induction cannot prove properties of infinite objects; it just proves properties of an infinite number of finite objects, which is not the same thing at all!

An example will clarify this issue. Suppose we define a function

$$reverse \ :: \ [a] \to [a]$$

which takes a list and returns a new list with the same elements, but in reverse order. For example, $reverse \ [1,2,3] \ = \ [3,2,1]$. Now, we want to state a theorem which says that if we reverse a list twice, we get the same list back. The following equation is one attempt to say that:

$$reverse \ (reverse \ xs) \ = \ xs$$

Alternatively, we might use extensional equality of functions, and just write

$$reverse \ . \ reverse \ = \ id.$$

It is straightforward to prove the first equation using induction, and the second equation follows immediately using the extensional definition of functional equality.

Theorem 27. *reverse . reverse = id*

Unfortunately, this theorem is untrue!

To see the problem, let's consider a concrete example. We will choose xs to be $[1..]$, which is the Haskell notation for the infinite list $[1,2,3,...]$. Now consider the following two expressions:

- $head \ (reverse \ (reverse \ [1..]))$

- $head \ [1..]$

Now the first of these expressions will go into an infinite loop, because the second (outermost) application of reverse needs to find the last element of its argument before it can return anything, and it will never find the last element of an infinite list. The second expression, however, does *not* go into an infinite loop; it returns the result 1 immediately. Yet, according to the dubious equations stated above, we should be able to replace $reverse \ (reverse \ [1..])$ by $[1..]$ without changing the value! What has gone wrong?

The problem is that mathematical induction only establishes that the theorem is valid for every element of the set that is connected to the base case by a finite number of steps. *It does not establish that the theorem is true for infinite lists, which are not reachable from the base case in a finite number of steps.* This means that all of our theorems over lists that were proved using induction have actually been proved only for finite lists. Note that there are an infinite number of lists of finite length; thus the induction is proving that a property holds for an infinite number of values, but it does not establish whether the property holds for values that are infinite in size.

There is a related point about natural numbers that sometimes confuses people. When we use induction over natural numbers, using 0 as the base case and $P(n) \rightarrow P(n+1)$ for the inductive case, we have established that the theorem holds for every natural number, and there is an infinite number of naturals. However, this does *not* prove that the theorem holds for infinity itself. There is an infinite number of naturals, but infinity is not itself a natural number.

Exercise 14. State the requirements on finite length that the proof of P imposes on the arguments of concat, where P is defined as

$$P(n) \equiv \text{concat xss} = \text{foldr } (+\!\!+) \; [] \; \text{xss}$$

Exercise 15. Check that Theorem 27 holds for the argument $[1, 2, 3]$.

Exercise 16. Prove the following theorem, using induction:

$$\textit{reverse } (xs+\!\!+ys) = \textit{reverse } ys+\!\!+\textit{reverse } xs$$

Then decide whether this theorem happens to be true for infinite lists like $[1..]$. Try to give a good argument for your conclusion, but you don't have to prove it.

Exercise 17. Use induction to prove Theorem 27. $\textit{reverse } (\textit{reverse } xs) = xs$.

Exercise 18. Explain why Theorem 27 does not hold for infinite lists.

4.9   Suggestions for Further Reading

Concrete Mathematics, by Graham, Knuth, and Patashnik [15] covers the more advanced mathematical techniques used in the analysis of algorithms. They include a number of problems on induction and also cover in depth the related topic of recurrences.

Many mathematics books contain more advanced examples of induction proofs. An entire chapter is devoted to induction in Engel's book, *Problem-Solving Strategies* [12], which is a good general source book for mathematical problems.

The textbooks on Haskell cited in Chapter 1 give examples of inductive proofs about recursive programs. The Bird-Meertens calculus [5] develops an extensive theory of programming, including many good applications of induction.

4.10   Review Exercises

Exercise 19. Assume that xss is a finite list of type [[a]], that it is of length
n, and that xs is a finite list and an arbitrary element of xss. Prove that
length (concat xss) = sum (map length xss).

Exercise 20. Prove that or defined over an argument that has an arbitrary
number of elements delivers the value True if True occurs as one of the
elements of its argument.

Exercise 21. Prove that and defined over an argument that has an arbitrary
number of elements delivers the value True if all of the elements in its
argument are True.

Exercise 22. Assume there is a function called max that delivers the larger of
its two arguments.
   max x y = x   *if* x >= y
   and
   max x y = y   *if* y >= x

Write a function maximum that, given a non-empty sequence of values
whose sizes can be compared (that is, values from a type of class Ord),
delivers the largest value in the sequence.

Exercise 23. Assume that the list xs is of type Ord a => [a], and that x is
an arbitrary element of xs.
Given the definition of maximum, defined as

maximum :: [Ord] -> Ord

maximum xs = foldr (max) y ys
             where xs = y:ys

prove that maximum has the following property:

(maximum xs) >= x

Exercise 24. Write a function that, given a sequence containing only non-
empty sequences, delivers the sequence made up of the first elements of
each of those non-empty sequences.

Exercise 25. Prove the equation concat = foldr (++) []. Assume that the
lists are finite, so that list induction can be used.

Exercise 26. Define an and operator using && and foldr.

Exercise 27. Given a list xs of type Bool, prove that

$$\text{and } ([\text{False}] \mathbin{+\!\!+} \text{xs}) = \text{False}.$$

Chapter 5

Trees

One of the most important data structures used in practical programming is the *tree*. Trees can be used to represent information with a hierarchical structure, they can be processed by recursive functions with a simple structure, and they are well suited for mathematical proofs. The study of trees illustrates connections between several of the main themes of this book, and offers many opportunities for exploiting formal mathematics in practical programming.

People use the idea of hierarchy for many purposes. For example, books are often organized as a sequence of chapters, each of which is a sequence of sections, which may have subsections, and so on. Corporations may be organized as collections of business units, each of which may have several departments. Departments, in turn, may have multiple sections, and so on. Biological taxonomy divides living organisms into kingdoms (plant, animal, etc.), then divides these categories into phylums (animals are vertebrate or invertebrate), divides phylums into classes (mammals, for example), and so on down through more and more narrow categories (order, family, genus, species, and subspecies). Software is organized as a collection of modules, any of which may be made up of several submodules, down to whatever level of refinement the designers find appropriate. At some level, modules are expressed in basic units such as objects, methods, or procedures. In other words, hierarchies provide an effective way to organize information.

Trees provide a model expressing the idea of hierarchy. Because they are formal, mathematical objects, trees provide a basis for precise reasoning about hierarchies.

5.1 Components of a Tree

Informally, trees are described in terms of a diagrammatic representation. In this sense, a tree is a collection of nodes connected by lines in a pattern that has no loops in it. That is, the pattern of connecting-lines does not admit a path along the lines that gets back to its starting point without retracing any steps.

There is not just one kind of tree; as we will see in the next section, there are many variations in the details of a tree structure. For now, we will discuss a semiformal definition of one particular kind of tree.

Definition 1. A *tree* is either an *empty tree* or it is a *node* together with a sequence of trees.

The definition is inductive because the term being defined, "tree", is used in its own definition. The starting point for the inductive definition is the empty tree. The definition does not say what an empty tree is; this is left as an undefined term, and the existence of the empty tree is accepted as an axiom. The term "node" is not defined, either, and the existence of nodes for building trees is also taken as an axiom. Later, when we use trees to represent specific mathematical entities, we will say exactly what entities comprise the set of nodes from which the trees are constructed.

Definition 2. The node portion of a non-empty tree is the *root* of the tree. The sequence-of-trees portion of a non-empty tree is called the *children* of the tree (and each individual member of the sequence is called a child of the tree).

Definition 3. A non-empty tree whose associated sequence of trees is empty is called a *leaf.*

A leaf, by itself, is the simplest kind of non-empty tree. In such a tree, the root is a leaf. In a more complicated tree (i.e., one consisting of a node with children) the root is not a leaf.

Figure 5.1 displays a tree as a diagram. The tree in this diagram is a node, together with two trees, one on the left and one on the right. These are, of course, subtrees of the complete tree represented by the diagram. Working our way down, we see that the left subtree consists of a node, together with two trees (one of which is a leaf). We see that the right subtree consists of a node, together with three trees, one on the left, one in the middle, and one on the right. The one on the left is a leaf, and the one on the right is a leaf. The one in the middle is not a leaf, but its subtrees are leaves. The tree, as a whole, has seven leaves and five interior nodes, one of which is the root.

Definition 4. A tree *s* is said to be a *subtree* of a tree *t* if either *s* is the same as *t*, or if *t* is non-empty and *s* is a subtree of one of the children of *t*.

The definition of the term "subtree" is also inductive. It is defined with respect to a given tree, and the basis of the inductive definition is the case when the given tree and the subtree are the same. The inductive case permits the subtree to be either one of the "immediate" constituents of the given tree or a subtree of one of those immediate constituents.

Definition 5. A tree *s* is said to be a *leaf* of a tree *t* if *s* is a subtree of *t* and *s* is a leaf.

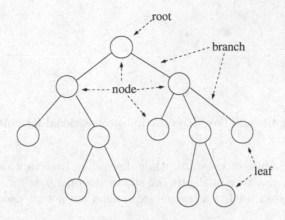

Figure 5.1: A tree diagram

Definition 6. A node n is said to *occur in* a tree t, written $n \in t$, if t consists of a node m together with a sequence of trees $[s_1, s_2, \ldots]$ and either n is the same as m or n occurs in one of the children of t.

The definition of the term "occurs in" is inductive because the term "occurs in" is used in its own definition. The basis of this inductive definition is the case in which the node occurs at the root. In the inductive case, the node occurs in one of the subtrees. The definition of "occurs in" makes it clear that no nodes occur in an empty tree because the tree in which a node occurs must consist of a node together with a sequence of trees.

Definition 7. A node n is said to be an *interior node* in the tree t if n occurs in t and there is a sequence of trees $[s_1, s_2, \ldots]$ such that n together with $[s_1, s_2, \ldots]$ is a subtree of t.

Trees usually contain additional data attached to the nodes and leaves. The tree structure (comprising the nodes and leaves) provides an organisation for the data values, making the information easier to use than if it were just kept in an array or list. Sometimes tree diagrams are drawn showing just the data values attached to nodes, but omitting the nodes themselves. For example, consider a tree used to represent linguistic constructs, such as the structure of a sentence in English or a statement in a programming language. Figure 5.2 shows the diagram of a tree representing the arithmetic expression $(3 * 4) + ((5 * 6)/8)$. The root of the tree is the $+$ operation, and the subtrees denote expressions describing the arguments to be added. The leaves of the tree are the numbers appearing in the expression. The value of the expression can be calculated by working from the leaves up to the root, calculating the intermediate values corresponding to each operator.

Most programming language interpreters and compilers use trees to represent the structure of the entire program. The tree representation makes explicit

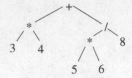

Figure 5.2: Tree diagram of a propositional formula

the essential structure of the data, while discarding irrelevant details. For example, the tree structure in Figure 5.2 shows the operands for each operator, and the parentheses used in a textual expression are not needed.

5.2   Representing Trees in Haskell

In traditional programming languages, you need to use low level programming with pointers or classes in order to process a tree. Haskell allows you to define tree structures directly and offers good support for writing programs that manipulate trees. This is especially important as there is not just one kind of tree; it is common to define new tree types that are tailored specifically for a particular application. Specifying trees formally in Haskell ensures that the data structures are expressed precisely, and it also makes algorithms on trees executable on a computer.

The special case of a tree type where every node must have exactly two children is called a *binary tree*. As stated earlier, in general a tree node may have any number of children. Binary trees are commonly used in practical computing applications.

To illustrate how to define tree structures in Haskell, we will define a binary tree type that has an integer data value attached to each node. Each kind of data structure defined in Haskell must have a unique data type; since we are considering a binary tree with integer data in each node, this new type will be called `BinTreeInt`.

The Haskell definition is based closely on the informal inductive definition of trees given in the previous section. The definition begins with `data BinTreeInt`, saying that a new datatype is being defined. There are are two kinds of `BinTreeInt`: empty ones (i.e., leaves) and nonempty ones (i.e., nodes with attached data). A leaf of the tree is denoted by the constructor `Leaf`, which takes no operands. A node of the tree is constructed by applying the constructor `Node` to the data value (of type `Integer`) and to two trees (which must have the same type, `BinTreeInt`).

```
data BinTreeInt
  = Leaf
  | Node Integer BinTreeInt BinTreeInt
```

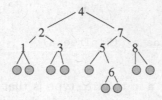

Figure 5.3: Tree diagram corresponding to the definition `tree3`

We can define values of the tree type using the constructors, which act just like functions. The following definitions declare that `tree1` has type `BinTreeInt` and consists just of a leaf:

```
tree1 :: BinTreeInt
tree1 = Leaf
```

To build a tree with some nodes and data, just apply `Node` to an integer and to expressions denoting the subtree, for example:

```
tree2 :: BinTreeInt
tree2 = Node 23 Leaf Leaf
```

Larger trees can also be defined straightforwardly. In the following example, the subtree operands are themselves expressions, so they need to be enclosed in parentheses. It is also a good idea to use indentation to make the tree structure clearer. Figure 5.3 shows the diagram corresponding to `tree3`.

```
tree3 :: BinTreeInt
tree3 =
  Node 4
    (Node 2
      (Node 1 Leaf Leaf)
      (Node 3 Leaf Leaf))
    (Node 7
      (Node 5
        Leaf
        (Node 6 Leaf Leaf))
      (Node 8 Leaf Leaf))
```

One limitation of the `BinTreeInt` type is that it requires the data attached to a node to be an integer. In practice, you are likely to have other types of data to be attached to nodes. Haskell allows you to define a *polymorphic* tree type, where the data attached to the nodes has some type a. The resulting tree has type `BinTree a`, which can be read as "binary tree with values of type a". Another useful refinement is the last line of the definition, `deriving Show`, which tells the compiler to generate code to print the tree automatically.

```
data BinTree a
  = BinLeaf
  | BinNode a (BinTree a) (BinTree a)
  deriving Show
```

The point of including a in the tree type is that it allows *any* type a to be used for the data attached to nodes, but it also requires that *all* the nodes have attached data of the same type. Here are some examples of correct tree definitions:

```
tree4 :: BinTree String
tree4 = BinNode "cat" BinLeaf (BinNode "dog" BinLeaf BinLeaf)

tree5 :: BinTree (Integer,Bool)
tree5 = BinNode (23,False)
          BinLeaf
          (BinNode (49,True) BinLeaf BinLeaf)

tree6 :: BinTree Int
tree6 = BinNode 4
          (BinNode 2
            (BinNode 1 BinLeaf BinLeaf)
            (BinNode 3 BinLeaf BinLeaf))
          (BinNode 6
            (BinNode 5 BinLeaf BinLeaf)
            (BinNode 7 BinLeaf BinLeaf))
```

The following definition produces a type error, because the nodes have attached data values of different types. The compiler will give an error message saying that Char, the type of one of the nodes, doesn't match Bool, the type of the other nodes.

```
treeBad = BinNode 'c'
            (BinNode True BinLeaf BinLeaf)
            (BinNode False BinLeaf BinLeaf)
```

Exercise 1. Define a Haskell datatype Tree1 for a tree that contains a character and an integer in each node, along with exactly three subtrees.

Exercise 2. Define a Haskell datatype Tree2 for a tree that contains an integer in each node, and that allows each node to have any number of subtrees.

5.3   Processing Trees with Recursion

Just as a for-loop is the natural programming technique for processing an array, recursion is the natural programming technique for processing trees. Recursion fits well with trees, because each subtree is a complete tree in its own right, so subtrees can be handled by recursive function applications. This section presents several typical algorithms on trees and shows how to implement them with recursive functions.

5.3.1 Tree Traversal

A natural task is to visit each node in the tree in order to process the data in that node, building a list of the results. An algorithm that works through a tree in this fashion is called a *tree traversal*. A traversal needs to decide what order to visit the nodes, and there are three specific traversal orders for binary trees that are commonly used in algorithms:

- *Preorder traversal:* first visit the root, then traverse the left subtree, then traverse the right subtree. A preorder traversal of the tree in Figure 5.3 produces $[4, 2, 1, 3, 7, 5, 6, 8]$.

- *Inorder traversal:* first visit the left subtree, then the root, then the right subtree. An inorder traversal of the tree in Figure 5.3 produces $[1, 2, 3, 4, 5, 6, 7, 8]$.

- *Postorder traversal:* Traverse the left subtree, then the right subtree, and finally visit the root. A postorder traversal of the tree in Figure 5.3 produces $[1, 3, 2, 6, 5, 8, 7, 4]$.

Functions can be defined to perform each of the traversals using recursion. The traversal functions take an argument which is a tree of type `BinTree a`, and they produce a flat list with type `[a]` consisting of the data values found attached to the nodes. All three of the traversal functions will produce a list containing all the data values in a tree; the only difference among them is the order of items in the list. The definitions of these functions are similar to the informal definitions of the traversal algorithms.

```
inorder :: BinTree a -> [a]
inorder BinLeaf = []
inorder (BinNode x t1 t2) = inorder t1 ++ [x] ++ inorder t2

preorder :: BinTree a -> [a]
preorder BinLeaf = []
preorder (BinNode x t1 t2) = [x] ++ preorder t1 ++ preorder t2

postorder :: BinTree a -> [a]
postorder BinLeaf = []
postorder (BinNode x t1 t2) =
  postorder t1 ++ postorder t2 ++ [x]
```

The traversal functions produce an empty list for an empty tree, and they use the ++ operator to construct lists for nonempty trees. The subtrees are handled by recursion, and the data value x attached to the current node is also placed in the result list. The best way to understand the recursion is by equational reasoning. (Some books approach recursion by considering stacks, return addresses, and other low level details, but those topics just confuse the issue and are appropriate only for compiler writers.) Here, for example, is the inorder traversal of of a small tree:

```
inorder (BinNode 4 (BinNode 8 BinLeaf BinLeaf) BinLeaf)
= inorder (BinNode 8 BinLeaf BinLeaf) ++ [4] ++ inorder BinLeaf
= (inorder BinLeaf ++ [8] ++ inorder BinLeaf) ++ [4] ++ []
= ([] ++ [8] ++ []) ++ [4] ++ []
= [8, 4]
```

All of these traversal functions convert a tree to a list containing the data attached to the tree's nodes. The inorder traversal produces a list with the data in the same order that would appear if you read the tree from left to right. In many applications this is the most natural way to represent the tree's data as a list. Because of this, the inorder function is often named flatten.

Exercise 3. Calculate the inorder traversal of tree3.

Exercise 4. Suppose that a tree has type BinTree a, and we have a function f :: a -> b. Write a new traversal function inorderf :: (a->b) -> BinTree a -> [b] that traverses the tree using inorder, but it applies f to the data value in each node before placing the result in the list. For example, inorder tree6 produces $[1, 2, 3, 4, 5, 6, 7]$, but inorderf (2*) tree6 produces $[2, 4, 6, 8, 10, 12, 14]$.

5.3.2   Processing Tree Structure

There are several functions that measure the size of a tree, or affect its shape. This section introduces some of the basic ones; more advanced operations are described in Chapter 12, which discusses AVL trees.

The reflect function takes a binary tree and returns its mirror image, where everything is reversed left-to-right.

```
reflect :: BinTree a -> BinTree a
reflect BinLeaf = BinLeaf
reflect (BinNode n l r) = BinNode n (reflect r) (reflect l)
```

Some of the most important properties of trees are concerned with the numbers of nodes in the two branches and with the heights of trees and subtrees. The time required by many algorithms depends on the heights of trees, so the science of algorithmics is often concerned with these properties.

The height of a tree is the distance between its root and its deepest leaf. An empty tree, consisting only of a leaf, has height 0. The height of a nonempty tree is the height of its taller subtree, plus one to account for the root node. Thus the height satisfies the following equations:

```
height :: BinTree a -> Integer
height BinLeaf = 0
height (BinNode x t1 t2) = 1 + max (height t1) (height t2)
```

The size of a tree is the number of nodes that it contains. This is measured simply by adding up the sizes of the subtrees, plus one for the root node.

```
size :: BinTreeInt -> Int
size Leaf = 0
size (Node x t1 t2) = 1 + size t1 + size t2
```

The size of a tree of type `BinTree a` tells you how many data values of type `a` are represented in the tree, and it is also related to the amount of computer memory required to represent the tree. The height of a tree is related to the tree's shape. At one extreme, a tree like `tree1` below, where all the left nodes are leaves, has a height that is the same as its size. At the other extreme, a tree like `tree2`, where the data is distributed evenly throughout the tree, will have a smaller height given the same amount of data. Such a tree is said to be *balanced*.

```
tree7, tree8 :: BinTree Integer

tree7 = BinNode 1
          BinLeaf
          (BinNode 2
             BinLeaf
             (BinNode 3 BinLeaf BinLeaf))

tree8 = BinNode 1
            (BinNode 2 BinLeaf BinLeaf)
            (BinNode 3 BinLeaf BinLeaf)
```

We can give a formal definition of balanced trees by writing down the following equations, which also define an executable function. The function application `balanced t` returns `True` if the tree `t` is perfectly balanced, and `False` otherwise.

```
balanced :: BinTree a -> Bool
balanced BinLeaf = True
balanced (BinNode x t1 t2) =
  balanced t1 && balanced t2 && (height t1 == height t2)
```

Exercise 5. Define two trees of size seven, one with the largest possible height and the other with the smallest possible height.

Exercise 6. Suppose that the last equation of the function `balanced` were changed to the following: `balanced (BinNode x t1 t2) = balanced t1 && balanced t2`. Give an example showing that the modified function returns `True` for an unbalanced tree.

Exercise 7. Suppose that the last equation of the function `balanced` were changed to the following: `balanced (BinNode x t1 t2) = height t1 == height t2`. Give an example showing that the modified function returns `True` for an unbalanced tree.

5.3.3   Evaluating Expression Trees

Software that is working with text written in some language óften uses trees to represent documents in the language. Trees express the essential structure of the tex while omitting unimportant details. Examples include programs that manipulate natural language, as well as compilers and interpreters for programming languages.

Consider a simple expression language, consisting of integer constants, additions, and multiplications. A document in the language can be expressed with a tree type:

```
data Exp
  = Const Integer
  | Add Exp Exp
  | Mult Exp Exp
```

A simple programming language interpreter can now be written as a tree traversal. The function takes an expression tree and returns the value that it denotes.

```
eval :: Exp -> Integer
eval (Const n) = n
eval (Add e1 e2) = eval e1 + eval e2
eval (Mult e1 e2) = eval e1 * eval e2
```

5.3.4   Binary Search Trees

Suppose you have a large set of pairs of keys (with some type `a`) and corresponding values (with some type `b`). For example, a key might be a person's name and the value their age. A crucial problem in computing with databases is to find the value corresponding to a given key.

A straightforward approach, called a *linear search*, is to store the data as a list of pairs, so that the database has type `[(a,b)]`. The search algorithm, `linSearch`, is given a key of type `a` and a database, and it works through the list sequentially until the right pair is found. The key type `a` must be in the equality class (i.e., it has to be a type that can be compared with `==`) because we have to be able to compare two keys to determine whether they are the same. It is possible that the key might not appear in the database, so we return the result as a `Maybe` type. Thus a failed search returns `Nothing` and a successful search returns `Just y`.

```
linSearch :: Eq a => a -> [(a,b)] -> Maybe b
linSearch k [] = Nothing
linSearch k ((x,y):xs) =
  if k==x
    then Just y
    else linSearch k xs
```

The linear search is simple; its only problem is efficiency. Since databases may contain millions of key-value pairs, it is important to perform the search quickly. For random data, it is reasonable to assume that an average linear search will find the pair at the middle of the list—half of the searches will find the data sooner, half will find it later. Therefore the time required for the search is proportional to the length of the list. If a bank with a million customers were to use a linear search, the average search time would be half a million iterations.

A much faster approach is the binary search algorithm, and a particularly good way to implement this is with a *binary search tree*. The idea is to store the data in such a way that a particular key-value pair can be found quickly, without examining half of the database.

A binary search tree is a tree of type `BinTree (a, b)`, where `a` is the key type and `b` is the value type. A binary search tree must also satisfy an additional property: if the key in a node has the value k, then all keys in the left subtree must be less than k, and all the keys in the right subtree must be greater than k. This property must hold throughout the entire tree, not just for the top node. If a tree of type `BinTree (a, b)` lacks this property, then it is still a perfectly good tree, but it is not a binary search tree. Since we need to compare keys for ordering, not just equality, the key type `a` must be in the `Ord` class (ensuring that `<` and `>` can be used).

The `bstSearch` algorithm first compares the key it is searching for with the key `x` of the root node. If there is a match, then the search has finished successfully. Otherwise, if `key < x` then it is guaranteed that the answer will be found in the left subtree, or not at all. Therefore the algorithm just continues with a recursive search down the left subtree. However, if `key > x` then the search continues down the right subtree.

```
bstSearch :: Ord a => a -> BinTree (a,b) -> Maybe b
bstSearch key BinLeaf = Nothing
```

```
bstSearch key (BinNode (x,y) t1 t2) =
  if key == x
    then Just y
    else if key < x
            then bstSearch key t1
            else bstSearch key t2
```

The average time required for a search is proportional to the height of the tree. With good luck, the database tree will be reasonably balanced, so that about half the data is in the left subtree and half in the right subtree. In this case, every step of the search algorithm discards half of the remaining data, so that the average search time for a database of size n is about $\log n$, compared with about $n/2$ for the linear search. For a large database, the difference is huge: with one million items, the average linSearch time is 500,000 while the average bstSearch time is 20. For a database containing a billion items, the number of steps in the search increases by only fifty percent, to 30 steps, even though the size of the database has increased by a factor of a thousand. The search time gets better and better, relatively speaking, as the size of the database increases. This property is crucial for large scale practical applications.

In practical applications, we start with an empty binary search tree (i.e., BinLeaf), and construct the database by inserting the key-value pairs one by one. The insert function takes a new pair, and an existing binary search tree, and it returns a new tree that contains the additional data. The function is defined carefully so that it creates a valid binary search tree (provided that its argument is valid). The function definition has a recursive structure similar to that of the search.

```
insert :: Ord a => (a,b) -> BinTree (a,b) -> BinTree (a,b)
insert (key,d) BinLeaf = BinNode (key,d) BinLeaf BinLeaf
insert (key,d) (BinNode (x,y) t1 t2) =
  if key == x
    then BinNode (key,d) t1 t2
    else if key < x
            then BinNode (x,y) (insert (key,d) t1) t2
            else BinNode (x,y) t1 (insert (key,d) t2)
```

There is an interesting point to notice about binary search trees. Sometimes the properties that a data structure must satisfy are specified completely by its type. For example, a linear search algorithm can be applied to list of key-value pairs as long as the list has the right type, [(a,b)]. In contrast, the binary search tree must be of the right type, but it must also satisfy the ordering constraint. Many data structures are like this, with some additional properties required beyond just the type.

The distinction is important, because the type of the data structure can be checked by the compiler, whereas the additional properties cannot. The type system of a programming language provides a way to describe required properties that can be checked at compile time; if the type checking indicates no problem, then it is *guaranteed* that the program will never generate any data that lacks the necessary properties. For example, we said above that the linear search algorithm requires its database to have a certain type. When the compiler checks the type of the program and finds no type errors, this constitutes a formal proof that the program will never generate a database for the linear search which is in the wrong form, and could thus cause a runtime error. The type system is unable, however, to prove that a binary search tree built using the `insert` function satisfies the required properties. Therefore the programmer needs to do the proof manually.

There is much more to say about the execution times of these algorithms. Books on the analysis of algorithms give a detailed discussion, but we will make just a few comments. In the first place, the times for individual searches will vary. Consider the linear search: with good luck, the desired data will be found at the very beginning of the list, and with bad luck it will appear at the end. It is not the case that every search takes $n/2$ time, for a database of size n. Instead, the average time over a large number of searches can be expected to be $n/2$. The correct way to calculate the average execution time of an algorithm is to add up the times required by every possible case, weighted by the probability of that case occurring.

The efficiency of the `bstSearch` algorithm is extremely sensitive to the shape of the tree. If the tree is perfectly balanced, the search time is $\log n$, but it is possible for the tree to be completely unbalanced. This causes the `bstSearch` to behave much like a linear search, requiring linear time.

In large scale practical applications, it isn't good enough to hope that the tree will be balanced through good fortune. A better approach is to change the `insert` function so that it guarantees a reasonably balanced result. This leads to a tradeoff: we can make the search faster at the cost of a slower insertion. Ideally, we would like algorithms that give good results for searches and insertions, regardless of the data values. These issues are considered in more depth in Chapter 12.

Exercise 8. Define a function `mapTree` that takes a function and applies it to every node in the tree, returning a new tree of results. The type should be `mapTree :: (a->b) -> BinTree a -> BinTree b`. This function is analogous to `map`, which operates over lists.

Exercise 9. Write `concatTree`, a function that takes a tree of lists and concatenates the lists in order from left to right. For example,

```
concatTree (Node [2] (Node [3,4] Tip Tip)
                     (Node [5] Tip Tip))
    ==> [3,4,2,5]
```

Exercise 10. Write `zipTree`, a function that takes two trees and pairs each of the corresponding elements in a list. Your function should return Nothing if the two trees do not have the same shape. For example,

```
zipTree (Node 2 (Node 1 Tip Tip) (Node 3 Tip Tip))
        (Node 5 (Node 4 Tip Tip) (Node 6 Tip Tip))
==> Just [(1,4),(2,5),(3,6)]
```

Exercise 11. Write zipWithTree, a function that is like zipWith except that it takes trees instead of lists. The first argument is a function of type a->b->c, the second argument is a tree with elements of type a, and the third argument is a tree with elements of type b. The function returns a list with type [c].

5.4   Induction on Trees

There are many useful theorems about the properties of trees, and the behaviour of functions on trees. To prove them, we can use induction, but the principle of induction needs to be made slightly more general for tree structures.

The general idea behind tree induction is similar to list induction. The base case is used for empty trees (or leaves), and the induction case is used for nodes. The principle of tree induction is stated below for binary trees, but it can be generalised to other types of tree as well.

Theorem 28 (Principle of induction on binary trees). Let BinTree a be a binary tree type as defined above, and let $P(t)$ be a proposition on trees. Suppose the following two requirements hold:

- Base case: P(BinLeaf)

- Induction case: For all t1 and t2 of type BinTree a, and all x :: a, suppose that the proposition holds for a tree consisting of a node, the value a, and the subtrees t1 and t2, provided that the proposition holds for t1 and t2. Using the notation of the chapter on propositional logic, this can be written formally as $P(\text{t1}) \land P(\text{t2}) \rightarrow P(\text{BinNode x t1 t2})$.

Then $\forall t$ :: BinTree a . $P(t)$; thus the proposition holds for all trees of finite size.

Carrying out a proof using tree induction is generally no harder than list induction. A number of examples are given below. In all cases, we will use trees of type BinTree a, for an arbitrary type a.

5.4.1   Repeated Reflection Theorem

Recall the reflect function, which reverses the order of the data in a binary tree. The following theorem says that if you reflect a tree twice, you get the same tree back.

Theorem 29. Let `t :: BinTree a`. Then `reflect (reflect t) = t`.

Proof. The proposition $P(t)$ that we need to prove says `reflect (reflect t)` = t. The theorem is proved by induction over t. The base case is `BinLeaf`:

```
reflect (reflect BinLeaf)
  = reflect BinLeaf                                    { reflect.1 }
  = BinLeaf                                            { reflect.1 }
```

For the inductive case, let `t1, t2 :: BinTree a` be trees, and assume $P(\text{t1})$ and $P(\text{t2})$. These are the inductive hypotheses. The aim is to prove that the proposition holds for a larger tree $P(\text{Node x t1 t2})$.

```
reflect (reflect (Node x t1 t2))
  = reflect (Node x (reflect t1) (reflect t2))    { reflect.2 }
  = Node x (reflect (reflect t1))                 { reflect.2 }
           (reflect (reflect t2))
  = Node x t1 t2                                   { hypothesis }
```

Now we have proved the base case, and we have proved the logical implication required for the induction case—that is, *if* the proposition holds for t1 and t2, then it must *also* hold for `Node x t1 t2`. Thus, by the principle of tree induction, the theorem holds for all finite trees. □

5.4.2 Reflection and Reversing

Reflecting a tree changes the order of data values, in a similar way to the reversal of a list. It seems that `reflect` and `reverse` are somehow doing the same thing, but these functions are definitely not equal.

Whenever we find functions that seem intuitively to be doing related things, it's useful to state this relationship precisely as an equation, rather than just a vague phrase in English. The resulting theorem gives a deeper understanding, and it may also be useful in practice. A powerful problem solving technique is to notice that a problem you have can be translated into a related notation for which you already have a solution.

To state precisely the relationship between reflecting a tree and reversing a list, we need to use an explicit translation between trees and lists. The `inorder` function is exactly what we need.

Theorem 30. Let `t :: BinTree a` be an arbitrary binary tree of finite size. Then `inorder (reflect t) = reverse (inorder t)`.

The proof requires two lemmas that describe properties of `reverse` and `(++)`. We leave the proofs of the lemmas as an exercise.

```
reverse xs ++ [x] = reverse ([x] ++ xs)
reverse (xs++ys) = reverse ys ++ reverse xs
```

Proof. Base case.

```
inorder (reflect BinLeaf)
  = inorder BinLeaf              { reflect.1 }
  = []                          { inorder.1 }
  = reverse []                  { reverse.1 }
  = reverse (inorder BinLeaf)   { inorder.1 }
```

Induction case. Let `t1`, `t2 ::  BinTree` a be arbitrary trees, and let `x ::` a be an arbitrary data value. Assume the inductive hypotheses:

```
inorder (reflect t1) = reverse (inorder t1)
inorder (reflect t2) = reverse (inorder t2)
```

Then

```
inorder (reflect (BinNode x t1 t2))
    = { reflect.2 }
inorder (BinNode x (reflect t2) (reflect t1))
    = { inorder.2 }
inorder (reflect t2) ++ [x] ++ inorder (reflect t1)
    = { hypotheses }
reverse (inorder t2) ++ [x] ++ reverse (inorder t1)
    = { reverse lemma 1 }
reverse ([x] ++ inorder t2) ++ reverse (inorder t1)
    = { reverse lemma 2 }
reverse (inorder t1 ++ [x] ++ inorder t2)
    = { inorder.2 }
reverse (inorder (BinNode x t1 t2))
```

With the base case and induction cases proved, the theorem holds by tree induction.   □

5.4.3   The Height of a Balanced Tree

If a binary tree is balanced then its shape is determined, and the number of nodes is determined by the height. The following theorem states this relationship precisely.

Theorem 31. Let $h =$ height t. If balanced t, then size $t = 2^h - 1$.

Proof. The proposition we want to prove is balanced $t \;\to\;$ size $t = 2^h - 1$. The proof is an induction over the tree structure. For the base case, we need to prove that the theorem holds for a leaf.

```
balanced BinLeaf  =  True
```
$h =$ height BinLeaf $= 0$
size BinLeaf $= 0$
$2^h - 1 = 0$

For the inductive case, let $t =$ Node x l r, and let $hl =$ height l and $hr =$ height r. Assume $P(1)$ and $P(r)$; the aim is to prove $P(t)$. There are two cases to consider. If t is not balanced, then the implication balanced $t \rightarrow P(t)$ is vacuously true. If t is balanced, however, then the implication is true if and only if $P(t)$ is true. Therefore we need to prove $P(t)$ given the following three assumptions: (1) $P(1)$ (inductive hypothesis), (2) $P(r)$ (inductive hypothesis), and balanced t (premise of implication to be proved).

$$
\begin{aligned}
h &= \text{height (Node x l r)} \\
&= 1 + \max \text{ (height l) (height r)} && \{\ height.2\ \} \\
&= 1 + \text{height l} && \{\ assumption\ \} \\
&= 1 + hl && \{\ def\ hl\ \}
\end{aligned}
$$

$$
\begin{aligned}
\text{size } t \\
&= \text{size (Node x l r)} && \{\ def\ t\ \} \\
&= 1 + \text{size l } + \text{ size r} && \{\ size.2\ \} \\
&= 1 + 2^{hl} - 1 + 2^{hr} - 1 && \{\ hypothesis\ \} \\
&= 2^{hl} + 2^{hr} - 1 && \{\ arithmetic\ \} \\
&= 2^{hl} + 2^{hl} - 1 && \{\ hl = hr\ \} \\
&= 2 \times 2^{hl} - 1 && \{\ algebra\ \} \\
&= 2^{hl+1} - 1 && \{\ algebra\ \} \\
&= 2^{h} - 1 && \{\ def\ h\ \}
\end{aligned}
$$

The base and inductive cases have been proved, so the theorem holds by the principle of tree induction. □

5.4.4 Length of a Flattened Tree

The following theorem says that if you flatten a tree, then the length of the resulting list is the same as the number of nodes in the tree.

Theorem 32. Let t :: BinTree a be any finite binary tree. Then length (inorder t) = size t.

Proof. The proof is a tree induction over t.
   Base case.

```
length (inorder BinLeaf)
  = length []
  = 0
  = size BinLeaf
```

   Induction case. Assume the induction hypotheses:

```
length (inorder t1) = size t1
length (inorder t2) = size t2
```

Then

```
length (inorder (BinNode x t1 2t))
  = length (inorder t1 ++ [x] ++ inorder t2)
  = length (inorder t1) + length [x] + length (inorder t2)
  = size t1 + 1 + size t2
  = size (Node x t1 t2)
```

Therefore the theorem holds by tree induction.   □

5.5   Improving Execution Time

A function definition consists of a type, along with a set of equations. These
equations serve two purposes:

- They give a mathematical specification of required properties. Since they
  are mathematical equations (not assignment statements), they can be
  used for formal proofs using equational reasoning.

- They serve as an executable computer program. The computer executes
  the program by a sequence of substitutions and simplifications. In effect,
  the Haskell compiler translates a set of equations into a machine language program that simplifies expressions using automated equational
  reasoning.

In general, inductive equations specifying properties of a function can be
successfully interpreted to compute the value of the function, given a particular argument, provided the equations have three essential characteristics. The
equations must be (1) consistent with properties of the function being defined,
(2) cover all relevant cases, and (3) supply simpler arguments to invocations
of the function that appear on the right-hand side than were supplied on the
left-hand side. What we mean by "simpler" is not easy to say in fully general terms. However, it amounts to making sure the amount of computation
involved in each invocation of the function on the right-hand side of the equation is significantly smaller than the amount of computation required by the
invocation on the left-hand side of the equation.

Consider, for example, the equations that define `inorder`:

```
inorder :: BinTree a -> [a]
inorder BinLeaf = []
inorder (BinNode x t1 t2) = inorder t1 ++ [x] ++ inorder t2
```

The trees supplied as arguments in the invocations of `inorder` on the right-
hand side of equation {`inorder.2`} have heights that are at least one smaller
than the height of the tree supplied as an argument. Because the height of a
tree must be at least zero, and the height is reduced by at least one in each
level of recursion, it eventually gets down to zero. Therefore the application
`inorder` t will terminate, provided that the tree is finite.

It makes sense to ask questions about the resources required for the specified computation. For example, we might ask how much time it would take to compute `inorder t` for some tree `t`.

It is a delightful fact that we can derive equations to answer this question about time directly from the equations specifying properties of the function—that is, the equations that our computational interpretation uses to carry out the work. Let `time e` denote the number of steps in the computation represented by the formula `e`. Our aim is to learn something about `time (inorder t)` for an arbitrary finite tree `t`. We will just estimate the time, by counting the number of basic operations, but will not seek an exact and precise analysis.

Assume that `time (inorder BinLeaf)` is zero. This is a harmless simplifying assumption; if the argument to `inorder` is a leaf, the machine will take a small amount of time to notice that fact and return. In practice, using a modern Haskell compiler, only a few machine instructions are needed. Since the time is so small, we ignore it.

Executing the second equation of `inorder` requires two recursive applications, and a concatenation (++). Some time is also needed to set up the equation (noticing that the argument is a node, doing the recursive calls, and returning); this requires time 1 (where the exact measurement unit is unspecified; it depends on how efficient the compiler is, how fast your computer is, etc.)

A minor optimisation to the function is to replace `[x] ++ inorder t2` by `x : inorder t2`. This is just a single equational reasoning step, and the compiler might even do this for us automatically. The time for the (:) operation is small (just a few machine instructions) and we will include that time in the one unit for the equation.

Before continuing, we need to know how many steps are involved in concatenation operations. This can be worked out from the equations in the definition of concatenation, using an analysis method like the one we are now using to work out the timing for `inorder`. For present purposes, we are going to skip that analysis and just state the result, which is this:

```
time (xs ++ ys) = length xs                    { time (++) }
```

That is, the time to perform a concatenation is proportional to the length of the first argument, and completely independent of the second. To understand why, consider that the equations defining (++) never look inside `ys`, but they perform a linear traversal over `xs`.

Armed with this information, we can continue with the time analysis of `inorder t`, for an arbitrary tree `t`.

```
time BinLeaf = 0
time (inorder (BinNode x t1 t2))
  = 1 + time (inorder t1 ++ [x] ++ inorder t2)
  = 1 + time (inorder t1) + time (inorder t2)
```

```
    + length (inorder t1)
= 1 + time (inorder t1) + time (inorder t2) + size t1
```

This result is not a simple formula that gives the execution time directly. Instead, it is a set of recursive equations. Such equations are commonly obtained in the performance analysis of algorithms, and they are known as *recurrence equations*. A recurrence equation is simply an inductive equation in which the values being equated are numbers.

There are many mathematical tricks that make it possible to solve various kinds of recurrence equations. To solve recurrence equations is to reduce them to equations that do not involve a recurrence; that is, equations in which none of the terms on the right-hand side refer to the function being defined on the left-hand side. Books on the analysis of algorithms often follow a systematic procedure: first, the algorithm is studied in order to derive a set of recurrence equations that describe its performance, and then a variety of mathematical techniques are used to solve the recurrence equation.

The study of solution methods for recurrence equations is a big topic, and we are not going to delve into it here. Instead, we will glean information from recurrence equations in ways that depend on circumstances. That is, we will rely on ad hoc analysis to derive information from recurrence equations.

In this case, it is the (`size t1`) term on the right-hand side of the recurrence equation that we want to focus on. Suppose that the tree is badly unbalanced in the sense that its left subtree contains all the nodes and its right subtree is empty. Furthermore, suppose this badly unbalanced condition persists all the way down to the leaves. That is, all the right subtrees are empty. In this case, the recurrence equations for time, height, and size specialize to the following form.

```
  time (inorder (BinNode x t1 BinLeaf))
   = 1 + time (inorder t1) + time (inorder BinLeaf) + size t1
   = 1 + time (inorder t1) + 0 + size t1
   = 1 + time (inorder t1) + size t1
  height (BinNode x t1 BinLeaf)
   = 1 + max (height t1) (height t2)
   = 1 + max (height t1) 0
   = 1 + height t1
  size (BinNode x t1 BinLeaf)
   = 1 + size t1 + size BinLeaf
   = 1 + size t1
```

The equations for the height and size are identical, so we can conclude that, in the case where all the right-hand subtrees are empty, the height of a tree is the same as the number of nodes in the tree. Therefore, the number of recurrence steps for `time` needed to reach the empty tree case is just the number of nodes in the tree being flattened. At each deeper level, the size term is one less than it was at the previous level. So, working down to the empty tree level amounts to adding all the integers starting with $n = size\ t1$ and ending with zero, plus the number of levels (since the number 1 is added at each level). Thus the result is

$$n + (n-1) + \ldots + 0.$$

As we proved already in Chapter 4, $\sum_{i=0}^{n} i = n(n+1)/2$. From these observations, we deduce the following formula for the time required to flatten a tree in which all the right-hand subtrees are empty:

```
time(inorder (BinNode x t1 BinLeaf)) = n(n+1)/2 + n,
    where n = size t1
```

This result is not good. It means that the number of steps needed to flatten a tree is proportional to the square of the number of nodes in the tree. That is too much time. One might hope that the number of computational steps needed to flatten a tree would be proportional to the number of nodes in the tree. Of course, this is a very special case, because all of the right-hand subtrees are empty. But, the formulas suggest that whenever the tree tends to have most of its nodes in its left subtrees instead of its right subtrees, flattening is going to take a long time. In the next section, we will consider another set of inductive equations for the flattening function that lead to a flattening time proportional to the number of nodes in the tree.

5.6 Flattening Trees in Linear Time

The reason that the `inorder` function is so slow is that it recopies lists repeatedly as it concatenates the partial results together. The trick for reducing the time required to flatten a tree is to accumulate the result-list in a collection of partial computations that permit pasting the results together directly, avoiding the expensive concatenations.

Without knowing the exact definition of the improved function, we can still write some equations that express properties it should have. The unknown function—call it g—will be similar to `inorder`, except it will take an extra argument ks of data values to be concatenated to the end of its result. That is, g will not simply return its result; it will return the concatenation of its result to a further list provided by some other source. This extra list is called the *continuation*.

```
g :: BinTree a -> [a] -> [a]
g BinLeaf ks = ks
g (BinNode x t1 t2) ks =  g t1 (x : g t2 ks)
```

These equations surely do not express the first properties of a flattening function that a person would think of. However, they do express properties that a person would expect a flattening function to have, and they avoid use of concatenation. Moreover, they have the three characteristics required to turn inductive equations into full specifications for a function (consistency, coverage of cases, and reduced computation on right-hand sides).

The question is, how long would it take a computing system interpreting these equations to produce a flattened version of a tree? We can use the equations defining g as a starting point to derive its execution time, just as we did for `inorder` in the preceding section.

Our hope is that the number of computation steps required to flatten a tree is on the order of the number of nodes in the tree. It turns out we can verify this conjecture using the principle of induction for trees. To be precise, we are trying to prove the validity of the following equation:

```
time (g t ks) = size tr
```

Now that we have guessed the equations for g, we need to verify that it actually works correctly. We conjecture the following theorem.

Theorem 33. Let t :: BinTree a be an arbitrary finite tree. Then g t ks = inorder t ++ ks.

Proof. Base case.

```
g BinLeaf ks
  = ks
  = [] ++ ks
  = inorder BinLeaf ++ ks
```

Induction case. Assume that

```
g t1 ks1 = inorder t1 ++ ks1
g t2 ks2 = inorder t2 ++ ks2
```

Then

```
g (BinNode x t1 t2) ks
   = { g.2 }
g t1 (x : g t2 ks)
   = { hypothesis.2 }
g t1 (x : inorder t2 ++ ks)
   = { (++).2 }
g (t1 ([x] ++ inorder t2 ++ ks))
   = { hypothesis.1 }
inorder t1 ++ ([x] ++ inorder t2 ++ ks)
   = { (++) associative }
(inorder t1 ++ [x] ++ inorder t2) ++ ks
```

By tree induction, the theorem holds.   □

To summarise the situation: we haven't done a conventional optimisation of an inefficient program. Instead, we have conjectured that there should exist an efficient program, and we have guessed the form it should have. Finally, we proved that the more efficient program does indeed compute the correct answer.

Functions like g, which use a continuation, are common in practical applications. However, it is not exactly equivalent to `inorder`—its type is different! If this is a concern, we can always define a new function that hides the continuation:

```
inorderEfficient :: BinTree a -> [a]
inorderEfficient t = g t []
```

Now that the new function has been shown to be correct, we should also try to verify our guess that it is efficient. Our aim now is to prove the following.

Theorem 34. Let t :: BinTree a be an arbitrary finite tree. Then time (g t ks) = size t.

Proof. Induction over the tree.
Base case.

```
time(g BinLeaf ks)
  = 0                         { a reasonable assumption }
  = size BinLeaf              { size.1 }
```

Inductive case. Assume

```
time (g t1 ks1) = size t1
time (g t2 ks2) = size t2
```

According to the assumption, time (g t1 ks1) = time (g t1 []); that is, the time depends only on the tree argument, but not the continuation. (This point is crucial, and is the fundamental reason why this algorithm is efficient.) Then

```
time (g (BinNode x t1 t2) ks)
  = time (g t1 (x : g t2 ks))
  = time (g t1 []) + 1 + time (g t2 [])
  = size t1 + 1 + size t2
  = size (BinNode x t1 t2)
```

So the theorem holds by the principle of tree induction. □

In summary, we have verified that g is mathematically equivalent to `inorder`: it computes the same result, given an empty continuation. Furthermore, it requires time proportional to the number of nodes in the tree. Thus g can be used as a faster replacement for `inorder`.

Programmers who use equations to specify software, rather than imperative procedures, must pay considerable attention to the form of the properties they specify for the functions they want their software to compute. Different properties of the equations lead to different computations, some of which are more efficient than others. We have been mostly concerned with correctness properties of functions, rather than resource utilization, but the same framework for reasoning can be used to analyze both correctness and resources. Serious software designers have to pay attention to both aspects of the programs they write.

Exercise 12. Write `appendTree`, a function that takes a binary tree and a list, and appends the contents of the tree (traversed from left to right) to the front of the list. For example,

```
appendTree (BinNode 2 (BinNode 1 BinLeaf BinLeaf)
                      (BinNode 3 BinLeaf BinLeaf))
           [4,5]
```

evaluates to [1,2,3,4,5]. Try to find an efficient solution that minimises recopying.

Part II

Logic

Chapter 6

Propositional Logic

Logic provides a powerful tool for reasoning correctly about mathematics, algorithms, and computers. It is used extensively throughout computer science, and you need to understand its basic concepts in order to study many of the more advanced subjects in computing. Here are just a few examples, spanning the entire range of computing applications, from practical commercial software to esoteric theory:

- In *software engineering*, it is good practice to specify what a system should do before starting to code it. Logic is frequently used for software specifications.

- In *safety-critical applications*, it is essential to establish that a program is correct. Conventional debugging isn't enough—what we want is a proof of correctness. Formal logic is the foundation of program correctness proofs.

- In *information retrieval*, including Web search engines, logical propositions are used to specify the properties that should (or should not) be present in a piece of information in order for it to be considered relevant.

- In *artificial intelligence*, formal logic is sometimes used to simulate intelligent thought processes. People don't do their ordinary reasoning using mathematical logic, but logic is a convenient tool for implementing certain forms of reasoning.

- In *digital circuit design and computer architecture*, logic is the language used to describe the signal values that are produced by components. A common problem is that a first-draft circuit design written by an engineer is too slow, so it has to be transformed into an equivalent circuit that is more efficient. This process is often quite tricky, and logic provides the framework for doing it.

- In *database systems*, complex queries are built up from simpler components. It's essential to have a clear, precise way to express such queries, so that users of the database can be sure they're getting the right information. Logic is the key to expressing such queries.

- In *compiler construction*, the typechecking phase must determine whether the program being translated uses any variables or functions inconsistently. It turns out that the method for doing this is similar to the method for performing logical inference. As a result, algorithms that were designed originally to perform calculations in mathematical logic are now embedded in modern compilers.

- In *programming language design*, one of the most commonly used methods for specifying the meaning of a computer program is the *lambda calculus*, which is actually a formal logical system originally invented by a mathematician for purely theoretical research.

- In *computability theory*, logic is used both to specify abstract machine models and to reason about their capabilities. There has been an extremely close interaction between mathematical logicians and theoretical computer scientists in developing a variety of machine models.

In this chapter, we discuss the difficulties with informal logical reasoning in English, and we show how to avoid those difficulties with formal logic. There are several different kinds of formal logic, and for now we will consider just the simplest one, called *propositional logic*. After looking at the language of propositional logic, we will consider in detail three completely different mathematical systems for reasoning formally about propositions: truth tables, natural deduction, and Boolean algebra.

Truth tables define the meanings of the logical operators, and they can be used to calculate the values of expressions and prove that two propositions are logically equivalent. As truth tables directly express the underlying meaning of propositions, they are a *semantic* technique. Truth tables are easy to understand for small problems, but they become impossibly large for most realistic problems.

Natural deduction is a formalisation of the basic principles of reasoning. It provides a set of *inference rules* that specify exactly what new facts you are allowed to deduce from some given facts. There is no notion of the 'value' of propositions; everything in this system is encapsulated in the inference rules. Since the rules are based on the forms of propositions, and we don't work with truth values, inference is a purely *syntactic* approach to logic. Many recently developed techniques in programming language theory are based on more advanced logical systems that are related to natural deduction.

Boolean algebra is another syntactic formalisation of logic, using a set of equations—the laws of Boolean algebra—to specify that certain propositions are equal to each other. Boolean algebra is an axiomatic approach, similar to elementary algebra and geometry. It provides an effective set of laws for manipulating propositions, and is an essential tool for digital circuit design.

6.1 The Need for Formalism

Formal logic was first developed by the ancient Greeks, who wanted to be able to reason carefully about statements in natural language. They were fascinated by the idea of statements that are known to be true with absolutely no doubt whatsoever. However, they quickly realised that logical reasoning is difficult and unreliable when using a natural language like Greek (or English!). All sorts of ambiguities arise, and it's hard to keep sight of the main line of reasoning without getting confused. We will begin by looking at some of these difficulties and how to get around them using *propositional variables*.

Suppose a friend says 'The sun is shining and I feel happy'. At first sight the meaning is obvious, but when you think about the sentence more carefully it isn't so clear. Perhaps your friend likes sunny days and feels happy *because* the sun is shining. It sounds like there is some connection between the two parts of the sentence, so in this context *and* means *and therefore*. But this reasoning depends on our experience with bright weather and happiness, and it has nothing to do with the logical structure of the sentence. Now consider another example: 'Cats are furry and elephants are heavy'. This has exactly the same structure as the preceding example, but nobody would assume that elephants are heavy *because of* the furriness of cats. In this case, *and* means *and also*. The word *and* has several subtly different meanings, and we choose the appropriate meaning using our knowledge of the world. Unfortunately, this means we can't even rely on a simple word like *and* while reasoning in English.

Furthermore, there are many other problems in working out the precise meanings of English sentences. For example, 'The sun is shining' is true some days and false other days. The meaning of 'That cloud looks like a motor bike' depends on who says it and which cloud they are pointing at. The list of such problems seems to be endless, and you can read about them in books on linguistics and philosophy.

There is no way to solve all the ambiguities of English. Who would want to do that anyway? The subtle nuances in natural language are not necessarily bad: they lead to much of the richness and expressiveness of literature. Yet they certainly can get in the way of logical thinking.

Instead of attempting the impossible—totally reliable reasoning in natural language—we need to *separate* the logical structure of an argument from all the connotations of the English. We do this using *propositions*.

A proposition is just a symbolic variable whose value must be either True or Falseand which stands for an English statement that could be either true or false. The crucial point here is that a proposition *must* be either True or False; there is no room for shades of meaning or interpretation. Usually we'll use A, B, C, D, etc. as propositional variables, but any variable name would do. For example, we can define some propositional variables to stand for the following English statements:

$$A \ = \ \text{The sun is shining.}$$
$$B \ = \ \text{I feel happy.}$$
$$C \ = \ \text{Cats are furry.}$$
$$D \ = \ \text{Elephants are heavy.}$$

The next step is to translate a complete English sentence into a mathematical statement that contains nothing but propositional variables (A, B etc.) and logical operators (*and, or, not, implies*).

| The sun is shining and I feel happy. | $\implies$ | A and B |
| Cats are furry and elephants are heavy. | $\implies$ | C and D |

The translation step is absolutely crucial, because it removes all the ambiguities of English and all our knowledge and experience of the world (like sunlight bringing happiness), and it leaves us with nothing but propositional variables and logical operators.

Sometimes it isn't clear how to translate an English statement into propositional logic. What about the sentence, 'It is raining but Jim is happy'—does this mean the same thing as 'It is raining *and* Jim is happy' or does the use of *but* indicate a different meaning? Such questions fall outside the realm of mathematics, and you just have to figure out what the English means or ask for a clarification. At least there is only one time when we need to worry about the subtleties of natural language—during the translation process—and we can forget about it thereafter.

A proposition must be either true or false; it cannot be 'maybe' or 'sometimes' or 'yes, but...'. If you translate an opinion like 'Cats are better than dogs because they purr' into a propositional variable P, then within the mathematics you'll just have to accept P as being true or false, and you won't be able to get at anything *inside* P, such as the reason that cats are better than dogs (indeed, you can't even get inside P to find out that it is about cats and dogs, not about turtles and rabbits).

There are many statements that cannot be represented by propositions because they require context in order to make their meaning clear. For example, if we define A to represent 'That cloud looks like a motor bike', then it must be clearly established in the English *which* cloud looks like a motor bike, and *who* thinks that. There are more complex logical systems that incorporate time, but propositional logic doesn't do that.

6.2   The Basic Logical Operators

Logical operators correspond to English words like *and, or, not*, and *therefore*, providing a way to build complex propositions from simpler ones. This section defines the exact meaning of the logical operators and shows how to translate between English statements and mathematical propositions.

6.2.1   Logical And ($\wedge$)

The *logical and* operator corresponds to the English conjunction 'and': it is used to claim that two statements are both true. Sometimes it is called 'logical conjunction', and its mathematical symbol is $\wedge$.

The proposition $A \wedge B$ is simply a statement that A is True and also B is True. It doesn't have any subtle connotations; in particular, it doesn't mean that there is any connection between A and B.

It's vitally important to remember that people can say things that are untrue! If someone tells you '$A \wedge B$', you have to bear in mind two possibilities:

- Their statement was correct.

- They lied to you.

If their statement was true, then both A and B are indeed true. However, if they lied to you[1] then you don't actually know about the truth of A or B.

Because we can't simply accept every statement as being true, we need a way to calculate whether a statement is true based on its constituent parts. For example if you already know that A is true and B is true, then the statement 'A and B' is certainly true. But if you know that A is false (or that B is false), then you know that the statement '$A \wedge B$' is false.

The mathematical symbol for logical *and* is $\wedge$. This symbol is shorter than '*and*', and it is clearly a mathematical operator—there is no danger of confusing $\wedge$ with the various vague meanings of the English word 'and'.[2]

Think of *and* as an operator over logical propositions, just as + is an operator over numbers. We can define the meaning of *and* by considering whether '$A \wedge B$' is true for all possible values of A and all possible values of B. Such a listing is called a truth table, and here is the definition of logical *and*:

A	B	$A \wedge B$
False	False	False
False	True	False
True	False	False
True	True	True

---

[1] A Los Angeles car salesman famous for his flamboyant television advertisements once ran a commercial claiming, 'We lose money on every car we sell, but we make it up on volume.'

[2] In the elderly (but still popular!) programming language Cobol, you do arithmetic with statements like Add a to b giving x instead of the more usual x := a+b. The designers of Cobol believed that + is mathematical, and therefore difficult, while 'Add ... giving ...' is English, and therefore easy. The Cobol notation is at least partly readable by nonprogrammers, which may be valuable in commercial applications, but it becomes unwieldy for large scale calculations. Once you get used to it, mathematical notation is *easier* than English!

Each time False is an argument value, ∧ returns False. Only when both arguments are True does it return True.

Various notations are used for the two truth values. Here we have used the full names True and False, but those become tedious to write when there are a lot of entries in the table. Another common notation is to use T and F. The usual notation in digital circuit design is to use 1 for True and 0 for False. Apart from being concise, this has the great advantage that 0 and 1 are more easily distinguishable than T and F, making the truth tables more readable. You may use any of these notations. However, if you use 0 and 1 as your notation for the truth values, be sure to remember that these are not numbers. For the sake of comparison, here is how the preceding truth table looks like with the 0, 1 notation.

A	B	$A \wedge B$
0	0	0
0	1	0
1	0	0
1	1	1

Given any two propositions, you can build a bigger one by connecting them with ∧. For example, from the propositional variables A, B and C we can build an endless list of more complex propositions, including the following:

$$A \wedge B$$
$$A \wedge (B \wedge B)$$
$$(A \wedge B) \wedge C$$
$$(A \wedge B) \wedge (B \wedge C)$$

6.2.2   Inclusive Logical Or (∨)

The *logical or* operator corresponds to the most common usage of the English word *or*. It takes two arguments and returns True if either argument (or both) is True; otherwise it returns False. Other commonly used names for this operation are *logical disjunction* and *inclusive or*, and the symbol for the operation is ∨. Here is the truth table defining ∨:

P	Q	$P \vee Q$
False	False	False
False	True	True
True	False	True
True	True	True

The English word 'or' has several different meanings. The *inclusive or* function corresponds to the simplest of these: if $A \vee B$ is true, then perhaps A is true, perhaps B is true, perhaps both are true, but you know they can't both be false. However, this is *all* that $A \vee B$ means. It does not indicate any connection between A and B.

6.2.3 Exclusive Logical Or ($\otimes$)

In English, we often think of 'A or B' as meaning 'either A or B—one or the other, but not both', excluding the possibility that A is true at the same time as B. A typical example is 'It will be bright and sunny at the picnic, or I will go home.' This meaning of the word 'or' is called *exclusive or* because if one of the alternatives is true, this excludes the possibility that the other is also true. Often the 'either ... or' construction indicates that the exclusive or is intended; for example: 'Either you pay me, or I will sue you.' However, you can't always rely on the presence of 'either'. The symbol for exclusive or is $\otimes$. Its truth table differs from the $\vee$ truth table only in the last line.

P	Q	$P \otimes Q$
False	False	False
False	True	True
True	False	True
True	True	False

The English word 'or' sometimes means inclusive or and sometimes exclusive or, so you have to be particularly careful when translating English sentences containing 'or' into propositions.

In many applications of mathematical logic, inclusive or plays a central role while exclusive or isn't too important. In many books and papers, the word 'or' is taken to mean inclusive or, and the exclusive variety isn't mentioned at all. When logic is applied to digital circuit design, both kinds of or are used heavily.

6.2.4 Logical Not ($\neg$)

The English word 'not' is used to assert that a statement is false. It corresponds to the *logical not* (also called *logical negation*) operator, whose symbol is $\neg$. It takes one argument and returns the opposite truth value:

A	$\neg A$
False	True
True	False

6.2.5 Logical Implication ($\rightarrow$)

The logical implication operator $\rightarrow$ corresponds to conditional phrases in English. For example, an English sentence in the form 'If A is true, then B is true.' would be translated into the proposition $A \rightarrow B$. Logical implication is also closely related to the **if ... then ... then** construct in programming languages. As usual, the precise definition is given by a truth table:

A	B	$A \to B$
False	False	True
False	True	True
True	False	False
True	True	True

Suppose you know that a proposition $A \to B$ is true. This says that *if* A is true, then B *must also* be true. But that's all it means—in particular, it doesn't tell you whether A is actually true. A common pitfall in informal English debate is to jump to the conclusion that A is true, when the speaker has merely said 'if A is true, then B is also true.' In the following example, $A \to B$ does *not* tell you what the weather will be like today!

A	It is sunny today.
B	There will be a picnic.
$A \to B$	If it is sunny today, then there will be a picnic.

There is a subtle but crucial difference between $\to$ and the corresponding English sentences. The last English sentence in the example above suggests a cause-and-effect relationship between the weather and the picnic: the picnic might be cancelled *because of* the rain. In contrast, $A \to B$ says nothing at all about any connection between A and B; it means nothing more than what the truth table says. Consider the following example:

A	The moon orbits the earth.
B	The sun is hot.
$A \to B$	If the moon orbits the earth, then the sun is hot.

These statements are all true! Logically, this is identical to the previous example. Since A and B are both true, the definition of $\to$ says that $A \to B$ is also true. However, the English translation sounds strange: it is true that the moon orbits the earth, and the sun is surely hot, but the *reason* the sun is hot has nothing to do with the moon. The English sentence is misleading, and you wouldn't want to use it in a debate, but there isn't anything wrong with the logical proposition. *Logical implication says nothing about cause-and-effect relationships.*

Another point about implication sometimes causes confusion: many people find the first line of the implication truth table surprising. If A is false, then why should $A \to B$ be true, when B is actually untrue? Let's see how this might translate into English, picking an example where A and B are both false:

A	The sun is cold.
B	The moon is made of green cheese.
$A \to B$	If the sun is cold, then the moon is made of green cheese.

Here, A and B are both false, but the definition says that $A \to B$ is true. Yet why should the English translation of $A \to B$ be true?

Even though it may seem strange to you to define False → False to be True, you probably already understand the idea underlying the definition: it reflects a colourful way to express skepticism. Suppose, for example, that your friend claims 'Fifty people came to my party last night', but you're sure there were only twenty. You might retort, 'If fifty people went to your party, then I'm the king of China'. Your reply has the form $A \to$ False, where A means fifty people went to the party, because you aren't the king of China. Furthermore, *the statement you are making is true*. Thus you are asserting $A \to$ False = True, and a quick study of the truth table for → shows that A must be False.

English sentences contain all sorts of connotations, but logical propositions mean nothing more than what the defining truth tables say. The truth table for → is just a definition. It is meaningless to argue about whether the definition given above is 'correct'; definitions aren't either right or wrong, they are just definitions.

The pertinent question to ask is *why is it convenient to define → so that* False → False *is true?* Perhaps the most honest answer is just to say that logicians have been refining their definitions for hundreds of years, and based on their experience, they feel this is the best way to define →. However, here's a more computer-oriented way to think about it: consider the programming language statement **if** A **then** B. If A turns out to be true, then the statement B will be executed, but if A is false, then the machine won't even look at B. It makes no difference at all what B says. The definition of → captures this indifference to B in the cases where A is false.

6.2.6   Logical Equivalence (↔)

The logical equivalence operator ↔ is used to claim that two propositions have the same value—either both are true, or both are false. The proposition $A \leftrightarrow B$ might be translated into an English sentence like 'saying A is just the same as saying B.' The definition of ↔ is

A	B	$A \leftrightarrow B$
False	False	True
False	True	False
True	False	False
True	True	True

The statement $A \leftrightarrow B$ could also be expressed by writing $(A \to B) \wedge (B \to A)$. Sometimes ↔ is simply considered to be an abbreviation for conjunction of two ordinary implications, rather than a fundamental logical operator.

There are two distinct mathematical ways to claim that the propositions A and B have the same value. One way is to write the equation $A = B$, and the other way is to claim that the proposition $A \leftrightarrow B$ has the value True. Logical equivalence is similar to *but not the same as* ordinary equality. The difference between them is important, and we'll say more about it later.

6.3   The Language of Propositional Logic

Before spending a lot of effort arguing about a statement somebody claims to be true, it's prudent to make sure the statement actually makes sense and to check that all the participants in the debate agree about what it means. It might be fun to discuss 'International trade creates jobs', but there's no point wasting time on 'Quickly raspberries running green'. That one isn't either true or false—it's just gibberish.

A big advantage of using logic rather than English is that we can define the language of propositional logic formally, precisely, and unambiguously. This allows us to check that a proposition makes sense before expending any further work deciding whether it's true. Because there is no formal definition of natural languages, it's fundamentally impossible to do that for English. It isn't even possible in all cases to decide definitively whether a statement in a natural language is grammatical: English has no formal grammar.

A proposition that 'makes sense' is called a *well-formed formula*, which is often abbreviated WFF (pronounced 'woof'). Every WFF has a well-defined meaning, and you can work out whether it's true or false given the values of the propositional variables. For example, $(A \rightarrow (B \wedge (\neg A)))$ is a well-formed formula, so we can calculate what it means, and it's worth spending a little time to decide whether there are any possible values of A and B for which it's true. (It's True if A is False.) On the other hand, what about $\vee AB\neg C$? This isn't a WFF. It's just nonsense, so it has no truth value.

Many programming languages (including Fortran, Pascal, Ada, Haskell, and many others) provide a Boolean data type[3] for use with control statements such as **if** statements, **while** loops, and so on. A variable of type Boolean has just two values, True and False, so it corresponds exactly to a propositional variable. These programming languages also provide some or all of the $\wedge$, $\vee$, $\neg$, $\rightarrow$, and $\leftrightarrow$ operators. Saying that a logical proposition is well-formed is just like saying that a Boolean expression in a programming language is syntactically valid.

There are also many programming languages (C, for example) that lack a Boolean type, using numbers instead. This is a loose approach to syntax, quite different from the precise rules for WFFs in logic. It is sometimes convenient, but it also leads to problematical definitions about whether expressions like $\sqrt{\sin x}$ are true or false, and it prevents the compiler from producing helpful type error messages if you accidentally use a non-Boolean expression where only a Boolean would make sense.

6.3.1   The Syntax of Well-Formed Formulas

A term in propositional logic is sometimes called a *formula*, and a term that is constructed correctly, following all the syntax rules, is called a *well-formed formula*, abbreviated *WFF*.

---

[3]Different languages use different names for the Boolean type; common choices include Bool, Boolean, and Logical.

Soon we will study in detail how to reason about the meanings of WFFs. However, we will never consider the meaning of ill-formed formulas; these are simply considered to be meaningless. For example, the formula $(P \land (\neg P))$ is well-formed; this means that we can go on to consider its meaning (this WFF happens to be False). However, the formula $P \lor \to Q$ violates the syntax rules of the language, so we refuse to be drawn into debates about whether it is True or False. The situation is similar to that of programming languages: we can talk about the behaviour of a program that is syntactically correct (and the behaviour might be correct or incorrect), but we never talk about the behaviour of a program with syntax errors—the compiler would refuse to translate the program, so there is no run-time behaviour.

The set of well-formed formulas (WFFs) is defined by saying precisely how you can construct them.

- The constants False and True are WFFs.
  The only examples are: False,   True.

- Any propositional variable is a WFF.
  Examples: P, Q, R.

- If a is a WFF, then $(\neg a)$ is a WFF.
  Examples: $(\neg P)$, $(\neg Q)$, $(\neg(\neg P))$.

- If a and b are WFFs, then $(a \land b)$ is a WFF.
  Examples: $(P \land Q)$, $((\neg P) \land (P \land Q))$.

- If a and b are WFFs, then $(a \lor b)$ is a WFF.
  Examples: $(P \lor Q)$, $((P \land Q) \lor (\neg Q))$.

- If a and b are WFFs, then $(a \to b)$ is a WFF.
  Examples: $(P \to Q)$, $((P \land Q) \to (P \lor Q))$.

- If a and b are WFFs, then $(a \leftrightarrow b)$ is a WFF.
  Examples: $(P \leftrightarrow Q)$, $((P \land Q) \leftrightarrow (Q \land P))$.

- Any formula that cannot be constructed using these rules is *not* a WFF.
  Examples: $(P \lor \land Q)$, $P \to \neg$.

The rules may be used repeatedly to build nested formulas. This process is called *recursion* and is the subject of Chapters 3, 4, and 9. For example, here is a demonstration, using recursion, that $(P \to (Q \land R))$ is a WFF:

1. P, Q and R are propositional variables, so they are all WFFs.

2. Because Q and R are WFFs, $(Q \land R)$ is also a WFF.

3. Because P and $(Q \land R)$ are both WFFs, so is $(P \to (Q \land R))$.

6.3.2   Precedence of Logical Operators

Well-formed formulas are fully parenthesised, so there is no ambiguity in their interpretation. Often, however, it's more convenient to omit some of the parentheses for the sake of readability. For example, we would prefer to write $P \to \neg Q \wedge R$ rather than $(P \to ((\neg Q) \wedge R))$.

The syntax rules given below define what an expression means when some of the parentheses are omitted. These conventions are analogous to those of elementary algebra, as well as most programming languages, where there is a precedence rule that says $a + b \times c$ means $a + (b \times c)$ rather than $(a + b) \times c$. The syntax rules for propositional logic are straightforward:

1. The most tightly binding operator is $\neg$. For example, $\neg P \wedge Q$ means $(\neg P) \wedge Q$. Furthermore, $\neg\neg P$ means $\neg(\neg P)$.

2. The second highest precedence is the $\wedge$ operator. In expressions combining $\wedge$ and $\vee$, the $\wedge$ operations come first. For example, $P \vee Q \wedge R$ means $P \vee (Q \wedge R)$. If there are several $\wedge$ operations in a sequence, they are performed left to right; for example, $P \wedge Q \wedge R \wedge S$ means $(((P \wedge Q) \wedge R) \wedge S)$. This property is described by saying '$\wedge$ associates to the left.'

3. The $\vee$ operator has the next level of precedence, and it associates to the left. For example, $P \wedge Q \vee R \vee U \wedge V$ means $(((P \wedge Q) \vee R) \vee (U \wedge V))$. This example would be more readable if a few of the parentheses are removed: $(P \wedge Q) \vee R \vee (U \wedge V)$.

4. The $\to$ operator has the next lower level of precedence. For example, $P \wedge Q \to P \vee Q$ means $(P \wedge Q) \to (P \vee Q)$. The $\to$ operator associates to the right; thus $P \to Q \to R \to S$ means $(P \to (Q \to (R \to S)))$.

5. The $\leftrightarrow$ operator has the lowest level of precedence, and it associates to the right, but we recommend that you use parentheses rather than relying on the associativity.

In general it's a good idea to omit parentheses that are clearly redundant but to include parentheses that improve readability. Too much reliance on these syntax conventions leads to inscrutable expressions.

6.3.3   Object Language and Meta-Language

Propositional logic is a precisely defined language of well-formed formulas. We need another richer language to use when we are reasoning *about* well-formed formulas. This means we will be working simultaneously with two distinct languages that must be kept separate. The WFFs of propositional logic will be called the *object language* because the objects we will be talking about are sentences in propositional logic. The algebraic language of equations, substitutions, and justifications is essentially just ordinary mathematics; we will call it the *meta-language* because it 'surrounds' the object language and can be used to talk *about* propositions.

This distinction between object language and meta language is common in computer science. For example, there are many programming languages, and we need both to write programs *in* them and also to make statements *about* them.

We have already seen all the operators in the propositional logic object language; these are ¬, ∧, ∨, → and ↔. Later we will use several operators that belong to the meta-language; these are ⊨ (which will be used in Section 6.4 on truth tables); ⊢ (which will be used in Section 6.5 on natural deduction); and = (which will be used in Section 6.7 on Boolean algebra). After studying each of those three major systems, we will look briefly at some of the meta-logical properties of ⊨, ⊢, and = in Section 6.9 on meta-logic.

6.3.4   Computing with Boolean Expressions

It's quicker and more reliable to compute large logical expressions with a computer than doing it by hand. In Haskell, logical variables have type `Bool`, and there are two constants of type `Bool`, namely `True` and `False`. Haskell contains built-in operators for a few of the logical operations, and the software tools provided with this book define the rest:

- $\neg x$ is written in Haskell as `not x`;

- $a \wedge b$ is written either as `a && b` or as `a /\ b`;

- $a \vee b$ is written either as `a || b` or as `a \/ b`;

- $a \rightarrow b$ is written as `a ==> b`;

- $a \leftrightarrow b$ is written as `a <=> b`.

Exercise 1. Imagine that you are visiting Hitech City, where all the buses are driverless (because they are fully automated) and at every bus stop there is a computer that understands spoken English. The computer can respond to yes-or-no questions and knows the routes of all buses. You are trying to get to the airport and have arrived at a bus stop shortly before a bus waiting there is ready to leave. You know that either this bus or the next one is going to the airport, but not both. You only have time to ask the computer one question before the bus leaves, and you must take this bus or the next to avoid missing your flight.

Unfortunately, some of the computers in Hitech City have a flaw in their design. They understand spoken English and can answer yes-or-know questions, but something in their circuitry causes them to answer "no" when a properly functioning computer would have answered "yes" and, vice versa, to answer "yes" when a properly functioning computer would have answered "no". You do not know whether the computer at this bus stop functions properly or contains a twisted-logic circuit.

All the computers have names, and no two of those names are the same. The computer at your bus stop is named "Bob". Devise a question that you could ask Bob and, based on Bob's yes-or-no answer, decide correctly whether to take this bus or the next one to make your flight. Remember, you are allowed only one question, the question must be phrased in a way that requires a yes-or-no answer, and you will get only one answer, which will be either "yes" or "no". The answer will be the correct answer to your question if Bob is a properly functioning computer, and it will be the wrong answer to your question if Bob contains the twisted-logic circuit, but you have no way to tell which kind of computer Bob is.

Exercise 2. Check your understanding of *or*, *and*, and *not* by deciding what value each of these expressions has and then evaluating it with the computer (you will need to import the `Stdm` module provided on the web page for this book):

(a) `False /\ True`

(b) `True \/ (not True)`

(c) `not (False \/ True)`

(d) `(not (False /\ True)) \/ False`

(e) `(not True) ==> True`

(f) `True \/ False ==> True`

(g) `True ==> (True /\ False)`

(h) `False ==> False`

(i) `(not False) <=> True`

(j) `True <=> (False ==> False)`

(k) `False <=> (True /\ (False ==> True))`

(l) `(not (True \/ False)) <=> False /\ True`

6.4   Truth Tables: Semantic Reasoning

Truth tables provide an easy method for reasoning about propositions (as long as they don't contain too many variables). The truth table approach treats propositions as expressions to be evaluated. All the expressions have type Bool, the constants are True and False, and the variables $A, B, \ldots$ stand for unknowns that must be either True or False.

When you write out truth tables by hand, it's really too tedious to keep writing True and False. We recommend abbreviating True as 1 and False as 0;

as pointed out earlier, T and F are also common abbreviations but are harder to read. However, it should be stressed that 0 and 1 are just abbreviations for the logical constants, whose type is Bool. Truth tables don't contain numbers. Only a few truth tables will appear below, so we will forgo the abbreviations, and stick safely with True and False. (However, when truth tables are used for reasoning about digital circuits, it is standard practice to use 0 and 1.)

6.4.1 Truth Table Calculations and Proofs

We have already been using truth tables to define the logical operators, so the format should be clear by now. Nothing could be simpler than using a truth table to evaluate a propositional expression: you just calculate the proposition for all possible values of the variables, write down the results in a table, and see what happens. Generally, it's useful to give several of the intermediate results in the truth table, not just the final value of the entire proposition. For example, let's consider the proposition $((A \to B) \land \neg B) \to \neg A$ and find out when it's true. On the left side of the table we list all the possible values of A and B, on the right side we give the value of the full proposition, and in the middle (separated by double vertical lines) we give the values of the subexpressions.

A	B	$A \to B$	$\neg B$	$(A \to B) \land \neg B$	$\neg A$	$((A \to B) \land \neg B) \to \neg A$
False	False	True	True	True	True	True
False	True	True	False	False	True	True
True	False	False	True	False	False	True
True	True	True	False	False	False	True

Many propositions may be either True or False, depending on the values of their variables; all of the propositions between the double vertical lines in the table above are like that. Special names are given to propositions where this is not the case:

Definition 8. A *tautology* is a proposition that is always True, regardless of the values of its variables.

Definition 9. A *contradiction* is a proposition that is always False, regardless of the values of its variables.

You can find out whether a proposition is a tautology or a contradiction (or neither) by writing down its truth table. If a column contains nothing but True the proposition is a tautology; if there is nothing but False it's a contradiction, and if there is a mixture of True and False the proposition is neither. For example, $P \lor \neg P$ is a tautology, but $P \land \neg P$ is a contradiction, and the following truth table proves it:

P	$\neg P$	$P \vee \neg P$	$P \wedge \neg P$
False	True	True	False
True	False	True	False

There is a special notation for expressing statements about propositions. As it's used to make statements *about* propositions, this notation belongs to the meta-language, and it uses the meta-operator $\models$, which is often pronounced 'double-turnstile' (to distinguish it from $\vdash$, which you'll see later, and which is pronounced 'turnstile').

Definition 10. The notation $P_1, P_2, \ldots, P_n \models Q$ means that if all of the propositions $P_1, P_2, \ldots, P_n$ are true, then the proposition Q is also true.

The $\models$ meta-operator makes a statement about the actual meanings of the propositions; it's concerned with which propositions are True and which are False. Truth tables can be used to prove statements containing $\models$. The meaning of a proposition is called its *semantics*. The set of basic truth values (True and False), along with a method for calculating the meaning of any well-formed formula, is called a *model of the logical system*.

There may be any number n of propositions $P_1, P_2, \ldots, P_n$ in the list of assumptions. If $n = 1$ then the statement looks like $P \models Q$. A particularly important case is when $n = 0$, so there are no assumptions at all, and the statement becomes simply $\models Q$. This means that Q is always true, regardless of the values of the propositional variables inside it; in other words, Q is a tautology.

6.4.2 Limitations of Truth Tables

The truth table method is straightforward to use; it just requires some brute-force calculation. You don't need any insight into *why* the proposition you're working on is true (or false, or sometimes true). This is both a strength and a weakness of the method. It's comforting to know that you can reliably crank out a proof, given enough time. However, a big advantage of the other two logical systems we will study, natural deduction and Boolean algebra, is that they give much better insight into *why* a theorem is true.

The truth table method requires one line for each combination of variable values. If there are k variables, this means there are 2^k lines in the table. For one variable, you get $2^1 = 2$ lines (for example, the truth table for the $\neg$ operator). For two variables, there are $2^2 = 4$ lines (like the definitions of $\vee$, $\wedge$, $\rightarrow$ and $\leftrightarrow$). For three variables there are $2^3 = 8$ lines, which is about as much as anybody can handle.

Since the number of lines in the table grows exponentially in the number of variables, the truth table method becomes unwieldy for most interesting problems. In Chapter 13 we will prove a theorem whose truth table has 2^{129} lines. That number is larger than the number of atoms in the known universe, so you'll be relieved to know that we will skip the truth table and use more powerful methods.

6.4.3 Computing Truth Tables

The software tools for this book contain functions that make it easy to compute truth tables automatically; see the online documentation.

Exercise 3. Use the truth table functions to determine which of the following formulas are tautologies.

(a) $(\text{True} \wedge P) \vee Q$

(b) $(P \vee Q) \to (P \wedge Q)$

(c) $(P \wedge Q) \to (P \vee Q)$

(d) $(P \vee Q) \to (Q \vee P)$

(e) $((P \vee Q) \wedge (P \vee R)) \leftrightarrow (P \wedge (Q \vee R))$

Exercise 4. In this exercise, and several more that follow it, you are given an expression. Verify that the expression is a WFF by analyzing its constituents at all levels down to the atomic primitives. Build a truth table for the formula with one column for the formula itself (this should be the rightmost column), and one column for each constituent of the formula (at any level in the analysis).

Each row in the table will correspond to a distinct combination of values of the atomic primitives in the formula, and the rows will exhaust all possible such combinations. Each column will begin with a heading, which is a WFF (the full WFF or one of its constituents), and will be filled with the value of the WFF that corresponds to the combination of values of the primitives represented by a particular row.

A WFF is said to be *satisfiable* if it is True for some values of its propositional variables. All WFFs fall into exactly one of three categories: tautology, contradiction, or satisfiable but not tautology. Based on the results in the truth table, place the WFF in one of these categories.

$$(\text{True} \wedge P) \vee Q$$

Exercise 5.

$$(P \vee Q) \wedge (P \vee R) \leftrightarrow P \wedge (Q \vee R)$$

Exercise 6.

$$((P \wedge \neg Q) \vee (Q \wedge \neg P)) \to \neg(P \leftrightarrow Q)$$

Exercise 7.

$$(P \to Q) \wedge (P \to \neg Q)$$

Exercise 8.

$$(P \to Q) \wedge (\neg P \to Q)$$

Exercise 9.

$$(P \to Q) \leftrightarrow (\neg Q \to \neg P)$$

6.5   Natural Deduction: Inference Reasoning

Natural deduction is a formal logical system that allows you to reason directly with logical propositions using *inference*, without having to substitute truth values for variables or evaluate expressions. Natural deduction provides a solid theoretical foundation for logic, and it helps focus attention on *why* logical propositions are true or false. Furthermore, natural deduction is well suited for automatic proof checking by computer, and it has (along with some closely related systems) a variety of applications in computer science.

In normal English usage, the verb *infer* means to reason about some statements in order to reach a conclusion. This is an informal process, and it can run into problems due to the ambiguities of English, but the intent is to establish that the conclusion is definitely true.

Logical inference means reasoning *formally* about a set of statements in order to decide, beyond all shadow of doubt, what is true. In order to make the reasoning absolutely clear cut, we need to do several things:

- The set of statements we're reasoning about must be defined. This is called the *object language*, and a typical choice would be propositional expressions.

- The methods for inferring new facts from given information must be specified precisely. These are called the *inference rules*.

- The form of argument must be defined precisely, so that if anybody claims to have an argument that proves a fact, we can determine whether it actually is a valid argument. This defines a *meta-language* in which proofs of statements in the object language can be written. Every step in the reasoning must be justified by an inference rule.

There are several standard ways to write down formal proofs. In this book we'll use the form which is most commonly used in computer science.

Definition 11. The notation $P_1, P_2, \ldots, P_n \vdash Q$ is called a *sequent*, and it means that if all of the propositions $P_1, P_2, \ldots, P_n$ are known, then the proposition Q can be inferred formally using the inference rules of natural deduction.

We have seen two similar notations: with truth tables, we had meta-logical statements with the $\models$ operator, like $P \models Q \to P$. This statement means that if P is True, then the proposition $Q \to P$ is also True. It says nothing about how we know that to be the case. In contrast, the notation $P \vdash Q \to P$, which will be used in this section, means *there is a proof* of $Q \to P$, and the proof assumes P. Both $\models$ and $\vdash$ are used to state theorems; the distinction is that $\models$ is concerned with the ultimate truth of the propositions, while $\vdash$ is concerned with whether we have a proof. We will return to the relationship between $\models$ and $\vdash$ in Section 6.9.

In formal logic, you can't just make intuitive arguments and hope they are convincing. Every step of reasoning you make has to be backed up by an inference rule. Intuitively, an inference rule says, '*If you know that Statement1 and Statement2 are established (either assumed or proven), then you may infer Statement3,*' and furthermore, the inference constitutes a proof. *Statement1* and *Statement2* are called the *assumptions* of the inference rule, and *Statement3* is called the *conclusion*. (There may be any number of assumptions—there don't have to be just two of them.) Here's an example of an inference rule about the ∧ operator, written informally in English:

> If you know some proposition a, and also the proposition b, then you are allowed to infer the proposition $a \wedge b$.

In this example, there are two assumptions—a and b—and the conclusion is $a \wedge b$. Note that we have expressed this inference using *meta-variables* a and b. These meta-variables could stand for any WFF; thus a might be P, $Q \to R$, etc. The variables P, Q etc, are propositional variables, belonging to the object language, and their values are True or False. We will adopt the following convention:

- Meta-Variables belong to the meta-language and are written as lower-case letters $a, b, c, \ldots$. The value of a meta-variable is a WFF. For example a might have the value $P \wedge Q$.

- Propositional variables belong to the object language and are written as upper-case letters $A, B, \ldots, P, Q, R, \ldots$. The value of a propositional variable is either True or False.

Formally, an inference rule is expressed by writing down the assumptions above a horizontal line, and writing the conclusion below the line:

$$\frac{\text{Statement1} \qquad \text{Statement2}}{\text{Statement3}}$$

This says that if you can somehow establish the truth of Statement1 and Statement2, then Statement3 is guaranteed (by the inference rule) to be true. The inference about ∧ would be written formally as

$$\frac{a \qquad b}{a \wedge b} \; .$$

An inference rule works in only one direction—for example, if you have established $a \wedge b$, you *cannot* use the rule above to infer a. A different inference rule, which we will see later, would be needed.

Figure 6.1 summarises all the inference rules of propositional logic. In the next several sections we will work systematically through them all. It would be a good idea to refer back frequently to Figure 6.1.

$$\frac{a \quad b}{a \wedge b}\{\wedge I\} \qquad \frac{a \wedge b}{a}\{\wedge E_L\} \qquad \frac{a \wedge b}{b}\{\wedge E_R\}$$

$$\frac{a}{a \vee b}\{\vee I_L\} \qquad \frac{b}{a \vee b}\{\vee I_R\} \qquad \frac{a \vee b \quad a \vdash c \quad b \vdash c}{c}\{\vee E\}$$

$$\frac{a \vdash b}{a \to b}\{\to I\} \qquad \frac{a \quad a \to b}{b}\{\to E\}$$

$$\frac{a}{a}\{ID\} \qquad \frac{\mathsf{False}}{a}\{CTR\} \qquad \frac{\neg a \vdash \mathsf{False}}{a}\{RAA\}$$

Figure 6.1: Inference Rules of Propositional Logic

Many of the rules fall into two categories. The *introduction rules* have a conclusion into which a new logical operator has been introduced; they serve to build up more complex expressions from simpler ones. In contrast, the *elimination rules* require an assumption that contains a logical operator which is eliminated from the conclusion; these are used to break down complex expressions into simpler ones. Introduction rules tell you what you need to know in order to introduce a logical operator; elimination rules tell you what you can infer from a proposition containing a particular operator.

6.5.1   Definitions of True, ¬, and ↔

Natural deduction works with a very minimal set of basic operators. In fact, the only primitive built-in objects are the constant False, and the three operators ∧, ∨ and →. Everything else is an abbreviation! It's particularly intriguing that False is the only primitive logic value in the natural deduction system.

Definition 12. The constant True and the operators ¬ and ↔ are abbreviations defined as follows:

$$
\begin{aligned}
\mathsf{True} &= \mathsf{False} \to \mathsf{False} \\
\neg a &= a \to \mathsf{False} \\
a \leftrightarrow b &= (a \to b) \wedge (b \to a)
\end{aligned}
$$

You can check all of these definitions semantically, using truth tables. In natural deduction, however, we will manipulate them only using the inference rules, and it will gradually become clear that the abbreviations work perfectly with the inference rules. The definition of ↔ should be clear, but let's look at ¬ and True more closely.

Notice that ¬False is defined to be False → False, which happens to be the definition of True. In other words, ¬False = True. Going the other direction, ¬True becomes True → False when the ¬ abbreviation is expanded out, and that becomes (False → False) → False when the True is expanded out. Later we will see an inference rule (called *Reductio ad Absurdum*) which will allow us to infer False from (False → False) → False. The result is that ¬True = False and ¬False = True.

We will write propositions with True, ¬ and ↔ just as usual, but sometimes to make a proof go through it will be necessary to expand out the abbreviations and work just with the primitives.

6.5.2 And Introduction {∧*I*}

The *And Introduction* inference rule says that if you know that some proposition *a* is true, and you also know that *b* is true, then you may infer that the proposition *a* ∧ *b* is true. As the name 'And Introduction' suggests, the rule specifies what you have to know in order to infer a proposition with a new occurrence of ∧.

When we write an inference, the horizontal line will be annotated with the name of the inference rule that was used. The abbreviation {∧*I*} stands for 'And Introduction', so the rule is written as follows:

$$\frac{a \qquad b}{a \wedge b}\{\wedge I\}$$

We will now work through several examples, starting with a theorem saying that the conclusion *P* ∧ *Q* can be inferred from the assumptions *P* and *Q*.

Theorem 35. $P, Q \vdash P \wedge Q$

Proof. The theorem is proved by just one application of the {∧*I*} inference; it's the simplest possible example of how to use the {∧*I*} inference rule.

$$\frac{P \qquad Q}{P \wedge Q}\{\wedge I\}$$

□

Notice that the theorem above involves two specific propositional variables, *P* and *Q*. The {∧*I*} rule does not require any particular propositions. It uses the meta-variables *a* and *b*, which can stand for any well-formed formula. For example, the following theorem has the same structure as the previous one, and is also proved with just one application of the {∧*I*} rule, but this time the meta-variable *a* stands for *R* → *S* and *b* stands for ¬*P*.

Theorem 36. $(R \to S), \neg P \vdash (R \to S) \wedge \neg P$

Proof.

$$\frac{(R \to S) \quad \neg P}{(R \to S) \land \neg P}\{\land I\}$$

$\square$

Usually you need several steps to prove a theorem, and each step requires explicit use of an inference rule. You can build up a proof in one diagram. When a subproof results in a conclusion that serves as an assumption for a larger proof, the entire subproof diagram appears above the line for the main inference in the large proof. Here is an example:

Theorem 37. $P, Q, R \vdash (P \land Q) \land R$

Proof. The main inference here has two assumptions: $P \land Q$ and R. However, the first assumption $P \land Q$ is not one of the assumptions of the entire theorem; it is the conclusion of another $\{\land I\}$ inference with assumptions P and Q. The entire proof is written in a single diagram. Notice how the conclusion $P \land Q$ of the first inference is sitting in exactly the right place above the longer line (the line belonging to the main inference).

$$\frac{\dfrac{P \quad Q}{P \land Q}\{\land I\} \quad R}{(P \land Q) \land R}\{\land I\}$$

$\square$

Inference proofs have a natural tree structure: assumptions of the theorem are like the leaves of a tree, and subproofs are like the nodes (forks) of a tree. The method we are using here for writing proofs makes this tree structure explicit.

There is an alternative format for writing logical inference proofs, where each inference is written on a numbered line. When the conclusion of one inference is required as an assumption to another one, this is indicated by explicit references to the line numbers. This format looks like a flat list of statements (much like an assembly language program), while the tree format we are using has a nested structure (much like a program in a block structured language).

Logical inference has many important applications in computer science. The tree format is normally used for computing applications, so we will use that notation in this book in order to help prepare you for more advanced studies. A clear exposition of the line-number format, which won't appear in this book, appears in Lemmon's book *Beginning Logic* [21].

Exercise 10. Prove $P, Q, R \vdash P \land (Q \land R)$.

Exercise 11. Consider the following two propositions:

$$x = A \wedge (B \wedge (C \wedge D))$$
$$y = (A \wedge B) \wedge (C \wedge D)$$

Describe the shapes of the proofs for x and y, Assuming A, B, C, and D. Suppose each proposition has 2^n propositional variables. What then would be the heights of the proof trees?

6.5.3 And Elimination $\{\wedge E_L\}$, $\{\wedge E_R\}$

There are two inference rules that allow the elimination of an And operation. These rules say that if $a \wedge b$ is known to be true, then a must be true, and also b must be true. The 'And Elimination Left' $\{\wedge E_L\}$ rule retains the left argument of $a \wedge b$ in the result, while the 'And Elimination Right' $\{\wedge E_R\}$ rule retains the right argument.

$$\frac{a \wedge b}{a}\{\wedge E_L\} \qquad \frac{a \wedge b}{b}\{\wedge E_R\}$$

Here is a simple example using And Elimination:

Theorem 38. $P, Q \wedge R \vdash P \wedge Q$

Proof. Two inferences are required. First the 'Left' form of the And Elimination rule $\{\wedge E_L\}$ is used to obtain the intermediate result Q, and the And Introduction rule is then used to combine this with P to obtain the result $P \wedge Q$.

$$\cfrac{P \qquad \cfrac{Q \wedge R}{Q}\{\wedge E_L\}}{P \wedge Q}\{\wedge I\}$$

□

The next example contains several applications of And Elimination, including both the Left and Right forms of the rule. As these two examples show, the three rules $\{\wedge I\}$, $\{\wedge E_L\}$ and $\{\wedge E_R\}$ can be used systematically to deduce consequences of logical conjunctions.

Theorem 39. $(P \wedge Q) \wedge R \vdash R \wedge Q$

Proof.

$$\cfrac{\cfrac{(P \wedge Q) \wedge R}{R}\{\wedge E_R\} \qquad \cfrac{\cfrac{(P \wedge Q) \wedge R}{P \wedge Q}\{\wedge E_L\}}{Q}\{\wedge E_R\}}{R \wedge Q}\{\wedge I\}$$

□

One of the most fundamental properties of the $\wedge$ operator is that it is *commutative*: the proposition $P \wedge Q$ is equivalent to $Q \wedge P$. The following theorem establishes this.

Theorem 40. $P \wedge Q \vdash Q \wedge P$

Proof. The idea behind this proof is to infer Q and P—in that order, with Q to the left of P—above the main line, so $\{\wedge I\}$ can be used to infer $Q \wedge P$. The intermediate results Q and P are obtained by And Elimination.

$$\cfrac{\cfrac{P \wedge Q}{Q}\{\wedge E_R\} \qquad \cfrac{P \wedge Q}{P}\{\wedge E_L\}}{Q \wedge P}\{\wedge I\}$$

$\square$

Inference proof trees can become quite large, but you can always work through them systematically, one inference at a time, until you see how all the parts of the proof fit together. You can check the individual inferences in any order you like, either top-down or bottom-up or any other order you like. We will give one more example of a theorem proved with the And Introduction and Elimination rules, with a larger proof.

Theorem 41. For any well formed propositions a, b and c,

$$a \wedge (b \wedge c) \vdash (a \wedge b) \wedge c$$

Proof.

$$\cfrac{\cfrac{\cfrac{a \wedge (b \wedge c)}{a}\{\wedge E_L\} \quad \cfrac{\cfrac{a \wedge (b \wedge c)}{b \wedge c}\{\wedge E_R\}}{b}\{\wedge E_L\}}{a \wedge b}\{\wedge I\} \quad \cfrac{\cfrac{a \wedge (b \wedge c)}{b \wedge c}\{\wedge E_R\}}{c}\{\wedge E_R\}}{(a \wedge b) \wedge c}\{\wedge I\}$$

$\square$

Exercise 12. Prove $(P \wedge Q) \wedge R \vdash P \wedge (Q \wedge R)$.

6.5.4 Imply Elimination $\{\rightarrow E\}$

As we work through the inference rules, you should refer back frequently to Figure 6.1. This will help build up a coherent picture of the complete system of natural deduction. Although the rules relating to $\vee$ come next in the figure, we will first study the rules for $\rightarrow$, which are slightly simpler.

The Imply Elimination rule $\{\rightarrow E\}$ says that if you know a is true, and also that a implies b, then you can infer b. The traditional Latin name for the rule is *Modus Ponens*.

$$\frac{a \qquad a \to b}{b}\{\to E\}$$

The following theorem provides a simple example of the application of $\{\to E\}$.

Theorem 42. $Q \land P, \ P \land Q \to R \vdash R$.

Proof.

$$\frac{\dfrac{Q \land P}{P}\{\land E_R\} \qquad \dfrac{Q \land P}{Q}\{\land E_L\}}{\dfrac{P \land Q}{}\{\land I\} \qquad P \land Q \to R}{R}\{\to E\}$$

□

Often we have chains of implication of the form $a \to b$, $b \to c$ and so on. The following theorem says that given a and these linked implications, you can infer c.

Theorem 43. For all propositions a, b and c, $a, a \to b, b \to c \vdash c$.

Proof.

$$\frac{\dfrac{a \quad a \to b}{b}\{\to E\} \qquad b \to c}{c}\{\to E\}$$

□

Exercise 13. Prove $P, \ P \to Q, \ (P \land Q) \to (R \land S) \vdash S$.

Exercise 14. Prove $P \to Q, \ R \to S, \ P \land R \vdash S \land R$.

6.5.5 Imply Introduction $\{\to I\}$

The Imply Introduction rule $\{\to I\}$ says that, in order to infer the logical implication $a \to b$, you must have a proof of b using a as an assumption.

$$\frac{a \vdash b}{a \to b}\{\to I\}$$

We will first give a simple example of Imply Introduction and then discuss the important issue of keeping track of the assumptions that have been made.

Theorem 44. $\vdash (P \wedge Q) \to Q$.

Proof. First consider the sequent $P \wedge Q \vdash Q$, which is proved by the following And Elimination inference:

$$\frac{P \wedge Q}{Q}\{\wedge E_R\}$$

Now we can use the sequent (that is, the theorem established by the inference) and the $\{\to I\}$ rule:

$$\frac{P \wedge Q \vdash Q}{P \wedge Q \to Q}\{\to I\}$$

$\square$

It is crucially important to be careful about what we are assuming. In fact, the reason for having the sequent notation with the $\vdash$ operator is to write down the assumptions of a theorem just as precisely as the conclusion.

In the $\{\wedge E_R\}$ inference we assumed $P \wedge Q$; without that assumption we could not have inferred Q. Therefore this assumption appears in the sequent to the left of the $\vdash$. However, the entire sequent $P \wedge Q \vdash Q$ does not rely on any further assumptions; it is independently true. Therefore we can put it above the line of the $\{\to I\}$ rule 'for free,' without actually making any assumptions. Since nothing at all needed to be assumed to support the application of the $\{\to I\}$ rule, we end up with a theorem that doesn't require any assumptions. The sequent that expresses the theorem, $\vdash P \wedge Q \to Q$, has nothing to the left of the $\vdash$ operator.

It is customary to write the entire proof as a single tree, where a complete proof tree appears above the line of the $\{\to I\}$ rule. That allows us to prove $\vdash P \wedge Q \to Q$ with just one diagram:

$$\frac{\dfrac{P \wedge Q}{Q}\{\wedge E_R\}}{P \wedge Q \to Q}\{\to I\}$$

From this diagram, it looks like there is an assumption $P \wedge Q$. However, that was a temporary, local assumption whose only purpose was to establish $P \wedge Q \vdash Q$. Once that result is obtained the assumption $P \wedge Q$ can be thrown away. An assumption that is made temporarily only in order to establish a sequent, and which is then thrown away, is said to be *discharged*. A discharged assumption does *not* need to appear to the left of the $\vdash$ of the main theorem. In our example, the proposition $P \wedge Q \to Q$ is *always* true, and it doesn't matter whether $P \wedge Q$ is true or false.

In big proof trees it may be tricky to keep track of which assumptions have been discharged and which have not. We will indicate the discharged assumptions by putting a box around them. (A more common notation is to draw a line through the discharged assumption, but for certain propositions that leads to a highly unreadable result.) Following this convention, the proof becomes:

$$\cfrac{\cfrac{\boxed{P \wedge Q}}{Q}\{\wedge E_R\}}{P \wedge Q \to Q}\{\to I\}$$

Recall the proof of Theorem 43, which used the chain of implications $a \to b$ and $b \to c$ to infer c, given also that a is true. A more satisfactory theorem would just focus on the chain of implications, without relying on a actually being true. The following theorem gives this purified chain property: it says that if $a \to b$ and $b \to c$, then $a \to c$.

Theorem 45 (Implication chain rule). For all propositions a, b and c,

$$a \to b, \; b \to c \vdash a \to c$$

Proof. Because we are proving an implication $a \to c$, we need to use the $\{\to I\}$ rule—there is no other way to introduce the $\to$ operator! That rule requires a proof of c given the assumption a, which is essentially the same proof tree used before in Theorem 43. The important point, however, is that the assumption a is discharged when we apply $\{\to I\}$. The other two assumptions, ($a \to b$ and $b \to c$), are not discharged. Consequently a doesn't appear to the left of the $\vdash$ in the theorem; instead it appears to the left of the $\to$ in the *conclusion* of the theorem.

$$\cfrac{\cfrac{\cfrac{\boxed{a} \quad a \to b}{b}\{\to E\} \quad b \to c}{c}\{\to E\}}{a \to c}\{\to I\}$$

$\square$

Sometimes in large proofs it can be confusing to see just where an assumption has been discharged. You may have an assumption P with a box around it, but there could be all sorts of implications $P \to \cdots$ which came from somewhere else. In such cases it's probably clearer to build up the proof in stages, with several separate proof tree diagrams.

A corollary of this implication chain rule is the following theorem, which says that if $a \to b$ but you know that b is false, than a must also be false. This is an important theorem which is widely used to prove other theorems, and its traditional Latin name is *Modus Tollens*.

Theorem 46 (*Modus Tollens*). For all propositions a and b,

$$a \to b, \; \neg b \vdash \neg a$$

Proof. First we need to expand out the abbreviations, using the definition that $\neg a$ means $a \rightarrow$ False. This results in the following sequent to be proved: $a \rightarrow b$, $b \rightarrow$ False $\vdash a \rightarrow$ False. This is an instance of Theorem 45.   □

Exercise 15. Prove $P \vdash Q \rightarrow P \wedge Q$.

Exercise 16. Prove $\vdash P \wedge Q \rightarrow Q \wedge P$.

6.5.6   Or Introduction $\{\vee I_L\}$, $\{\vee I_R\}$

Like all the introduction rules, Or Introduction specifies what you have to establish in order to infer a proposition containing a new $\vee$ operator. If the proposition a is true, then both $a \vee b$ and $b \vee a$ must also be true (you can see this by checking the truth table definition of $\vee$). Or Introduction comes in two forms, Left and Right.

$$\frac{a}{a \vee b}\{\vee I_L\} \qquad \frac{b}{a \vee b}\{\vee I_R\}$$

Theorem 47. $P \wedge Q \vdash P \vee Q$

Proof. The proof requires the use of Or Introduction. There are two equally valid ways to organise the proof. One method begins by establishing P and then uses $\{\vee I_L\}$ to put the P to the left of $\vee$:

$$\frac{\dfrac{P \wedge Q}{P}\{\wedge E_L\}}{P \vee Q}\{\vee I_L\}$$

An alternative proof first establishes Q and then uses $\{\vee I_R\}$ to put the Q to the right of $\vee$:

$$\frac{\dfrac{P \wedge Q}{Q}\{\wedge E_R\}}{P \vee Q}\{\vee I_R\}$$

Normally, of course, you would choose one of these proofs randomly; there is no reason to give both.   □

Exercise 17. Prove $P \rightarrow$ False $\vee P$.

Exercise 18. Prove P, $Q \vdash (P \wedge Q) \vee (Q \vee R)$.

6.5.7   Or Elimination $\{\vee E\}$

The Or Elimination rule specifies what you can conclude if you know that a proposition of the form $a \vee b$ is true. We can't conclude anything about either a or b, because either of those might be false even if $a \vee b$ is true. However, suppose we know $a \vee b$ is true, and also suppose there is some conclusion c that can be inferred from a and can also be inferred from b. Then c must also be true.

$$\frac{a \vee b \qquad a \vdash c \qquad b \vdash c}{c} \{\vee E\}$$

Or Elimination is a formal version of *proof by case analysis*. It amounts to the following argument: 'There are two cases: (1) if a is true, then c holds; (2) if b is true then c holds. Therefore c is true'. Here is an example:

Theorem 48. $(P \wedge Q) \vee (P \wedge R) \vdash P$

Proof. There are two proofs above the line. The first proof assumes $P \wedge Q$ in order to infer P. However, that inference is all that we need; $P \wedge Q$ is discharged and is not an assumption of the main theorem. For the same reason, $P \wedge R$ is also discharged. The only undischarged assumption is $(P \wedge Q) \vee (P \vee R)$, which therefore must appear to the left of the $\vdash$ in the main theorem.

$$\frac{(P \wedge Q) \vee (P \wedge R) \qquad \dfrac{\boxed{P \wedge Q}}{P} \{\wedge E_L\} \qquad \dfrac{\boxed{P \wedge R}}{P} \{\wedge E_L\}}{P} \{\vee E\}$$

□

Finally, here is a slightly more complex example:

Theorem 49. $(a \wedge b) \vee (a \wedge c) \vdash b \vee c$

Proof.

$$\frac{(a \wedge b) \vee (a \wedge c) \qquad \dfrac{\dfrac{\boxed{a \wedge b}}{b} \{\wedge E_R\}}{b \vee c} \{\vee I_L\} \qquad \dfrac{\dfrac{\boxed{a \wedge c}}{c} \{\wedge E_R\}}{b \vee c} \{\vee I_R\}}{b \vee c} \{\vee E\}$$

□

6.5.8 Identity {*ID*}

The Identity rule {*ID*} says, rather obviously, that if you know a is true, then you know a is true.

$$\boxed{\dfrac{a}{a}\{ID\}}$$

Although the Identity rule seems trivial, it is also necessary. Without it, for example, there would be no way to prove the following theorem, which we would certainly want to be true.

Theorem 50. $P \vdash P$

Proof.

$$\frac{P}{P}\{ID\}$$

□

An interesting consequence of the Identity rule is that True is true, as the following theorem states.

Theorem 51. $\vdash$ True

Proof. Recall that True is an abbreviation for False $\rightarrow$ False. We need to infer this implication using the {$\rightarrow I$} rule, and that in turn requires a proof of False given the assumption False. This can be done with the {*ID*} rule; it could also be done with the Contradiction rule, which we'll study shortly.

$$\frac{\dfrac{\boxed{\text{False}}}{\text{False}}\{ID\}}{\text{False} \rightarrow \text{False}}\{\rightarrow I\}$$

Notice that we had to assume False in order to prove this theorem, but fortunately that assumption was discharged when we inferred False $\rightarrow$ False. The assumption of False was just temporary, and it doesn't appear as an assumption of the theorem. That's a relief; it would never do if we had to assume that False is true in order to prove that True is true!   □

6.5.9 Contradiction {*CTR*}

The Contradiction rule says that you can infer *anything at all* given the assumption that False is true.

$$\boxed{\dfrac{\mathsf{False}}{a}\{CTR\}}$$

In effect, this rule says that False is untrue, and it expresses that fact purely through the mechanism of logical inference. It would be disappointing if we had to describe the fundamental falseness of False by making a meaningless statement outside the system, such as 'False is wrong.' After all, the whole point of natural deduction is to describe the process of logical reasoning formally, using a small set of clearly specified inference rules. It would also be a bad idea to try to define False as equal to ¬True. Since True is already defined to be ¬False, that would be a meaningless and useless circular definition.

The Contradiction rule describes the untruthfulness of False indirectly, by saying that everything would become provable if False is ever assumed or inferred.

Theorem 52. $P, \neg P \vdash Q$

Proof. Recall that $\neg P$ is just an abbreviation for $P \rightarrow$ False. That means we can use the $\{\rightarrow E\}$ rule to infer False, and once that happens we can use $\{CTR\}$ to support any conclusion we feel like—even Q, which isn't even mentioned in the theorem's assumptions!

$$\dfrac{\dfrac{P \qquad P \rightarrow \mathsf{False}}{\mathsf{False}}\{\rightarrow E\}}{Q}\{CTR\}$$

$\square$

The Identity and Contradiction rules often turn up in larger proofs. A typical example occurs in the following theorem, which states an important property of the logical Or operator ∨. This theorem says that if $a \vee b$ is true, but a is false, then b has to be true. It should be intuitively obvious, but the proof is subtle and merits careful study.

Theorem 53. For all propositions a and b,

$$a \vee b, \neg a \vdash b$$

Proof. As usual, the $\neg a$ abbreviation should be expanded to $a \rightarrow$ False. Because we're given $a \vee b$, the basic structure will be an Or Elimination, and there will be two smaller proof trees corresponding to the two cases for $a \vee b$. In the first case, when a is assumed temporarily, we obtain a contradiction with the theorem's assumption of $\neg a$: False is inferred, from which we can infer anything else (and here we want b). In the second case, when b is assumed temporarily, the desired result of b is an immediate consequence of the $\{ID\}$ rule.

$$
\cfrac{
  \cfrac{
    \boxed{a} \quad a \to \text{False}
  }{\text{False}} \{\to E\}
}{
  \cfrac{a \vee b \qquad\qquad b \qquad\qquad \cfrac{\boxed{b}}{b}\{ID\}}{b}\{CTR\}
}
$$

$$\{\vee E\}$$

Note that there are two undischarged assumptions, $a \vee b$ and $a \to \text{False}$, and there are two temporary assumptions that are discharged by the $\{\vee E\}$ rule. $\square$

6.5.10   *Reductio ad Absurdum* $\{RAA\}$

The *Reductio ad Absurdum* (reduce to absurdity) rule says that if you can infer False from an assumption $\neg a$, then a must be true. This rule underpins the *proof by contradiction* strategy: if you want to prove a, first assume the contradiction $\neg a$ and infer False; the $\{RAA\}$ rule then allows you to infer a.

$$
\boxed{\cfrac{\neg a \vdash \text{False}}{a}\{RAA\}}
$$

Theorem 54 (Double negation). $\neg\neg a \vdash a$

Proof. Our strategy is to use a proof by contradiction. That is, if we can assume $\neg a$ and infer False, the $\{RAA\}$ rule would then yield the inference a. Because we are given $\neg\neg a$, the contradiction will have the following general form:

$$
\cfrac{\neg a \qquad \neg\neg a}{\text{False}}
$$

To make this go through, we need to replace the abbreviations by their full defined values. Recall that $\neg a$ is an abbreviation for $a \to \text{False}$, so $\neg\neg a$ actually means $(a \to \text{False}) \to \text{False}$. Once we expand out these definitions, the inference becomes much clearer: it is just a simple application of $\{\to E\}$:

$$
\cfrac{
  \cfrac{\boxed{a \to \text{False}} \qquad (a \to \text{False}) \to \text{False}}{\text{False}}\{\to E\}
}{a}\{RAA\}
$$

Both requirements for the $\{RAA\}$ rule have now been provided, and the $\{RAA\}$ gives the final result a. $\square$

6.5.11   Inferring the Operator Truth Tables

The inference rules of natural deduction are intended to serve as a formal foundation for logical reasoning. This raises two fundamental questions: *Are the inference rules actually powerful enough? Will they ever allow us to make mistakes?* These are profound and difficult questions. Modern research in mathematical logic addresses such questions and has led to some astonishing results. In this section we start with an easy version of the first question: *Are the inference rules powerful enough to calculate the truth tables of the logical operators?*

Answering this question will provide some good practice in using the inference rules. A deeper point is that it's philosophically satisfying to know that the inference rules provide a complete foundation for propositional logic. Nothing else is required: the truth tables given earlier in the chapter are the most intuitive place to start learning logic, and we treated them like definitions at the beginning of the chapter, but it isn't necessary to accept them as the fundamental definitions of the operators. If we take instead the inference rules as the foundation of mathematical logic, then truth tables are no longer definitions; they merely summarise a lot of calculations using the rules.

To illustrate the idea, let's use the inference rules to calculate the value of $\text{True} \wedge \text{False}$. To answer this fully, we need to prove that $\text{True} \wedge \text{False}$ is logically equivalent to False, and also that it is *not* logically equivalent to True. Recall that False is a primitive constant, but True is defined as $\text{False} \rightarrow \text{False}$. First, here is a proof of the sequent $\vdash \text{True} \wedge \text{False} \rightarrow \text{False}$:

$$\frac{\dfrac{(\text{False} \rightarrow \text{False}) \wedge \text{False}}{\text{False}} \{\wedge E_R\}}{((\text{False} \rightarrow \text{False}) \wedge \text{False}) \rightarrow \text{False}} \{\rightarrow I\}$$

Next, we prove the sequent $\vdash \text{False} \rightarrow ((\text{False} \rightarrow \text{False}) \wedge \text{False})$:

$$\frac{\dfrac{\text{False}}{(\text{False} \rightarrow \text{False}) \wedge \text{False}} \{CTR\}}{\text{False} \rightarrow ((\text{False} \rightarrow \text{False}) \wedge \text{False})} \{\rightarrow I\}$$

Putting these results together, and reintroducing the abbreviations, we get

$$\text{True} \wedge \text{False} \leftrightarrow \text{False}$$

We have thereby calculated one of the lines of the truth table definition of $\wedge$. The complete truth tables for all the logical operators can be inferred using similar calculations.

Exercise 19. Use the inference rules to calculate the value of $\text{True} \wedge \text{True}$.

Exercise 20. Use the inference rules to calculate the value of $\text{True} \vee \text{False}$.

Exercise 21. Notice that in the proof of $\vdash$ True$\wedge$False $\rightarrow$ False we used $\{\wedge E_R\}$ to obtain False from (False $\rightarrow$ False) $\wedge$ False, and everything worked fine. However, we could have used $\{\wedge E_L\}$ instead to infer False $\rightarrow$ False, which is True. What would happen if that choice is made? Would it result in calculating the wrong value of True $\wedge$ False? Is it possible to show that True $\wedge$ False is *not* logically equivalent to True?

6.6   Proof Checking by Computer

One of the great benefits of formal logic is the possibility of using computer software to check proofs automatically. Informal arguments in English are full of imprecision and ambiguity, so there is no hope of writing a computer program to determine whether one is valid. People don't always agree as to whether an argument holds water! Formal proofs, however, are totally precise and unambiguous. Formal arguments may become long and involved, but computers are good at long and involved calculations.

Formal proofs are intended to provide the utmost confidence that a theorem is correct. The language of propositional logic, along with the natural deduction inference system, are solid foundations for accurate reasoning. However, everyone makes mistakes, and even a tiny error would make a large proof completely untrustworthy. To get the full benefit of formal logic, we need computers to help with the proofs.

A *proof checker* is a computer program that reads in a theorem and a proof, and determines whether the proof is valid and actually establishes the theorem. A *theorem prover* is a computer program that reads in a theorem and attempts to generate a proof from scratch. The generic term *proof tools* refers to any computer software that helps with formal proofs, including proof checkers and theorem provers.

The advantage of a theorem prover is that it can sometimes save the user a lot of work. However, theorem provers don't always succeed, since there are plenty of true theorems whose proofs are too difficult for them to find. A theorem prover may stop with the message 'I cannot prove it', or it may go into an infinite loop and never stop at all. The main advantage of a proof checker is that it can *always* determine whether a purported proof is valid or invalid. Proof checkers are also suitable when you want to write a proof by hand, and you also want to be sure it's correct.

Proof tools are the subject of active current research, but they are not limited to research: they are also becoming practical for real applications. There are a number of 'industrial-strength' proof tools for various logical systems (see Section 6.10 for references). Proof tools have already been applied successfully to some very difficult and important real-world problems. One example is the proof that the floating point hardware in the Intel Pentium Pro processor is correct [25]. As proof checkers become more widely used, it will become increasingly important for computing professionals to have a good working un-

Theorem 55. $\vdash Q \to ((P \wedge R) \to (R \wedge Q))$

```
example_theorem :: Theorem
example_theorem =
  Theorem
    []
    (Imp Q (Imp (And P R) (And R Q)))
```

Figure 6.2: Theorem 55 and Its Haskell Representation

derstanding of logic.

We'll now look at a simple proof checking system implemented in Haskell, which is part of the software that accompanies this book. You can obtain the software and its documentation from this book's web page; see the Preface for the web address.

Studying the proof checker will introduce you to a technology that is likely to become increasingly important in the near future. You can also use the software to check your own proofs; it's nice to be sure your exercises are correct before handing them in! Furthermore, you may find it interesting to study the implementation of the checker.

Documentation of the proof checker is available on the book web page. The program is written in Haskell. Although it uses some advanced features of Haskell that aren't covered in Chapter 1, most of it isn't too difficult to understand. Haskell is well suited for writing proof checkers; indeed, most of the industrial strength proof tools are implemented in functional programming languages, .

6.6.1 Example of Proof Checking

Before getting into picky details, let's start by checking a real theorem:

Theorem 55. $\vdash Q \to ((P \wedge R) \to (R \wedge Q))$

Proof. See Figure 6.3. □

In order to process this theorem with a Haskell program, we have to represent it in a form that can be typed into a file. Figure 6.2 gives the theorem in both forms, with the mathematical notation and the proof checker's notation shown side by side. The value named `example_theorem` has type `Theorem`, and it consists of three parts: (1) a constructor `Theorem` that starts the data structure; (2) a list of assumptions (i.e., propositions to the left of the ⊢), which is `[]` for this example; and (3) the proposition to be proved. As you can see, the computer-readable proposition is expressed with prefix operators; the name `And` represents ∧, while the name `Imp` represents →.

Figure 6.3 shows the proof of Theorem 55. The conventional mathematical notation is given alongside the Haskell representation, so you can compare them easily. The Haskell representation of the proof is a value named proof1, and its type is Proof. Notice that the mathematical notation has a tree structure, and the proof is also represented as a tree structure in Haskell. The only difference is that the mathematical notation uses lines and positioning on the page to show the structure, while the Haskell notation uses punctuation. The computer representation uses prefix operators to indicate inferences; for example, the name AndI indicates the beginning of an And-Introduction $\{\wedge I\}$ inference. The proof is a value named proof1, and it has type Proof.

The proof checker is a Haskell function named check_proof which takes two arguments: a Theorem and a Proof. The function checks to see whether the proof is valid, and also whether it serves to establish the truth of the theorem. You might expect the checker to return a result of type Bool, where True means the proof is valid and False means the proof is incorrect. However, it's much better for the checker to give a detailed explanation if something is wrong with the proof, and to do that it needs to perform some Input/Output. Because of this, the function returns a Haskell value of type IO () instead of a simple Bool. You do *not* need to know how the Input/Output works. The function's type signature is:

```
check_proof :: Theorem -> Proof -> IO ()
```

The proof given in Figure 6.3 is valid: all of the inferences are sound; the conclusion of the main inference matches the result to the right of the ⊢ in the theorem, and there are no undischarged assumptions that fail to appear to the left of the ⊢ in the statement of the theorem.

To check the proof, we must first start an interactive session with Haskell and load the software tools (see the Appendix for instructions on how to do this). Here is the output produced by running the proof checker on the example:

```
> check_proof example_theorem proof1
The proof is valid
```

In order to see what happens when something goes wrong with a proof, let's introduce a small mistake into the previous example. Figure 6.4 gives the modified proof, along with its Haskell representation. The error is that the two subproofs above the line of the And Introduction step now appear in the wrong order: on the left is the assumption Q, and on the right is the proof of R. Because of this, the $\{\wedge I\}$ rule infers the proposition $Q \wedge R$, but we still have $R \wedge Q$ below the line. Here is the result of running the incorrect proof through the proof checker:

```
> check_proof example_theorem proof2
Invalid And-Introduction: the conclusion
    And R Q
```

$$\cfrac{\cfrac{\boxed{P \wedge R}}{R}\{\wedge E_R\} \qquad \boxed{Q}}{\cfrac{\cfrac{R \wedge Q}{(P \wedge R) \to (R \wedge Q)}\{\to I\}}{Q \to ((P \wedge R) \to (R \wedge Q))}\{\to I\}}\{\wedge I\}$$

```
proof1 :: Proof
proof1 =
  ImpI
    (ImpI
      (AndI
        ((AndER
            (Assume (And P R))
            R),
          Assume Q)
        (And R Q))
      (Imp (And P R) (And R Q)))
    (Imp Q (Imp (And P R) (And R Q)))
```

Figure 6.3: A Valid Proof of Theorem 55 and Its Haskell Representation

```
    must be the logical And of the assumption
        Q
    with the assumption
        R
  The proof is invalid
```

To use the proof checker on your own, you will need to know how to represent propositions (WFFs) and how to represent proofs. The following sections describe these issues briefly, but you should also read the online documentation on the book's web page.

Exercise 22. Suppose we simply replace $R \wedge Q$ below the $\{\wedge I\}$ line with $Q \wedge R$. This fixes the `Invalid And-Introduction` error, but it introduces another error into the proof.

  (a) Edit `proof2` to reflect this change; call the result `proof3`.

  (b) Decide exactly what is wrong with `proof3`.

  (c) Run the proof checker on `proof3`, and see whether it reports the same error that you predicted.

$$\cfrac{\cfrac{\boxed{Q}\quad \cfrac{\boxed{P \wedge R}}{R}\{\wedge E_R\}}{R \wedge Q}\{\wedge I\}}{\cfrac{(P \wedge R) \to (R \wedge Q)}{Q \to ((P \wedge R) \to (R \wedge Q))}\{\to I\}}\{\to I\} \qquad \longleftarrow \textbf{\textit{Wrong!}}$$

```
proof2 :: Proof
proof2 =
  ImpI
   (ImpI
     (AndI
        (Assume Q,
        (AndER
           (Assume (And P R))
           R))
        (And R Q))
     (Imp (And P R) (And R Q)))
   (Imp Q (Imp (And P R) (And R Q)))
```

Figure 6.4: An Invalid Proof of Theorem 55 and Its Haskell Representation

6.6.2   Representation of WFFs

The well-formed formulas of propositional logic can be represented in Haskell as an algebraic data type. This representation allows us to use the compiler to check that a formula is indeed well-formed, and it also provides a way to express terms that appear in logical proofs—later in this chapter we will exploit that in order to support a program that checks logical proofs for correctness.

The Boolean constants are represented by FALSE and TRUE. Intuitively, these correspond in meaning to the familiar Boolean constants False and True. Use False (or True) when you want to write Boolean expressions to be evaluated; use FALSE (or TRUE) when you want to write a WFF and reason about it. We will come back to this subtle but important distinction at the end of the section.

The traditional names used for propositional variables are upper-case letters $P, Q, R \ldots$. We allow every upper-case letter to be used as a propositional variable, except that F and T are disallowed because someone reading the code might wonder whether these are supposed to be variables or the constants *false* and *true*. In addition, you can make any string into a propositional variable by writing Pvar "name".

The logical expression $P \wedge Q$ is written And P Q. If the arguments to the And are themselves logical expressions, they should be surrounded by parentheses, for example: And (And P Q) R. In a similar way, $P \vee Q$ is written Or P Q, and $P \to Q$ is written Imp P Q, and the negation $\neg P$ is written Not P. In all cases, arguments that are not simple logical variables should be enclosed in parentheses.

WFFs are defined directly as a Haskell data type `Prop`. It is common in mathematics to define structures recursively, as WFFs were defined in the previous section. Haskell's algebraic data types allow such standard mathematical definitions to be turned directly into a computer program. The definition is

```
data Prop
  = FALSE
  | TRUE
  | A | B | C | D | E | G | H | I | J | K | L | M
  | N | O | P | Q | R | S | U | V | W | X | Y | Z
  | Pvar String
  | And Prop Prop
  | Or Prop Prop
  | Not Prop
  | Imp Prop Prop
  | Equ Prop Prop
  deriving (Eq,Show)
```

Exercise 23. Define each of the following well-formed formulas as a Haskell value of type `Prop`.

(a) P

(b) $Q \lor \text{False}$

(c) $Q \to (P \to (P \land Q))$

(d) $P \land (\neg Q)$

(e) $\neg P \to Q$

(f) $(P \land \neg Q) \lor (\neg P \land Q) \to (P \lor Q)$

Exercise 24. Translate each of the following Haskell expressions into the conventional mathematical notation.

(a) `And P Q`

(b) `Imply (Not P) (Or R S)`

(c) `Equ (Imply P Q) (Or (Not P) Q)`

6.6.3 Representing Proofs

A proof is represented by another Haskell algebraic data type. Technically, a proof is a data structure that contains a proposition along with a formal argument for the truth of the proposition. There is a separate constructor for every kind of proof:

- `Assume Prop`. The simplest way to establish that a proposition is true is to assume it. No further justification is necessary, and any proposition at all may be assumed. See `proof1` above for an example. Note that assumptions may be discharged by the Imply Introduction rule $\{\to I\}$, so there is a limited and well-defined scope for all assumptions. Unfortunately, however, it isn't easy to see what the scope of an assumption is just by looking at it; you need to study what it means to discharge.

- `AndI (Proof,Proof) Prop`. The And Introduction inference rule $\{\wedge I\}$ requires two separate proofs above the line, as well as the claimed conclusion (which is a proposition).

- `AndEL Proof Prop`. The And Elimination Left inference rule $\{\wedge E_L\}$ has just one proof above the line; like all the inference rules, it has exactly one conclusion, which is a proposition.

- `AndER Proof Prop`. The And Elimination Right rule $\{\wedge E_R\}$. This is the Right version of And Elimination $\{\wedge E_R\}$

- `OrIL Proof Prop`. The Or Introduction Left rule $\{\vee I_L\}$;

- `OrIR Proof Prop`. The Or Introduction Right rule $\{\vee I_R\}$;

- `OrE (Proof,Proof,Proof) Prop`. The Or Elimination rule $\{\vee E\}$ requires three proofs above the line. The first one is a proof whose conclusion has the form $a \vee b$, the second is a proof of some conclusion c given a, and the third is a proof of the same conclusion c given b. The conclusion of the rule must be the proposition c.

- `ImpI Proof Prop`. The Imply Introduction rule $\{\to I\}$.

- `ImpE (Proof,Proof) Prop`. The Imply Elimination rule $\{\to E\}$.

- `ID Proof Prop`. The Identity rule $\{ID\}$.

- `CTR Proof Prop`. The Contradiction rule $\{CTR\}$.

- `RAA Proof Prop`. The *Reductio ad Absurdum* rule $\{RAA\}$.

The best way to understand how to represent proofs is to look at some examples, and then try your own. Start by studying `proof1` and `proof2` above.

The representation of a theorem is very simple: it just contains a list of assumptions of type `[Prop]` and a single conclusion of type `Prop`.

```
data Theorem = Theorem [Prop] Prop
```

To represent a theorem of the form

$$a_1, a_2, a_3 \vdash c,$$

you would write

Theorem [a_1, a_2, a_3] c.

The proof checker function takes two arguments: a theorem of type `Theorem` and a proof of type `Proof`. The easiest way to use it is to give a name to the theorem and the proof, write a defining equation for each, and save the definitions in a file. Don't try to type in proofs interactively! Follow the defining equations above for `example_theorem` and `proof1` as a model.

There are many things that can go wrong with a proof. The proof checker tries to give the best messages possible, but sometimes it is hard to debug a proof. Actually, proof debugging has much in common with program debugging. See the book's web page for hints on debugging proofs.

6.7 Boolean Algebra: Equational Reasoning

We have already looked at two major approaches to propositional logic: the *semantic* approach with truth tables, and the *syntactic* approach with the inference rules of natural deduction. We now look at the third major system, Boolean algebra, which is an *axiomatic* approach to logic.

The earliest attempts to develop a formal logic system were based on inference. The most famous of these was Aristotle's Theory of Syllogisms, which profoundly affected the entire development of logic and philosophy for more than two thousand years.

During the past several centuries, however, a completely different style of mathematical reasoning appeared: algebra. Algebraic techniques were enormously successful for reasoning about numbers, polynomials, and functions. One of the most appealing benefits of algebra is that many problems can be expressed as an equation involving an unknown quantity x that you would like to determine; you can then use algebraic laws to manipulate the equation systematically in order to solve for x.

A natural question is: can the power of algebraic techniques be applied to other areas of mathematics? Perhaps the most famous such application is Descartes' analytical geometry. George Boole, a nineteenth century British mathematician, saw that algebraic methods might also be applied to make formal logical reasoning easier, and he attempted to create such a system. A successful modern algebraic approach to logic has been developed and is named in his honour.

Boolean algebra is a form of *equational reasoning*. There are two crucial ideas: (1) you show that two values are the same by building up *chains of equalities*, and (2) you can *substitute equals for equals* in order to add a new link to the chain.

A *chain of equalities* relies on the fact that if you know $a = b$ and also $b = c$, then you can deduce formally that $a = c$. For example, the following chain of equations allows us to conclude that $a = e$:

$$a = b$$
$$= c$$
$$= d$$
$$= e$$

Substituting equals for equals means that if you know $x = y$, and if you have a big expression that contains x, then you can replace x by y without changing the value of the big expression. For example, suppose you're given that $x = 2 + p$ and $y = 5 \times x + 3$. Then you replace x by $2 + p$, resulting in $y = 5 \times (2 + p) + 3$.

There is a minor but important point to observe in the example above: you have to use parentheses properly to ensure that the value you substitute into the expression sticks together as one value. In this example, we had to put parentheses around the value $2 + p$, because otherwise we would have written the incorrect equation $y = 5 \times 2 + p + 3$, which has a quite different meaning.

When we build a chain of equations using Boolean algebra, it's good practice to give a justification for each step in the chain. The justifications help a reader to understand the proof, and they also make it easier to check that the proof is actually valid. A standard way to write chains of equations is to start each line (except the first) with an $=$ sign, followed by the next expression in our chain, followed by the justification which explains how we obtained it from the previous expression. You'll see plenty of examples as we work through the laws of Boolean algebra.

6.7.1   The Laws of Boolean Algebra

Modern Boolean algebra is based on a set of equations that describe the basic algebraic properties of propositions. These equations are called *laws*; a law is a proposition that is always true, for every possible assignment of truth values to the logical variables.

The laws of Boolean Algebra are analogous to ordinary algebraic laws. For example, elementary algebra has commutative equations for addition and multiplication, $x + y = y + x$ and $x \times y = y \times x$. There are analogous commutative laws for $\vee$ and $\wedge$, saying that $x \vee y = y \vee x$ and $x \wedge y = y \wedge x$. It can be enlightening to compare the laws of Boolean algebra with those of elementary algebra, but don't get carried away: there are differences as well as similarities.

There is one particularly dangerous trap. In many ways, logical And ($\wedge$) behaves like multiplication ($\times$) while logical Or ($\vee$) behaves like addition ($+$). In fact, these similarities have tempted many people to use the $+$ symbol for Or and the $\times$ symbol (or $\cdot$) for And. George Boole carried similar analogies very far—much too far—in his original work.

However, $\wedge$ does not behave like $\times$ in all respects, and $\vee$ does not behave like $+$ in all respects (see, for example, Section 6.7.4). Reading too much significance into the similarities between laws on numeric and Boolean operations can lead you astray.

The essence of algebra is *not* that there are fundamental addition and multiplication operators that appear everywhere. The essential idea is that we can use equations to state axioms on a set of operators, and then use equational reasoning to explore the properties of the resulting system. Some algebraic systems have addition and multiplication operators, and some algebraic systems don't. Boolean algebra doesn't.

Table 6.1 summarises the laws of Boolean Algebra, and we'll discuss them in more detail in the following sections. If we were mainly concerned with the foundations of algebra, our aim would be to take the smallest possible set of equations as axioms and to derive other ones as theorems. However, we are more concerned here with the practical application of Boolean algebra in computer science, so Table 6.1 gives a richer set of laws that are easier to use for practical calculation than a minimal set of axioms would be.

6.7.2 Operations with Constants

These simple laws describe how $\wedge$ and $\vee$ interact with the Boolean constants True and False.

$$
\begin{array}{rcll}
a \wedge \text{False} & = & \text{False} & \{\wedge \text{ null}\} \\
a \vee \text{True} & = & \text{True} & \{\vee \text{ null}\} \\
a \wedge \text{True} & = & a & \{\wedge \text{ identity}\} \\
a \vee \text{False} & = & a & \{\vee \text{ identity}\}
\end{array}
$$

Often it's possible to simplify Boolean expressions with equational reasoning using the constant laws. If you already know what the final simplified result will be, then the equational reasoning serves as a proof of the equation. Here, for example, is a simplification of the expression $(P \wedge \text{True}) \vee \text{False}$. Alternatively, the following reasoning is a proof of the equation $(P \wedge \text{True}) \vee \text{False} = P$.

$$
\begin{array}{ll}
(P \wedge \text{True}) \vee \text{False} & \\
= P \wedge \text{True} & \{\vee \text{ identity}\} \\
= P & \{\wedge \text{ identity}\}
\end{array}
$$

Note the form of the proof. We are trying to prove an equation, and the proof consists of a chain of equations. The chain begins with the left-hand side of the theorem and ends with the right-hand side of the theorem. Each step of the chain is justified by one of the laws of Boolean algebra, and the name of the law is written to the right.

Exercise 25. Simplify $(P \wedge \text{False}) \vee (Q \wedge \text{True})$.

Exercise 26. Prove the equation $(P \wedge \text{False}) \wedge \text{True} = \text{False}$.

Table 6.1: Laws of Boolean Algebra

$$
\begin{array}{rcl l}
a \wedge \text{False} & = & \text{False} & \{\wedge \text{ null}\} \\
a \vee \text{True} & = & \text{True} & \{\vee \text{ null}\} \\
a \wedge \text{True} & = & a & \{\wedge \text{ identity}\} \\
a \vee \text{False} & = & a & \{\vee \text{ identity}\} \\
\\
a & \to & a \vee b & \{\text{disjunctive implication}\} \\
a \wedge b & \to & a & \{\text{conjunctive implication}\} \\
a \wedge a & = & a & \{\wedge \text{ idempotent}\} \\
a \vee a & = & a & \{\vee \text{ idempotent}\} \\
a \wedge b & = & b \wedge a & \{\wedge \text{ commutative}\} \\
a \vee b & = & b \vee a & \{\vee \text{ commutative}\} \\
(a \wedge b) \wedge c & = & a \wedge (b \wedge c) & \{\wedge \text{ associative}\} \\
(a \vee b) \vee c & = & a \vee (b \vee c) & \{\vee \text{ associative}\} \\
\\
a \wedge (b \vee c) & = & (a \wedge b) \vee (a \wedge c) & \{\wedge \text{ distributes over } \vee\} \\
a \vee (b \wedge c) & = & (a \vee b) \wedge (a \vee c) & \{\vee \text{ distributes over } \wedge\} \\
\neg(a \wedge b) & = & \neg a \vee \neg b & \{\text{DeMorgan's law}\} \\
\neg(a \vee b) & = & \neg a \wedge \neg b & \{\text{DeMorgan's law}\} \\
\\
\neg \text{True} & = & \text{False} & \{\text{negate True}\} \\
\neg \text{False} & = & \text{True} & \{\text{negate False}\} \\
a \wedge \neg a & = & \text{False} & \{\wedge \text{ complement}\} \\
a \vee \neg a & = & \text{True} & \{\vee \text{ complement}\} \\
\neg(\neg a) & = & a & \{\text{double negation}\} \\
\\
a \wedge (a \to b) & \to & b & \{\textit{Modus Ponens}\} \\
(a \to b) \wedge \neg b & \to & \neg a & \{\textit{Modus Tollens}\} \\
(a \vee b) \wedge \neg a & \to & b & \{\text{disjunctive syllogism}\} \\
(a \to b) \wedge (b \to c) & \to & a \to c & \{\text{implication chain}\} \\
(a \to b) \wedge (c \to d) & \to & (a \wedge c) \to (b \wedge d) & \{\text{implication combination}\} \\
(a \wedge b) \to c & = & a \to (b \to c) & \{\text{Currying}\} \\
a \to b & = & \neg a \vee b & \{\text{implication}\} \\
a \to b & = & \neg b \to \neg a & \{\text{contrapositive}\} \\
(a \to b) \wedge (a \to \neg b) & = & \neg a & \{\text{absurdity}\} \\
\\
a \leftrightarrow b & = & (a \to b) \wedge (b \to a) & \{\text{equivalence}\}
\end{array}
$$

6.7.3 Basic Properties of $\wedge$ and $\vee$

The following laws describe the basic properties of the $\wedge$ and $\vee$ operators. An *idempotent* property allows you to collapse expressions like $a \wedge a \wedge a$ down to just a. Commutativity means that the order of the operands can be reversed, and associativity means that the grouping of parentheses can be changed without affecting the meaning.

$$
\begin{array}{rcll}
a & \rightarrow & a \vee b & \{\text{disjunctive implication}\} \\
a \wedge b & \rightarrow & a & \{\text{conjunctive implication}\} \\
a \wedge a & = & a & \{\wedge \text{ idempotent}\} \\
a \vee a & = & a & \{\vee \text{ idempotent}\} \\
a \wedge b & = & b \wedge a & \{\wedge \text{ commutative}\} \\
a \vee b & = & b \vee a & \{\vee \text{ commutative}\} \\
(a \wedge b) \wedge c & = & a \wedge (b \wedge c) & \{\wedge \text{ associative}\} \\
(a \vee b) \vee c & = & a \vee (b \vee c) & \{\vee \text{ associative}\}
\end{array}
$$

Commutative operators take two operands, but the order doesn't matter. The commutative properties are often needed to put an expression into a form where you can use another of the identities. Although we don't have a law saying that $\text{False} \wedge a = \text{False}$, the commutativity of $\wedge$ can be applied to rearrange an expression so that the law we do have, $a \wedge \text{False} = \text{False}$, becomes usable. As an example, here is a proof of the equation $(\text{False} \wedge P) \vee Q = Q$:

$$
\begin{array}{ll}
(\text{False} \wedge P) \vee Q & \\
= (P \wedge \text{False}) \vee Q & \{\wedge \text{ commutative}\} \\
= \text{False} \vee Q & \{\wedge \text{ null}\} \\
= Q \vee \text{False} & \{\vee \text{ commutative}\} \\
= Q & \{\vee \text{ identity}\}
\end{array}
$$

An associative operator gives the same result regardless of grouping. For example, ordinary addition is associative, so $2 + (3 + 4) = (2 + 3) + 4$. In a similar way, $\wedge$ and $\vee$ are both associative.

Since the parentheses don't actually affect the result in an expression where an associative operator is repeated, you can safely omit them. For example, we commonly write expressions like $2 + x + y$, without insisting on $2 + (x + y)$ or $(2 + x) + y$. The same thing happens in Boolean algebra: the propositions $P \wedge Q \wedge R$ and $A \vee B \vee C \vee D$ are unambigous because the $\wedge$ and $\vee$ operators are associative, and it makes no difference what order the operations are performed. When you mix different operators, however, parentheses are important: $P \wedge (Q \vee R)$ is not the same thing as $(P \wedge Q) \vee R$.

Exercise 27. Prove $(P \wedge ((Q \vee R) \vee Q)) \wedge S = S \wedge ((R \vee Q) \wedge P)$.

Exercise 28. Prove $P \wedge (Q \wedge (R \wedge S)) = ((P \wedge Q) \wedge R) \wedge S$.

6.7.4   Distributive and DeMorgan's Laws

The laws in this section describe some important properties of expressions that contain both the $\vee$ and $\wedge$ operators.

$$
\begin{array}{rcll}
a \wedge (b \vee c) & = & (a \wedge b) \vee (a \wedge c) & \{\wedge \text{ distributes over } \vee\} \\
a \vee (b \wedge c) & = & (a \vee b) \wedge (a \vee c) & \{\vee \text{ distributes over } \wedge\} \\
\neg(a \wedge b) & = & \neg a \vee \neg b & \{\text{DeMorgan's law}\} \\
\neg(a \vee b) & = & \neg a \wedge \neg b & \{\text{DeMorgan's law}\}
\end{array}
$$

The distributive laws are analogous to the way multiplication distributes over addition in elementary algebra: $a \times (b+c) = (a \times b) + (a \times c)$. However, it is *not* the case that addition distributes over multiplication, because $a + (b \times c) \neq (a + b) \times (a + c)$. Here is a significant reason that you should not think of $\vee$ and $\wedge$ as being addition and multiplication.

There is an intuitive reading for both of DeMorgan's laws. The proposition $\neg(a \wedge b)$ says 'a and b aren't *both* true'. An equivalent way to say this is 'either a or b must be false,' which corresponds to $\neg a \vee \neg b$.

Exercise 29. Give an intuitive explanation of the second DeMorgan's law.

6.7.5   Laws on Negation

The following laws state some simple properties of logical negation ($\neg$). We'll see some more subtle properties in the following section, where negation is mixed with implication.

$$
\begin{array}{rcll}
\neg\text{True} & = & \text{False} & \{\text{negate True}\} \\
\neg\text{False} & = & \text{True} & \{\text{negate False}\} \\
a \wedge \neg a & = & \text{False} & \{\wedge \text{ complement}\} \\
a \vee \neg a & = & \text{True} & \{\vee \text{ complement}\} \\
\neg(\neg a) & = & a & \{\text{double negation}\}
\end{array}
$$

The following example shows how equational reasoning can be used to simplify $P \wedge \neg(Q \vee P)$:

$$
\begin{array}{rll}
P \wedge \neg(Q \vee P) & & \\
= P \wedge (\neg Q \wedge \neg P) & & \{\text{DeMorgan's law}\} \\
= P \wedge (\neg P \wedge \neg Q) & & \{\wedge \text{ commutative}\} \\
= (P \wedge \neg P) \wedge \neg Q & & \{\wedge \text{ associative}\} \\
= \text{False} \wedge \neg Q & & \{\wedge \text{ complement}\} \\
= \neg Q \wedge \text{False} & & \{\wedge \text{ commutative}\} \\
= \text{False} & & \{\wedge \text{ null}\}
\end{array}
$$

6.7.6 Laws on Implication

The laws on implication are frequently useful for solving problems, and they are also subtle enough to warrant careful study—especially the ones that combine the $\rightarrow$ and $\neg$ operators. Note that some of these laws are implications, and others are equations. A good way to understand an implication law is to find a counterexample demonstrating that it would not be valid as an equation. For example, the conjunctive implication law says that the conjunction $a \wedge b$ implies a. However, it is *not* valid to write the implication in the other direction: $a \rightarrow (a \wedge b)$ is False when $a =$ True and $b =$ False.

$a \wedge (a \rightarrow b)$	$\rightarrow$	b	{*Modus Ponens*}
$(a \rightarrow b) \wedge \neg b$	$\rightarrow$	$\neg a$	{*Modus Tollens*}
$(a \vee b) \wedge \neg a$	$\rightarrow$	b	{disjunctive syllogism}
$(a \rightarrow b) \wedge (b \rightarrow c)$	$\rightarrow$	$a \rightarrow c$	{implication chain}
$(a \rightarrow b) \wedge (c \rightarrow d)$	$\rightarrow$	$(a \wedge c) \rightarrow (b \wedge d)$	{implication combination}
$(a \wedge b) \rightarrow c$	$=$	$a \rightarrow (b \rightarrow c)$	{Currying}
$a \rightarrow b$	$=$	$\neg a \vee b$	{implication}
$a \rightarrow b$	$=$	$\neg b \rightarrow \neg a$	{contrapositive}
$(a \rightarrow b) \wedge (a \rightarrow \neg b)$	$=$	$\neg a$	{absurdity}

Consider the Currying law, which is a logical form of Curried function arguments (which will be covered in Chapter 11). Suppose two conditions a and b are sufficient to ensure that c must be true. The Currying law says, in effect, that there are two equivalent ways to establish that a and b both hold: either we can require that $a \wedge b$ is true, or we can require that an implication on a is satisfied and also an implication on b. If either a or b is false, then $a \wedge b$ will be false, so the implication $(a \wedge b) \rightarrow c$ is vacuous. Furthermore at least one of the implications in $a \rightarrow (b \rightarrow c)$ will also be vacuous.

The second law, $a \rightarrow b = \neg a \vee b$, often provides the easiest way to prove implications in Boolean algebra. Notice that one side of the equation contains the $\rightarrow$ operator and the other doesn't; therefore you can use this equation to introduce a $\rightarrow$ where none was present before. Boolean algebra doesn't have any notion of inference, so you can't prove a proposition containing a $\rightarrow$ with an introduction rule.

The contrapositive law lets you turn around an implication. Suppose you know that $a \rightarrow b$; then if b is false it can't be the case that a is true.

The absurdity law is quite powerful, because it allows us to deduce the value of a just from implications on a. This is worth thinking about: you might expect that an implication $a \rightarrow b$ tells you something about b *if you know a*, but it can't tell you whether a is true. Suppose, however, we know both $a \rightarrow b$ and also $a \rightarrow \neg b$. Then a can't be true because $b \wedge \neg b$ can't be true.

6.7.7   Equivalence

Strictly speaking, we don't need the logical equivalence operator $\leftrightarrow$ at all. The proposition $a \leftrightarrow b$ is simply an abbreviation for $(a \to b) \wedge (b \to a)$, as stated by the following equation.

$$a \leftrightarrow b \;\; = \;\; (a \to b) \wedge (b \to a) \quad \{\text{equivalence}\}$$

Logical equivalence is essentially similar to equality, but there is a subtle distinction. The well-formed formula $P \wedge P \leftrightarrow P$ is a single proposition, which happens to have the value True. In contrast, the equation $P \wedge P = P$ is not a proposition. It is an equation, whose left and right-hand sides are propositions. The equation is a statement in the meta-language *about* propositions that are expressed in the object language. You can say that two propositions have the same value in either language; if you're in the meta-language, talking about propositions, use =, but if you're in the object language, trying to write propositions that express properties of other propositions, use $\leftrightarrow$.

Exercise 30. Prove the following statements using equational reasoning with the laws of Boolean algebra:

$$(A \vee B) \wedge B \leftrightarrow B$$

[$\wedge$ absorption]

Exercise 31.

$$((\neg A \wedge B) \vee (A \wedge \neg B)) \leftrightarrow (A \vee B) \wedge (\neg(A \wedge B))$$

[half adder]

Exercise 32.
$$\neg(A \wedge B) \leftrightarrow \neg A \vee \neg B$$

[Restriction: This is DeMorgan's first law. Do not use that law in the proof. You may use DeMorgan's other law.]

Exercise 33.

$$(A \vee B) \wedge (\neg A \vee C) \wedge (B \vee C) \leftrightarrow (A \vee B) \wedge (\neg A \vee C)$$

6.8   Logic in Computer Science

Logic and computer science are strongly connected subjects. Each one has a major influence on the other. The proof checking software described in Section 6.6 is a typical example of the application of computing to logic. In this section, we're concerned with the other direction: the benefits of logic to computing. There are far too many applications of logic to computing to mention them all here. This section describes just a few examples in order to help put the topics in this chapter into perspective. Section 6.10 gives references that will tell you more about these topics.

Formal correctness of software. Many large programming projects have been devastated by ineradicable bugs. There has been much discussion of the 'Software Crisis': how can we write software that works correctly? One approach to this problem is to use mathematics to specify what the software is supposed to do and to prove its correctness. There are many different methods for doing this, but ultimately they are all based on formal logical reasoning.

In general, it doesn't work very well to take a poorly-written program and try to prove it correct. Even if the program contains no bugs, the sloppy structure will make the logical correctness proof impossibly difficult. However, there has been considerable success in using the formal logical reasoning to help derive the software from a mathematical specification.

The Curry-Howard Isomorphism and type systems. Many modern programming languages—especially functional languages like Haskell—have powerful and expressive type systems. We need effective methods to help deal with type systems: to help programmers understand them, to help compiler writers implement them, and to help language designers to specify them.

There is a remarkable connection between the inference rules of logic and the typing rules of programming languages; in fact, they are essentially the same! This connection was observed in the 1950s by the logicians Curry and Howard, and has ultimately resulted in great improvements in programming languages.

Linear logic and access control. Often in computing, we are concerned with controlling the access to some resource. One example arises in functional programming, where array operations can be implemented more efficiently if the compiler can guarantee that the program observes certain constraints on the way the array is accessed.

There is a logical system, called linear logic, which keeps track of where each intermediate result in an inference proof is used. The inference rules of linear logic are careful to discharge every assumption exactly once, and intermediate results must also be discharged. The system is able to express certain kinds of resource utilisation through the inference rules. There is significant current research on the application of linear logic to language design and compilers.

Digital hardware design. Computers are built out of digital hardware. These circuits are very complex, containing enormous numbers of components. Discrete mathematics is an essential tool for designing digital circuits. Using the mathematics makes the design process easier and also makes it possible to prove that a circuit is correct. In Chapter 13 we will return to this application in more detail.

6.9   Meta-Logic

Meta-Logic is concerned with stepping outside the language of logic, so that we can make general statements about the properties of the logical system itself. Meta-logic enables us to talk *about* logic rather than just saying things *in* logic. Within the logical system, the only elements of our vocabulary are the propositional variables and logical operators; we can say things like $A \wedge B \rightarrow A \vee B$ but that's all. The purpose of meta-logic is to enable us to think about deeper questions.

At this point, we have covered three quite different methods for reasoning about propositions: truth tables, logical inference, and Boolean algebra. These methods have completely different styles:

- Truth tables enable us to calculate the values (or *meanings*) of propositions, given the values of their variables. The basic technique is calculation, and it results in a logical value (True or False). The meaning of an expression is called its *semantics*, so calculation with truth tables is a form of semantic reasoning.

- Inference rules enable us to prove theorems. The basic technique involves matching the structures of propositions with the structure of the formulas in the inference rules. The structure of an expression is called its *syntax*, so logical inference is a form of syntactic reasoning.

- Boolean algebra allows the use of equational reasoning to prove the equality of two expressions, or to calculate the values of expressions. It applies the power of algebra to the subject of logic.

Just as propositional logic needs a vocabulary for talking about truth values ($\wedge$, $\vee$ etc.), meta-logic needs a vocabulary for talking about logical reasoning itself. There are two fundamental operator symbols in meta-logic, $\models$ and $\vdash$, which correspond to semantic and syntactic reasoning respectively. We have already defined these operators. Recall that:

- $P_1, P_2, \ldots, P_n \vdash Q$ means that there is a proof which infers the conclusion Q from the assumptions $P_1, \ldots, P_n$ using the formal inference rules of natural deduction;

- $P_1, P_2, \ldots, P_n \models Q$ means that Q must be True if $P_1, \ldots, P_n$ are all True, but it says nothing about whether we have a proof, or indeed even whether a proof is possible.

The $\models$ and the $\vdash$ operators are describing two different notions of truth, and it's important to know whether they are in fact equivalent. Is it possible to prove a theorem that is false? Is there a true theorem for which no proof exists?

Definition 13. A formal system is *consistent* if the following statement is true for all well-formed formulas a and b:

$$\text{If } a \vdash b \text{ then } a \models b.$$

In other words, the system is consistent if each proposition provable using the inference rules is *actually* true.

Definition 14. A formal system is *complete* if the following statement is true for all well formed formulas a and b:

$$\text{If } a \models b \text{ then } a \vdash b.$$

In other words, the system is complete if the inference rules are powerful enough to prove every proposition which is true.

Fortunately, it turns out that propositional logic is both consistent and complete. This means that you can't prove false theorems and if a theorem is true it has a proof.

Theorem 56. Propositional logic is consistent and complete.

You can find the proof of Theorem 56 in some of the books suggested in the next section. It is interesting to note, however, that this is a (meta-logical) theorem that is proved mathematically; it isn't just an assumption.

There are many logical systems, and some of them are much richer and more expressive than propositional logic. In the next chapter we will look at a more powerful logical system called *predicate logic*, which is also consistent and complete.

It turns out that any logical system that is powerful enough to express ordinary arithmetic must be either inconsistent or incomplete. This means it's impossible to capture all of mathematics in a safe logical system. This result, the famous *Gödel's Theorem*, has had a profound impact on the philosophy of mathematics and is also relevant to theoretical computer science.

6.10 Suggestions for Further Reading

Logic plays an increasingly important role in computer science, and there are many books and papers where you can learn more about the connections between these two subjects. There is a regular international conference, *Logic in Computer Science* (LICS), devoted to current research. A number of recommendations for further reading are given below; the complete citations appear in the Bibliography.

- *Forever Undecided: A Puzzle Guide to Gödel*, by Raymond Smullyan [28]. This is a collection of puzzles set on an island of Knights and Knaves. Knights are always truthful, and knaves always lie. The tricky problem is that you can't tell whether someone is a knight or a knave simply by looking at them. A typical problem is to think of a question to ask someone that will enable you to figure out what you want to know, regardless of whether the person you ask happens to be a knight or a knave. Smullyan manages to capture the essence of several meta-logical problems, including the famous Incompleteness Theorem of Gödel, and to express deep problems in terms of entertaining and relatively elementary puzzles. This book is a classic of logic and philosophy. Don't miss it!

- *Gödel, Escher, Bach: An Eternal Golden Braid*, by Douglas R. Hofstadter [17] is 'A metaphorical fugue on minds and machines in the spirit of Lewis Carroll.' The themes of the book are drawn from mathematics, art, music, cognitive science, and computer science. Another unmissable classic.

- *How To Prove It: A Structured Approach*, by Daniel J. Velleman [33], is a systematic presentation of the standard methods for logical reasoning in carrying out proofs. This is a good source for hints on technique, and it contains lots of examples.

- *Computer-Aided Reasoning: An Approach*, by M. Kaufmann, P. Manolios, and J. Strother Moore [1], describes ACL2, a combination programming language and theorem prover that is used in commercial applications, including avionics and verification of VLSI circuits.

- *Introduction to HOL: A theorem proving environment for higher order logic*, by M. Gordon and T. Melham, describes a logical system that is more general than the one presented in this chapter, and also documents a software tool that helps to develop and check proofs. HOL has been used in many practical applications.

- *Logic for Mathematics and Computer Science*, by Stanley N. Burris [6], is a more advanced coverage of mathematical logic. It gives detailed presentations of some important proof techniques that are useful for automated systems. There is also a good explanation of some interesting historical topics, including syllogisms and Boole's original attempt to apply algebra to logic.

- *A Mathematical Introduction to Logic*, by Herbert B. Enderton [10], gives a standard presentation of logic from a mathematical perspective, including advanced topics such as models, soundness and completeness, and undecidability.

- *Type Theory and Functional Programming*, by Simon Thompson [31], gives a detailed development of the relationship between the inference rules of natural deduction and the rules used to define type systems for programming languages.

- *Logic and Declarative Language*, by Michael Downward [9], covers the relationship between logic and declarative languages, especially logic programming languages like Prolog.

- *Proofs and Types*, by Girard, Lafont and Taylor [13], presents the relationship between inference proofs and type systems at a research level.

6.11 Review Exercises

Exercise 34. Prove the following using inference rules; natural deduction and proof-checker notation:

$$A \wedge \neg A \vdash \mathsf{False}$$

Exercise 35.

$$A \vdash \neg(\neg A)$$

Exercise 36.

$$A, A \to B, B \to C, C \to D \vdash D$$

[Note: Extension of Implication Chain Rule, Theorem 11]

Exercise 37.

$$A \to B \vdash \neg B \to \neg A$$

[Note: This conclusion is called the contrapositive of the premise]

Exercise 38. For the following problems, omit the proof checker code.

$$A \to B, A \to \neg B \vdash \neg A$$

Exercise 39.

$$\neg A \vdash (A \to B) \wedge (A \to \neg B)$$

Exercise 40.

$$A \vee (B \vee C) \vdash (A \vee B) \vee C$$

Exercise 41. Prove $P, Q, R, S \vdash (P \wedge Q) \wedge (R \wedge S)$.

Exercise 42. Prove $P \to R \vdash P \wedge Q \to R$.

Exercise 43. Use the inference rules to calculate the value of True $\vee$ True.

Exercise 44. Use the inference rules to calculate the value of False → True. Many people find the truth table definition False → True = True to be highly counterintuitive. Back in Section 6.4, this was just an arbitrary definition. Now, however, you can see the real reason behind the truth table definition of →: it's necessary in order to make the semantic truth table definition behave consistently with the formal inference rules of propositional logic.

Exercise 45. Suppose that you were given the following code:

```
logicExpr1 :: Bool -> Bool -> Bool
logicExpr1 a b = a /\ b \/ a <=> a

logicExpr2 :: Bool -> Bool -> Bool
logicExpr2 a b = (a \/ b) /\ b <=> a /\ b
```

Each of these functions specifies a Boolean expression. What are the truth values of these expressions? How would you write a list comprehension that can calculate the values for you to check your work?

Exercise 46. Work out the values of these expressions, then check with a list comprehension:

```
logicExpr3 :: Bool -> Bool -> Bool -> Bool
logicExpr3 a b c
  = (a /\ b) \/ (a /\ c) ==> a \/ b

logicExpr4 :: Bool -> Bool -> Bool -> Bool
logicExpr4 a b c
  = (a /\ (b \/ c)) \/ (a \/ c) ==> a \/ c
```

Exercise 47. Using the `Logic` data type defined below, define a function `distribute` that rewrites an expression using the distributive law, and a function `deMorgan` that does the same for DeMorgan's law.

```
data Logic = A | B | C
           | And Logic Logic
           | Or  Logic Logic
           | Not Logic
           | Imply Logic Logic
           | Equiv Logic Logic
           deriving (Eq, Show)
```

Exercise 48. Use equational reasoning (Boolean algebra) to prove that

$$(C \wedge A \wedge B) \vee C = C \wedge (C \vee (A \wedge B)).$$

Exercise 49. Prove that

$$C \vee (A \wedge (B \vee C)) = ((C \vee A) \wedge C) \vee A \wedge B.$$

Chapter 7

Predicate Logic

It is frequently necessary to reason logically about statements of the form *everything has the property p* or *something has the property p*. One of the oldest and most famous pieces of logical reasoning, which was known to the ancient Greeks, is an example:

>All men are mortal. Socrates is a man. Therefore Socrates is mortal.

In general, propositional logic is not expressive enough to support such reasoning. We could define a proposition P to mean 'all men are mortal', but P is an atomic symbol—it has no internal structure—so we cannot do any formal reasoning that makes use of the meaning of 'all'.

Predicate logic, also called first order logic, is an extension to propositional logic that adds two *quantifiers* that allow statements like the examples above to be expressed. Everything in propositional logic is also in predicate logic: all the definitions, inference rules, theorems, algebraic laws, etc., still hold.

7.1 The Language of Predicate Logic

The formal language of predicate logic consists of propositional logic, augmented with variables, predicates, and quantifiers.

7.1.1 Predicates

A *predicate* is a statement that an object x has a certain property. Such statements may be either true or false. For example, the statement '$x > 5$' is a predicate, and its truth depends on the value of x. A predicate can be extended to several variables; for example, '$x > y$' is a predicate about x and y.

The conditional expressions used to control execution of computer programs are predicates. For example, the Haskell expression if x<0 then -x else x uses the predicate x<0 to make a decision.

In predicate logic, it is traditional to write predicates concisely in the form $F(x)$, where F is the predicate and x is the variable it is applied to. A predicate containing two variables could be written $G(x, y)$. A predicate is essentially a function that returns a Boolean result. Often in this book, we will use Haskell notation for function applications, which does not require parentheses: for example, $f\ x$ is the application of f to x. For predicate logic, we will use the traditional notation with parentheses, $F(x)$.

Definition 15. Any term in the form $F(x)$, where F is a predicate name and x is a variable name, is a well-formed formula. Similarly, $F(x_1, x_2, \ldots, x_k)$ is a well-formed formula; this is a predicate containing k variables.

When predicate logic is used to solve a reasoning problem, the first step is to translate from English (or mathematics) into the formal language of predicate logic. This means the predicates are defined; for example we might define:

$$\begin{aligned} F(x) &\equiv\ x > 0 \\ G(x, y) &\equiv\ x > y \end{aligned}$$

The *universe of discourse*, often simply called the *universe* or abbreviated U, is the set of possible values that the variables can have. Usually the universe is specified just once, at the beginning of a piece of logical reasoning, but this specification cannot be omitted. For example, consider the statement 'For every x there exists a y such that $x = 2 \times y$'. If the universe is the set of even integers, or the set of real numbers, then the statement is true. However, if the universe is the set of natural numbers then the statement is false (let x be any odd number). If the universe doesn't contain numbers at all, then the statement is not true or false; it is meaningless.

Several notational conventions are very common in predicate logic, although some authors do not follow them. These standard notations will, however, be used in this book. The universe is called U, and its constants are written as lower-case letters, typically c and p (to suggest a constant value, or a particular value). Variables are also lower-case letters, typically x, y, z. Predicates are upper-case letters F, G, H, ... For example, $F(x)$ is a valid expression in the language of predicate logic, and its intuitive meaning is 'the variable x has the property F'. Generic expressions are written with a lower-case predicate; for example $f(x)$ could stand for *any* predicate f applied to a variable x.

7.1.2   Quantifiers

There are two *quantifiers* in predicate logic; these are the special symbols $\forall$ and $\exists$.

Definition 16. If $F(x)$ is a well-formed formula containing the variable x, then $\forall x.\ F(x)$ is a well-formed formula called a *universal quantification*. This is a statement that *everything* in the universe has a certain property: 'For all

x in the universe, the predicate $F(x)$ holds'. An alternative reading is 'Every x has the property F'.

Universal quantifications are often used to state required properties. For example, if you want to say formally that a computer program will give the correct output for all inputs, you would use $\forall$. The upside-down A symbol is intended to remind you of *All*.

Example 1. Let U be the set of even numbers. Let $E(x)$ mean x is even. Then $\forall x.\ E(x)$ is a well-formed formula, and its value is true.

Example 2. Let U be the set of natural numbers. Let $E(x)$ mean x is even. Then $\forall x.\ E(x)$ is a well-formed formula, and its value is false.

Definition 17. If $F(x)$ is a well-formed formula containing the variable x, then $\exists x.\ F(x)$ is a well-formed formula called an *existential quantification*. This is a statement that *something* in the universe has a certain property: 'There exists an x in the universe for which the predicate $F(x)$ holds'. An alternative reading is 'Some x has the property F'.

Existential quantifications are used to state properties that must occur at least once. For example, we might want to state that a database contains a record for a certain person; this would be done with $\exists$. The backwards E symbol is reminiscent of *Exists*.

Example 3. Let U be the set of natural numbers. Let $F(x,y)$ be defined as $2 \times x = y$. Then $\exists x.F(x,6)$ is a well-formed formula; it says that there is a natural number x which gives 6 when doubled; 3 satisfies the predicate, so the formula is true. However, $\exists x.\ F(x,7)$ is false.

Quantified expressions can be nested. Let the universe be the set of integers, and define $F(x) = \exists y.\ x < y$; thus $F(x)$ means 'There is some number larger than x'. Now we can say that every integer has this property by stating $\forall x.F(x)$. An equivalent way to write this is $\forall x.\ (\exists y.\ x < y)$. The parentheses are not required, since there is no ambiguity in writing $\forall x.\ \exists y.\ x < y$. All the quantified variables must be members of the universe.

In the example above, both x and y are integers. However, it is often useful to have several variables that are members of *different* sets. For example, suppose we are reasoning about people who live in cities, and want to make statements like 'There is at least one person living in every city'. It is natural to define $L(x,y)$ to mean 'The person x lives in the city y', and the expression $\forall x.\exists y.L(y,x)$ then means 'Every city has somebody living in it'. But what is the universe?

The way to handle this problem is to define a separate set of possible values for each variable. For example, let $C = \{\text{London}, \text{Paris}, \text{Los Angeles}, \text{München}\}$ be the set of cities, and let $P = \{\text{Smith}, \text{Jones}, \cdots\}$ be the set of persons. Now we can let the universe contain *all* the possible variable values: $U =$

$C \cup P$. Quantified expressions need to restrict each variable to the set of relevant values, as it is no longer intended that a variable x could be *any* element of U. This is expressed by writing $\forall x \in S.\ F(x)$ or $\exists x \in S.\ F(x)$, which say that x must be an element of S (and therefore also a member of the universe). Now the statement 'There is at least one person living in every city' is written

$$\forall x \in C.\ \exists y \in P.\ L(y, x).$$

Universal quantification over an empty set is vacuously true, and existential quantification over an empty set is vacuously false. Often we require that the universe is non-empty, so quantifications over the universe are not automatically true or false.

7.1.3 Expanding Quantified Expressions

If the universe is finite (or if the variables are restricted to a finite set), expressions with quantifiers can be interpreted as ordinary terms in propositional logic. Suppose $U = \{c_1, c_2, \ldots, c_n\}$, where the size of the universe is n. Then quantified expressions can be expanded as follows:

$$\forall x.\ F(x) \quad = \quad F(c_1) \wedge F(c_2) \wedge \cdots \wedge F(c_n) \qquad (7.1)$$
$$\exists x.\ F(x) \quad = \quad F(c_1) \vee F(c_2) \vee \cdots \vee F(c_n) \qquad (7.2)$$

With a finite universe, therefore, the quantifiers are just syntactic abbreviations. With a small universe it is perfectly feasible to reason directly with the expanded expressions. In many computing applications the universe is finite but may contain millions of elements; in this case the quantifiers are needed to make logical reasoning practical, although they are not needed in principle.

If the variables are not restricted to a finite set, it is impossible even in principle to expand a quantified expression. It may be intuitively clear to write $F(c_1) \wedge F(c_2) \wedge F(c_3) \wedge \cdots$, but this is not a well-formed formula. Every well-formed formula has a finite size, although there is no bound on how large a formula may be. This means that in the presence of an infinite universe, quantifiers make the language of predicate logic more expressive than propositional logic.

The expansion formulas, Equations 7.1 and 7.2, are useful for computing with predicates.

Example 4. Let the universe $U = \{1, 2, 3\}$, and define the predicates *even* and *odd* as follows:

$$even\ x \quad \equiv \quad (x \bmod 2 = 0)$$
$$odd\ x \quad \equiv \quad (x \bmod 2 = 1)$$

Two quantified expressions will be expanded and evaluated using these definitions:

$\forall x.\ (even\ x \to \neg(odd\ x))$
$= (even\ 1 \to \neg(odd\ 1)) \land (even\ 2 \to \neg(odd\ 2)) \land (even\ 3 \to \neg(odd\ 3))$
$=$ (False $\to \neg$True) $\land$ (True $\to \neg$False) $\land$ (False $\to \neg$True)
$=$ True $\land$ True $\land$ True
$=$ True
$\exists x.\ (even\ x \land odd\ x)$
$= (even\ 1 \land odd\ 1) \lor (even\ 2 \land odd\ 2) \lor (even\ 3 \land odd\ 3)$
$=$ (False $\land$ True) $\lor$ (True $\land$ False) $\lor$ (False $\land$ True)
$=$ False $\lor$ False $\lor$ False
$=$ False

Example 5. Let $S = \{0, 2, 4, 6\}$ and $R = \{0, 1, 2, 3\}$. Then we can state that every element of S is twice some element of R as follows:

$$\forall x \in S.\ \exists y \in R.\ x = 2 \times y$$

This can be expanded into a quantifier-free expression in two steps. The first step is to expand the *outer* quantifier:

$$(\exists y \in R.\ 0 = 2 \times y)$$
$$\land\ (\exists y \in R.\ 2 = 2 \times y)$$
$$\land\ (\exists y \in R.\ 4 = 2 \times y)$$
$$\land\ (\exists y \in R.\ 6 = 2 \times y)$$

The second step is to expand all four of the remaining quantifiers:

$$\big((0 = 2 \times 0) \lor (0 = 2 \times 1) \lor (0 = 2 \times 2) \lor (0 = 2 \times 3)\big)$$
$$\land\ \big((2 = 2 \times 0) \lor (2 = 2 \times 1) \lor (2 = 2 \times 2) \lor (2 = 2 \times 3)\big)$$
$$\land\ \big((4 = 2 \times 0) \lor (4 = 2 \times 1) \lor (4 = 2 \times 2) \lor (4 = 2 \times 3)\big)$$
$$\land\ \big((6 = 2 \times 0) \lor (6 = 2 \times 1) \lor (6 = 2 \times 2) \lor (6 = 2 \times 3)\big)$$

Two short cuts have been taken in the notation here. (1) Since every quantified variable is restricted to a set, the universe was not stated explicitly; however we can define $U = S \cup R$. (2) Instead of defining $F(x, y)$ to mean $x = 2 \times y$ and writing $\forall x \in S.\ \exists y \in R.\ F(x, y)$, we simply wrote the expression $x = 2 \times y$ inside the expression. Both short cuts are frequently used in practice.

Exercise 1. Let the universe $U = \{1, 2, 3\}$. Expand the following expressions into propositional term (i.e., remove the quantifiers):

(a) $\forall x.\ F(x)$

(b) $\exists x.\ F(x)$

(c) $\exists x.\ \forall y.\ G(x, y)$

Exercise 2. Let the universe be the set of integers. Expand the following expression: $\forall x \in \{1, 2, 3, 4\}.\ \exists y \in \{5, 6\}.\ F(x, y)$

7.1.4   The Scope of Variable Bindings

Quantifiers *bind* variables by assigning them values from a universe. A dangling expression without explicit quantification, such as $x+2$, has no explicit variable binding. If such an expression appears, x is assumed implicitly to be an element of the universe, and the author should have told you explicitly somewhere what the universe is.

The extent of a variable binding is called its *scope*. For example, the scope of x in the expression $\exists x.\ F(x)$ is the subexpression $F(x)$. For $\forall x \in S.\ \exists y \in R.\ F(x,y)$, the scope of x is $\exists y \in R.\ F(x,y)$, and the scope of y is $F(x,y)$.

It is good practice to use parentheses to make expressions clear and readable. The expression

$$\forall x.\, p(x) \lor q(x)$$

is not clear: it can be read in two different ways. It could mean either

$$\forall x.\, (p(x) \lor q(x))$$

or

$$(\forall x.\, p(x)) \lor q(x).$$

It is probably best to use parentheses in case of doubt, but there is a convention that resolves unclear expressions: the quantifier extends over the smallest subexpression possible unless parentheses indicate otherwise. In other words, the scope of a variable binding is the smallest possible. So, in the assertion given above, the variable x in $q(x)$ is not bound by the $\forall$, so it must have been bound at some outer level (i.e., this expression has to be embedded inside a bigger one).

Often the same quantifier is used several times in a row to define several variables:

$$\forall x.\ \forall y.\ F(x,y)$$

It is common to write this in an abbreviated form, with just one use of the $\forall$ operator followed by several variables separated by commas. For example, the previous expression would be abbreviated as follows:

$$\forall x, y.\ F(x,y)$$

This abbreviation may be used for any number of variables, and it can also be used if the variables are restricted to be in a set, as long as they all have the same restriction. For example, the abbreviated expression

$$\forall x, y, z \in S.\ F(x,y,z)$$

is equivalent to the full expression

$$\forall x \in S.\ \forall y \in S.\ \forall z \in S.\ F(x,y,z).$$

7.1.5 Translating Between English and Logic

Sometimes it is straightforward to translate an English statement into logic. If an English statement has no internal structure that is relevant to the reasoning, it can be represented by an ordinary propositional variable:

A	$\equiv$	Elephants are big.
B	$\equiv$	Cats are furry.
C	$\equiv$	Cats are good pets.

An English statement built up with words like *and, or, not, therefore,* and so on, where the meaning corresponds to the logical operators, can be represented by a propositional expression.

$\neg A$	$\equiv$	Elephants are small.
$A \wedge B$	$\equiv$	Elephants are big and cats are furry.
$B \to C$	$\equiv$	If cats are furry then they make good pets.

Notice in the examples above that no use has been made of the internal structure of the English statements. (The sentence 'elephants are small' may appear to violate this, but it could just as easily have been written 'it is untrue that elephants are big', which corresponds exactly to $\neg A$.)

When general statements are made about classes of objects, then predicates and quantifiers are needed in order to draw conclusions. For example, suppose we try these definitions in propositional logic, without using predicates:

A	$\equiv$	Small animals are good pets.
C	$\equiv$	Cats are animals.
S	$\equiv$	Cats are small.

In ordinary conversation, it would be natural to conclude that cats are good pets, but this cannot be concluded with propositional logic. All we have are three propositions: A, C, and S are known, but nothing else, and the only conclusions that can be drawn are uninteresting ones like $A \wedge C$, $S \vee A$, and the like. The substantive conclusion, that cats are good pets, requires reasoning about the internal structure of the English statements. The solution is to use predicates to give a more refined translation of the sentences:

$A(x)$	$\equiv$	x is an animal.
$C(x)$	$\equiv$	x is a cat.
$S(x)$	$\equiv$	x is small.
$GP(x)$	$\equiv$	x is a good pet.

Now a much richer kind of English sentence can be translated into predicate logic:

$\forall x.\ C(x) \to A(x)$	$\equiv$	Cats are animals.
$\forall x.\ C(x) \to S(x)$	$\equiv$	Cats are small.
$\forall x.\ C(x) \to S(x) \wedge A(x)$	$\equiv$	Cats are small animals.
$\forall x.\ S(x) \wedge A(x) \to GP(x)$	$\equiv$	Small animals are good pets.

It is generally straightforward to translate from formal predicate logic into English, since you can just turn each logical operator directly into an English word or phrase. For example,

$$\forall x.\ S(x) \land A(x) \to GP(x)$$

could be translated into English literally:

(1) For every thing, if that thing is small and that thing is an animal, then that thing is a good pet.

This is graceless English, but at least it's comprehensible and correct. The style can be improved:

(2) Everything which is small and which is an animal is a good pet.

Even better would be:

(3) Small animals make good pets.

Such stylistic improvements in the English are optional. It is important to be sure that the effort to improve the literary style doesn't affect the meaning, but this is a question of proper usage of natural language, not of formal logic.

It is sometimes trickier to translate from English into formal logic, precisely because the English usually does not correspond obviously to the logical quantifiers and operators. Sentence (1) above can be translated straightforwardly into logic, sentence (3) is harder. The difficulty is not really in the logic; it is in figuring out exactly what the English sentence says.

Often the real difficulty in translating English into logic is in figuring out what the English says, or what the speaker meant to say. For example, many people make statements like 'All people are not rich'. What this statement *actually says* is

$$\forall x.\neg R(x),$$

where the universe is the set of people and $R(x)$ means 'x is rich'. What is usually meant, however, by such a statement is

$$\neg \forall x.\ R(x),$$

that is, it is not true that all people are rich (alternatively, not all people are rich). The intended meaning is equivalent to

$$\exists x.\ \neg R(x).$$

Such problems of ambiguity or incorrect grammar in English cannot be solved mathematically, but they do illustrate one of the benefits of mathematics: simply translating a problem from English into formal logic may expose confusion or misunderstanding.

Example 6. Consider the translation of the sentence 'Some birds can fly' into logic. Let the universe be a set that contains all birds (it is all right if it contains other things too, such as frogs and other animals). Let $B(x)$ mean 'x is a bird' and $F(x)$ mean 'x can fly'. Then 'Some birds can fly' is translated as

$$\exists x.\ B(x) \wedge F(x)$$

Warning! A common pitfall is to translate 'Some birds can fly' as

$$\exists x.\ B(x) \to F(x) \qquad \textit{Wrong translation!}$$

To see why this is wrong, let p be a frog that somehow got into the universe. Now $B(p)$ is false, so $B(p) \to F(p)$ is true (remember False $\to$ False = True). This is just saying 'If that frog were a bird then it would be able to fly', which is true; it doesn't mean the frog actually is a bird, or that it actually can fly. However, we have now found a value of x—namely the frog p—for which $B(x) \to F(x)$ is true, and that is enough to satisfy $\exists x.\ B(x) \to F(x)$, even if all the birds in the universe happen to be penguins (which cannot fly).

Exercise 3. Express the following statements formally, using the universe of natural numbers, and the predicates $E(x) \equiv x$ is even and $O(x) \equiv x$ is odd.

- There is an even number.
- Every number is either even or odd.
- No number is both even and odd.
- The sum of two odd numbers is even.
- The sum of an odd number and an even number is odd.

Exercise 4. Let the universe be the set of all animals, and define the following predicates:

$$
\begin{array}{lcl}
B(x) & \equiv & x \text{ is a bird.} \\
D(x) & \equiv & x \text{ is a dove.} \\
C(x) & \equiv & x \text{ is a chicken.} \\
P(x) & \equiv & x \text{ is a pig.} \\
F(x) & \equiv & x \text{ can fly.} \\
W(x) & \equiv & x \text{ has wings.} \\
M(x,y) & \equiv & x \text{ has more feathers than } y \text{ does.}
\end{array}
$$

Translate the following sentences into logic. There are generally several correct answers. Some of the English sentences are fairly close to logic, while others require more interpretation before they can be rendered in logic.

- Chickens are birds.
- Some doves can fly.

- Pigs are not birds.
- Some birds can fly, and some can't.
- An animal needs wings in order to fly.
- If a chicken can fly, then pigs have wings.
- Chickens have more feathers than pigs do.
- An animal with more feathers than any chicken can fly.

Exercise 5. Translate the following into English.

- $\forall x.\ (\exists y.\ \text{wantsToDanceWith}\ (x, y))$
- $\exists x.\ (\forall y.\ \text{wantsToPhone}\ (y, x))$
- $\exists x.\ (\text{tired}\ (x) \land \forall y.\ \text{helpsMoveHouse}\ (x, y))$

7.2   Computing with Quantifiers

As long as the universe is finite, a computer is useful for evaluating logical expressions with quantifiers. This provides a good way to check your understanding of expressions in predicate logic. Even more importantly, many software applications are expressed in terms of finite sets of data that are manipulated using predicate logic expressions.

The software tools file provides several Haskell functions that are helpful for computing with predicate logic, and this section explains how to use them. To keep things simple, we assume that the universe is a set of numbers represented as a list. In programming terminology, a *predicate* is a function that returns a Boolean value; this is the same meaning that predicate has in logic.

The function `forall` takes the universe, represented as a list of numbers, and a predicate. It applies the predicate to each value in the universe and returns the conjunction of the results:

```
forall :: [Int] -> (Int -> Bool) -> Bool
```

For example, `forall [1,2] (>5)` means $\forall x.\ x > 5$ where the universe $U = \{1, 2\}$. The implementation of `forall` simply expands the quantified expression, using Equation 7.1, and then evaluates it. The expansion uses the Haskell function `and ::  [Bool] -> Bool`; thus $F(c_1) \land F(c_2) \land F(c_3)$ would be expressed in Haskell as `and [f c1, f c2, f c3]`.

Exercise 6. Write the predicate logic expressions corresponding to the following Haskell expressions. Then decide whether the value is `True` or `False`, and evaluate using the computer. Note that `(== 2)` is a function that takes a number and compares it with 2, while `(< 4)` is a function that takes a number and returns `True` if it is less than 4.

```
forall [1,2,3] (== 2)
forall [1,2,3] (< 4)
```

Like `forall`, the function `exists` applies its second argument to all of the elements in its first argument:

```
exists :: [Int] -> (Int -> Bool) -> Bool
```

However, `exists` forms the disjunction of the result, using the Haskell function `or :: [Bool] -> Bool`.

Exercise 7. Again, rewrite the following in predicate logic, work out the values by hand and evaluate on the computer:

```
exists [0,1,2] (== 2)
exists [1,2,3] (> 5)
```

The functions `exists` and `forall` can be nested in the same way as quantifiers can be nested in predicate logic. It's convenient to express inner quantified formulas as separate functions.

Example 7. $\forall x \in \{1,2\}. \ (\exists y \in \{1,2\}. \ x = y)$ has an inner assertion that can be implemented as follows:

```
inner_fun :: Int -> Bool
inner_fun x = exists [1,2] (== x)
```

Now consider the evaluation of:

```
forall [1,2] inner_fun
```

The evaluation can be calculated step by step. The function **and** takes a list of Boolean values and combines them all using the $\wedge$ operation:

```
forall [1,2] inner_fun
= and [inner_fun 1, inner_fun 2]
= and [exists [1,2] (== 1),
       exists [1,2] (== 2)]
= and [or [1==1, 2==1],
       or [1==2, 2==2]]
= and [True, True]
= True
```

Example 8. Define:

```
inner_fun x = exists [1,2,3] (== x+2)
exists [1,2,3] inner_fun
```

Here is the evaluation:

```
  exists [1,2,3] inner_fun
= or [inner_fun 1, inner_fun 2, inner_fun 3]
= or [exists [1,2,3] (== 1+2),
      exists [1,2,3] (== 2+2),
      exists [1,2,3] (== 3+2)]
= or [or [1 == 1+2, 2 == 1+2, 3 == 1+2],
      or [1 == 2+2, 2 == 2+2, 3 == 2+2],
      or [1 == 3+2, 2 == 3+2, 3 == 3+2]]
= or [or [False, False, True],
      or [False, False, False],
      or [False, False, False]]
= or [True, False, False]
= True
```

An important distinction between mathematical quantification and the Haskell functions `exists` and `forall` is that quantification is defined over both finite and infinite universes, whereas these Haskell functions do not always terminate when applied to infinite universes.

Exercise 8. Define the predicate $p\ x\ y$ to mean $x = y+1$, and let the universe be $\{1, 2\}$. Calculate the value of each of the following expressions, and then check your solution using Haskell.

  (a) $\forall x.\ (\exists y.\ p(x, y))$

  (b) $\exists x, y.\ p(x, y)$

  (c) $\exists x.\ (\forall y.\ p(x, y))$

  (d) $\forall x, y.\ p(x, y)$

7.3   Logical Inference with Predicates

The inference rules for propositional logic can be extended to handle predicate logic as well. Four additional rules are required (Figure 7.1): an introduction rule and an elimination rule for both of the quantifiers $\forall$ and $\exists$.

A good way to understand the inference rules of predicate logic is to view them as generalisations of the corresponding rules of propositional logic. For example, there is a similarity between inferring $F(P) \wedge F(Q)$ in propositional logic, and inferring $\forall x.F(x)$ in predicate logic. If the universe is finite, then the predicate logic is not, in principle, even necessary. We could express $\forall x.F(x)$ by $F(p_1) \wedge F(p_2) \wedge \ldots \wedge F(p_n)$, where n is the size of the universe. If the universe is infinite, however, then the inference rules of predicate logic allow deductions that would be impossible using just the propositional logic rules.

We will always assume that the universe is non-empty. This is especially important with the rule for forall elimination, which is invalid when the universe of discourse is empty. Also, in the rule for exists elimination, the conclusion of

$$\frac{F(x) \quad \{x \text{ arbitrary}\}}{\forall x.F(x)}\{\forall I\} \qquad \frac{\forall x.F(x)}{F(p)}\{\forall E\}$$

$$\frac{F(p)}{\exists x.F(x)}\{\exists I\} \qquad \frac{\exists x.F(x) \quad F(x) \vdash A \quad \{x \text{ not free in } A\}}{A}\{\exists E\}$$

Figure 7.1: Inference Rules of Predicate Logic. In these rules, p must be an element of the universe.

the rule (which is also the conclusion of the sequent in the hypothesis of the rule) must be a proposition that does not depend in any way on the parameter of the universe of discourse. That is, the proposition A in the statement of the rule must not be a predicate that depends on x.

7.3.1 Universal Introduction $\{\forall I\}$

A standard proof technique, which is used frequently in mathematics and computer science, is to state and prove a property about an arbitrary value x, which is an element of the universe, and then to interpret this as a statement about *all* elements of the universe. A typical example is the following simple theorem about Haskell lists, which says that there are two equivalent methods for attaching a singleton x in front of a list xs. The definition of $(+\!+)$ is

```
(++) :: [a] -> [a] -> [a]
[] ++ ys = ys
(x:xs) ++ ys = x : (xs ++ ys)
```

Theorem 57. Let $x :: a$ and $xs :: [a]$. Then $x : xs = [x] +\!+ xs$.

It is important to realise that this theorem is stating a property about *arbitrary* x and xs. It really means that the property holds for *all* values of the variables, and this could be stated more formally with an explicit quantifier:

Theorem 58. $\forall x :: a. \ \forall xs :: [a]. \ x : xs = [x] +\!+ xs$

These two theorems have exactly the same meaning; the only difference is the style[1] in which they are expressed: the first is a little more like English, and the second is a little more formal. Both styles are common. For a theorem in the first style, the use of *arbitrary* variables means an implicit $\forall$ is meant. Now, consider the proof of this theorem; the following proof could be used for either the formal or the informal statement of the theorem:

---

[1]For a discussion about good style in mathematics, see the pointer to *Mathematical Writing* in Section 7.5.

Proof.

$$
\begin{aligned}
[x] +\!\!+ xs & \\
&= (x : [\,]) +\!\!+ xs && \text{def. of notation} \\
&= x : ([\,] +\!\!+ xs) && (+\!\!+).2 \\
&= x : xs && (+\!\!+).1
\end{aligned}
$$

$\square$

Again, there is a significant point about this proof: it consists of formal reasoning about one value x and one value xs, but these values are *arbitrary*, and the conclusion we reach at the end of the proof is that the theorem is true for *all* values of the variables. In other words, if we can prove a theorem for an arbitrary variable, then we infer that the theorem is true for *all* possible values of that variable.

These ideas are expressed formally by the following inference rule, which says that if the expression a (which may contain a variable x) can be proved for *arbitrary* x, then we may infer the proposition $\forall x.\ a$. Since this rule specifies what we need to know in order to infer an expression containing $\forall$, its name is $\{\forall I\}$.

$$
\frac{F(x) \quad \{x \text{ arbitrary}\}}{\forall x.F(x)}\{\forall I\}
$$

To clarify exactly what this rule means, we will compare two examples: one where it can be used, and one where it cannot. Let the universe be the set of natural numbers N, and let $E(x)$ be the proposition 'x is even'. First, consider the following theorem:

Theorem 59. $\vdash \forall x.\ E(x) \to \big(E(x) \lor \neg E(x)\big)$

The proof uses the $\forall$ introduction rule. The important point is that this inference does not depend on the particular value of p; thus the value of p is arbitrary, and the $\{\forall I\}$ rule allows us to infer $\forall x.F(x) \lor \neg F(x)$.

Proof.

$$
\frac{\dfrac{\dfrac{\boxed{E(p)}}{E(p) \lor \neg E(p)}\{\lor I_L\}}{E(p) \to E(p) \lor \neg E(p)}\{\to I\}}{\forall x.\ E(x) \to E(x) \lor \neg E(x)}\{\forall I\}
$$

$\square$

Now consider the following *incorrect* proof, which purports to show that all natural numbers are even:

$$\frac{\overline{E(2)}}{\forall x.E(x)}\{\forall I\} \quad \textbf{\textit{Wrong!}}$$

The theorem $E(2)$ is established for a particular value, 2. However, 2 is not *arbitrary*: the proof that 2 is even relies on its value, and we could not substitute 3 without invalidating it. Because we have not proved $E(x)$ for an arbitrary value, the requirements of the $\{\forall I\}$ inference rule are not satisfied, and we cannot conclude $\forall x.E(x)$.

The notion of an arbitrary variable is commonly used in semi-formal treatments of mathematical logic. In a completely formal presentation of logic—for example, in proof-checking software—the requirement that variables be arbitrary is replaced by a purely syntactic constraint.

The problem with talking about arbitrary variables is that arbitrary is an adjective, like blue or green. How can you look at a variable and decide whether it is or is not "arbitrary"? In short, you cannot, because arbitrariness is a property of the context in which the variable appears, and not the variable itself.

The syntactic constraint on the $\{\forall I\}$ rule is that that the variable x is not allowed to appear "free" (that is, unbound by a quantifier) in any undischarged hypothesis that was used to infer $F(x)$. For example, here is another wrong inference:

$$\frac{\dfrac{\overline{x=2}}{E(x)}}{\forall x.E(x)}\{\forall I\} \quad \textbf{\textit{Wrong!}}$$

In this case, we have assumed $x = 2$ and then inferred $E(x)$ (this would actually require several inference steps, or recourse to an auxiliary theorem, but it is shown here as a single inference). Now, we have $E(x)$ above the line, so it looks like we can use the $\{\forall I\}$ rule to infer $\forall x.E(x)$. However, the proof above the line of $E(x)$ was based on the assumption $x = 2$, and x appears free in that assumption. Furthermore, the assumption was not discharged. Consequently, the syntactic requirement for the $\{\forall I\}$ rule has not been satisfied, and the inference is not valid.

The syntactic constraint is really just a more formal way to state the same thing as the "arbitrary variable" requirement. You can explore these issues in more detail in the proof checker software (see the web page for this book).

7.3.2 Universal Elimination $\{\forall E\}$

The universal elimination rule says that if you have established $\forall x.F(x)$, and p is a particular element of the universe, then you can infer $F(p)$.

$$\frac{\forall x.F(x)}{F(p)}\{\forall E\}$$

The following theorem allows you to apply a universal implication, in the form $\forall x.F(x) \to G(x)$, to a particular proposition $F(p)$, and its proof illustrates the $\{\forall E\}$ inference rule.

Theorem 60. $F(p), \forall x.F(x) \to G(x) \vdash G(p)$

Proof.

$$\frac{F(p) \quad \dfrac{\forall x.F(x) \to G(x)}{F(p) \to G(p)}\{\forall E\}}{G(p)}\{\to E\}$$

$\square$

In Chapter 6 the implication chain theorem was proved; this says that from $a \to b$ and $b \to c$ you can infer $a \to c$. The $\{\forall I\}$ inference rule can be used to prove the corresponding theorem on universal implications: from $\forall x.F(x) \to G(x)$ and $\forall x.G(x) \to H(x)$, you can infer $\forall x.F(x) \to H(x)$. However, in order to use $\{\forall I\}$ we have to establish first, for an arbitrary p in the universe, that $F(p) \to H(p)$, and the proof of that proposition requires using the $\{\forall E\}$ rule twice to prove the particular propositions $F(p) \to G(p)$ and $G(p) \to H(p)$.

Theorem 61. $\forall x.F(x) \to G(x), \forall x.G(x) \to H(x) \vdash \forall x.F(x) \to H(x)$

Proof.

$$\frac{\dfrac{\dfrac{\forall x.F(x) \to G(x)}{F(p) \to G(p)}\{\forall E\} \quad \dfrac{\forall x.G(x) \to H(x)}{G(p) \to H(p)}\{\forall E\}}{F(p) \to H(p)}\text{Th. Imp Chain}}{\forall x.F(x) \to H(x)}\{\forall I\}$$

$\square$

The following theorem says that you can change the order in which the variables are bound in $\forall x. \forall y. F(x,y)$. This theorem is simple but extremely important.

Theorem 62. $\forall x. \forall y. F(x,y) \vdash \forall y. \forall x. F(x,y)$

Proof.

$$\frac{\dfrac{\dfrac{\dfrac{\forall x. \forall y. F(x,y)}{\forall y. F(p,y)}\{\forall E\}}{F(p,q)}\{\forall E\}}{\forall x. F(x,q)}\{\forall I\}}{\forall y. \forall x. F(x,y)}\{\forall I\}$$

□

This theorem says that if, for all x, a proposition P implies $f(x)$, then P implies $\forall x.f(x)$. This allows you to pull a proposition P, which does not use x, out of an implication bound by $\forall$.

Theorem 63. $\forall x.\ P \to f(x) \ \vdash\ P \to \forall x.\ f(x)$

Proof.

$$\cfrac{\cfrac{\cfrac{\forall x.\ P \to f(x)}{P \to f(c)}\ {\scriptstyle\{\forall E\}}\quad (c\ \text{arbitrary})}{\cfrac{f(c)}{\cfrac{\forall x.\ f(x)}{P \to \forall x.\ f(x)}{\scriptstyle\{\to I\}}}{\scriptstyle\{\forall I\}}}{}}{}$$

□

Exercise 9. Prove $\forall x.F(x), \forall x.F(x) \to G(x) \vdash \forall x.G(x)$.

7.3.3   Existential Introduction $\{\exists I\}$

The $\{\exists I\}$ rule says that if $f(p)$ has been established for a particular p, then you can infer $\exists x.f(x)$.

$$\boxed{\cfrac{f(p)}{\exists x.f(x)}{\scriptstyle\{\exists I\}}}$$

The following theorem says that if $F(x)$ holds for all elements of the universe, then it must hold for one of them. Recall that we require the universe of discourse to be non-empty; otherwise this theorem would not hold.

Theorem 64. $\forall x.F(x) \vdash \exists x.F(x)$

Proof.

$$\cfrac{\cfrac{\forall x.F(x)}{F(p)}{\scriptstyle\{\forall E\}}}{\exists x.F(x)}{\scriptstyle\{\exists I\}}$$

□

7.3.4 Existential Elimination $\{\exists E\}$

Recall the $\{\vee E\}$ inference rule of propositional logic; this says that if you know $a \vee b$, and also that c follows from a and c follows from b, then you can infer c.

If the universe is finite, then $\exists x.F(x)$ can be expressed in the form $F(p_1) \vee \ldots \vee F(p_n)$, where the universe is $\{p_1,\ldots,p_n\}$. We could extend the $\{\vee E\}$ rule so that if we know that $F(p_i)$ holds for some i, and furthermore that A must hold if $F(x)$ holds for *arbitrary* x, then A can be inferred.

The existential elimination rule $\{\exists E\}$ captures this idea, and it provides a much more convenient tool for reasoning than repeated applications of $\{\vee E\}$. Its fundamental importance, however, is that $\{\exists E\}$ may also be used if the universe is infinite. This means it is more powerful than $\{\vee E\}$, as that can be used only for an $\vee$ expression with a finite number of terms. (Recall that a proof must have a finite length.)

$$\frac{\exists x.F(x) \qquad F(x) \vdash A \quad \{x \text{ arbitrary}\}}{A}\{\exists E\}$$

The following theorem gives an example of $\{\exists E\}$. It says that if $P(x)$ always implies $Q(x)$, and also that $P(x)$ holds for some x, then $Q(x)$ also holds for some x.

Theorem 65. $\exists x . P(x),\ \forall x . P(x) \rightarrow Q(x) \vdash \exists x . Q(x)$

Proof.

$$\frac{\exists x . P(x) \qquad \dfrac{\boxed{P(c)} \quad \dfrac{\forall x . P(x) \rightarrow Q(x)}{P(c) \rightarrow Q(c)}\{\forall E\}}{\dfrac{Q(c)}{Q(c)}\{\rightarrow E\}}\{\exists E\}}{\exists x . Q(x)}\{\exists I\}$$

$\square$

The following theorem says that a $\forall$ directly inside an $\exists$ can be brought outside the $\exists$.

Theorem 66. $\exists x. \forall y. F(x,y) \vdash \forall y. \exists x. F(x,y)$

Proof.

$$\frac{\exists x. \forall y. F(x,y) \qquad \dfrac{\dfrac{\dfrac{\boxed{\forall y. F(p,y)}}{F(p,q)}\{\forall E\}}{\dfrac{\exists x. F(x,q)}{\forall y. \exists x. F(x,y)}\{\exists I\}}\{\forall I\}}{\forall y. \exists x. F(x,y)}}{\forall y. \exists x. F(x,y)}\{\exists E\}$$

Table 7.1: Algebraic Laws of Predicate Logic

$$\forall x.\ f(x) \quad \rightarrow \quad f(c) \tag{7.3}$$
$$f(c) \quad \rightarrow \quad \exists\, x.\ f(x) \tag{7.4}$$

$$\forall x.\,\neg\, f(x) \quad = \quad \neg\exists x.\ f(x) \tag{7.5}$$
$$\exists\, x.\,\neg\, f(x) \quad = \quad \neg\forall x.\ f(x) \tag{7.6}$$

Provided that x does not occur free in q:
$$(\forall x.\ f(x)) \wedge q \quad = \quad \forall x.\ (f(x) \wedge q) \tag{7.7}$$
$$(\forall x.\ f(x)) \vee q \quad = \quad \forall x.\ (f(x) \vee q) \tag{7.8}$$
$$(\exists x.\ f(x)) \wedge q \quad = \quad \exists x.\ (f(x) \wedge q) \tag{7.9}$$
$$(\exists x.\ f(x)) \vee q \quad = \quad \exists x.\ (f(x) \vee q) \tag{7.10}$$

$$(\forall x.\ f(x)) \wedge (\forall x.\ g(x)) \quad = \quad \forall x.\ (f(x) \wedge g(x)) \tag{7.11}$$
$$(\forall x.\ f(x)) \vee (\forall x.\ g(x)) \quad \rightarrow \quad \forall x.\ (f(x) \vee g(x)) \tag{7.12}$$
$$\exists x.\ (f(x) \wedge g(x)) \quad \rightarrow \quad (\exists x.\ f(x)) \wedge (\exists x.\ g(x)) \tag{7.13}$$
$$(\exists x.\ f(x)) \vee (\exists x.\ g(x)) \quad = \quad \exists x.\ (f(x) \vee g(x)) \tag{7.14}$$

□

Exercise 10. Prove $\exists x.\ \exists y.\ F(x,y) \vdash \exists y.\ \exists x.\ F(x,y)$.

Exercise 11. The converse of Theorem 66 is the following:

$$\forall y.\ \exists x.\ F(x,y) \vdash \exists x.\ \forall y.\ F(x,y) \quad \textbf{\textit{Wrong!}}$$

Give a counterexample that demonstrates that this statement is *not* valid.

Exercise 12. Prove $\forall x.(F(x) \wedge G(x)) \vdash (\forall x.F(x)) \wedge (\forall x.G(x))$.

7.4 Algebraic Laws of Predicate Logic

The previous section presented predicate logic as a natural deduction inference system. An alternative style of reasoning is based on a set of algebraic laws about propositions with predicates, listed in Table 7.1.

This is not the minimal possible set of laws; some of them correspond to inference rules, and others are provable as theorems. The focus in this section, however, is on practical calculations using the laws, rather than on theoretical foundations.

The following two laws express, in algebraic form, the $\{\forall E\}$ and $\{\exists I\}$ inference rules. As they correspond to inference rules, these laws are logical implications, not equations.

$$\forall x.\ f(x) \quad \rightarrow \quad f(c) \tag{7.3}$$
$$f(c) \quad \rightarrow \quad \exists x.\ f(x) \tag{7.4}$$

In both of these laws, x is bound by the quantifier, and it may be any element of the universe. The element c is any fixed element of the universe. Thus the first law says that if the predicate f holds for all elements of the universe, it must hold for a particular one c, and the second law says that if f holds for an *arbitrarily chosen* element c, then it must hold for all elements of the universe.

The following theorem combines these two laws and is often useful in proving other theorems. Its proof uses the line-by-line style, which is standard when reasoning about predicate logic with algebraic laws.

Theorem 67. $\forall x.\ f(x) \ \rightarrow \ \exists x.\ f(x)$

Proof.

$$\forall x.\ f(x)$$
$$\rightarrow \ f(c) \qquad \{7.3\}$$
$$\rightarrow \ \exists x.\ f(x) \quad \{7.4\}$$

$\square$

The next two laws state how the quantifiers combine with logical negation. The first one says that if $f(x)$ is always false, then it is never true; the second says that if $f(x)$ is ever untrue, then it is not always true.

$$\forall x.\, \neg f(x) \ = \ \neg \exists x.\, f(x) \tag{7.5}$$
$$\exists x.\, \neg f(x) \ = \ \neg \forall x.\, f(x) \tag{7.6}$$

The following four laws show how a predicate $f(x)$ combines with a proposition q that does not contain x. These are useful for bringing constant terms into or out of quantified expressions.

$$(\forall x.\ f(x)) \wedge q \ = \ \forall x.\, (f(x) \wedge q) \tag{7.7}$$
$$(\forall x.\ f(x)) \vee q \ = \ \forall x.\, (f(x) \vee q) \tag{7.8}$$
$$(\exists x.\ f(x)) \wedge q \ = \ \exists x.\, (f(x) \wedge q) \tag{7.9}$$
$$(\exists x.\ f(x)) \vee q \ = \ \exists x.\, (f(x) \vee q) \tag{7.10}$$

The final group of laws concerns the combination of quantifiers with $\wedge$ and $\vee$. It is important to note that two of them are equations (or double implications), whereas the other two are implications. Therefore they can be used in only one direction, and they must be used at the top level, not on subexpressions.

$$\forall x.\, f(x) \land \forall x.\, g(x) \;=\; \forall x.\, (f(x) \land g(x)) \qquad (7.11)$$
$$\forall x.\, f(x) \lor \forall x.\, g(x) \;\to\; \forall x.\, (f(x) \lor g(x)) \qquad (7.12)$$
$$\exists x.\, (f(x) \land g(x)) \;\to\; \exists x.\, f(x) \land \exists x.\, g(x) \qquad (7.13)$$
$$\exists x.\, f(x) \lor \exists x.\, g(x) \;=\; \exists x.\, (f(x) \lor g(x)) \qquad (7.14)$$

Example 9. The following equation can be proved algebraically:

$$\forall x.\, (f(x) \land \neg g(x)) \;=\; \forall x. f(x) \land \neg \exists x.\, g(x)$$

This is established through a sequence of steps. Each step should be justified by one of the algebraic laws, or by another equation that has already been proved. When the purpose is actually to prove a theorem, the justifications should be written explicitly. Often this kind of reasoning is used informally, like a straightforward algebraic calculation, and the formal justifications are sometimes omitted.

$$
\begin{aligned}
&\forall x.\, (f(x) \land \neg g(x)) \\
&= \forall x.\, f(x) \land \forall x.\, \neg g(x) \quad \{7.11\} \\
&= \forall x. f(x) \land \neg \exists x.\, g(x) \quad \{7.5\}
\end{aligned}
$$

Example 10. The following equation says that if $f(x)$ sometimes implies $g(x)$, and $f(x)$ is always true, then $g(x)$ is sometimes true.

$$\exists x.\, (f(x) \to g(x)) \,\land\, (\forall x.\, f(x)) \,\to\, \exists x.\, g(x)$$

The first step of the proof replaces the local variable x by y in the $\forall$ expression. This is not actually necessary, but it may help to avoid confusion; whenever the same variable is playing different roles in different expressions, and there seems to be a danger of getting them mixed up, it is safest just to change the local variable. In the next step, the $\forall$ expression is brought inside the $\exists$; in the following step it is now possible to pick a particular value for y: namely, the x bound by $\exists$.

$$
\begin{aligned}
&\exists x.\, (f(x) \to g(x)) \,\land\, (\forall x.\, f(x)) \\
&= \big(\exists x.\, (f(x) \to g(x))\big) \,\land\, \big(\forall y.\, f(y)\big) \quad \text{change of variable} \\
&= \exists x.\, \big((f(x) \to g(x)) \,\land\, (\forall y.\, f(y))\big) \quad \{7.9\} \\
&= \exists x.\, \big((f(x) \to g(x)) \,\land\, f(x)\big) \quad \{7.3\} \\
&\to \exists x.\, g(x) \quad \{\textit{Modus Ponens}\}
\end{aligned}
$$

7.5 Suggestions for Further Reading

The books on logic recommended at the end of Chapter 6 also cover predicate logic. Those citations are not repeated here; instead, two excellent books on style and elegance in mathematical proofs are suggested.

Mathematical Writing, by Knuth, Larrabee and Roberts [20], is filled with good advice about how to write mathematics in a style which is clear, rigorous and lively. This short book is based on a course on mathematical writing given by the authors at Stanford.

A *rigorous* proof is careful, and it covers scrupulously all the relevant aspects of the problem. Routine, straightforward issues may be treated lightly in a rigorous proof, but they really must be sound. In contrast, a *formal* proof takes no short cuts at all; it does *everything* using the rules of some formal system, such as the logical inference rules. Formal proofs are good for machine checking and generally fit well with computing applications. Proofs that are informal but rigorous should be written in a clear, elegant style that is convincing to a knowledgeable reader.

Proofs from THE BOOK, by Aigner and Ziegler [3], is an outstanding collection of elegant and rigorous proofs written in normal (but particularly good) mathematical style. It is worth looking at for its beauty, although some of its contents are rather advanced. The book was inspired by an idea of Paul Erdős, one of the leading mathematicians of the twentieth century. Erdős imagined The Book, which contains the most elegant proofs of the most interesting theorems. The Book doesn't actually exist; it is an ideal to which real people can only aspire, but it is nevertheless inspiring to mathematicians to find the best approximation to it that they can. *Proofs from THE BOOK* is such an effort, and it is likely to become a mathematical classic.

7.6   Review Exercises

Exercise 13. Suppose the universe contains 10 elements. How many times will F occur when $\forall x.\ \exists y.\forall z.\ F(x,y,z)$ is expanded into quantifier-free form? How large in general are expanded expressions?

Exercise 14. Prove $(\exists x.\ f(x)) \vee (\exists x.\ g(x)) \vdash \exists x.\ (f(x) \vee g(x))$.

Exercise 15. Prove $(\forall x.\ f(x)) \vee (\forall x.\ g(x)) \vdash \forall x.\ (f(x) \vee g(x))$.

Exercise 16. Prove the converse of Theorem 63.

Exercise 17. Find counterexamples that show that Laws 7.12 and 7.13, which are implications, would not be valid as equations.

Exercise 18. Prove the following implication:
$$\big(\forall x.\ f(x) \to h(x) \ \wedge\ \forall x.\ g(x) \to h(x)\big)$$
$$\to\quad \forall x.\ (f(x) \vee g(x) \to h(x))$$

Exercise 19. Define a predicate (with the natural numbers $0,1,2,\ldots$ as its universe) that expresses the notion that all of the elements that occur in either of the sequences supplied as operands to the append operator

(++) also occur as elements of the sequence it delivers. That is, the predicate states that under certain constraints on the number of elements in the sequence xs, any element that occurs in either the sequence xs or the sequence ys also occurs in the sequence xs ++ ys. Hint: Take 'x *occurs in* xs' to mean $\exists y.\exists ys.(xs = (y : ys) \wedge (x = y) \vee (x$ *occurs in* $ys))$. That is, the proposition 'x *occurs in* xs' always has the same value as the proposition $\exists y.\exists ys.(xs = (y : ys) \wedge (x = y) \vee (x$ *occurs in* $ys))$. Denote the predicate 'x occurs in xs' by the formula 'x $\in$ xs' (overloading the '$\in$' symbol used to denote set membership).

Part III

Set Theory

Chapter 8

Set Theory

Set theory is one of the most fundamental branches of mathematics. Many profound advances in mathematics over the past century have taken place in set theory, and there is a deep connection between set theory and logic. More importantly for computer science, it has turned out that the notation and terminology of elementary set theory is extremely useful for describing algorithms, and nearly every branch of computing uses sets from time to time.

This chapter introduces the concepts from set theory that you will need for computer science. Section 8.1 begins by describing what sets are and giving several notations for describing them. There are many useful operations that can be performed on sets, and these are presented in Section 8.2. In Section 8.3, we consider a particular kind of set that is well suited for computing applications: the finite sets with equality. Next, in Section 8.4 we study a variety of mathematical laws that describe properties of sets that are useful in computing. The chapter concludes with a summary of the notations and the main theorems of set theory.

8.1 Notations for Describing Sets

We will not define formally what a set is. The reason for this is that set theory was intended originally to serve as the foundation for all of mathematics: everything else in mathematics is to be defined—at least in principle—in terms of sets. As sets are the lowest level concept, there is nothing more primitive that could be used to define them formally.

Informally, a set is just a collection of objects called *members* or *elements*. You can think of a set as a group of members, or a collection, or a class, etc. However, these words are just synonyms—they don't constitute a precise definition of a set. One way to describe a set is to write down all its members inside braces { }. Here are some examples:

$$
\begin{aligned}
A &= \{\text{dog}, \text{cat}, \text{horse}\} \\
B &= \{\text{canary}, \text{eagle}\} \\
C &= \{0, 1, 2, 3, 4\} \\
D &= \{0, 1, \dots, 100\} \\
E &= \{\,\} \\
N &= \{0, 1, 2, 3, \dots\} \\
Z &= \{3, \{\text{dog}, 7\}, \text{horse}\}
\end{aligned}
$$

Any particular thing can appear only once in a set; this means that it makes sense to ask whether x is a member of a set S—the answer must be yes or no—but it doesn't make sense to ask how many times x occurs in S. It is bad notation to write a set with some element appearing several times, because the extra occurrences of the element are meaningless, and they might be confusing. You can always remove the redundant copies of an element without changing the set.

A set can have any number of elements. For example, A has three elements, E has zero elements, and N has an infinite number of elements.

It is common to use lower-case letters $(a, b, \dots)$ to refer to members of a set and to use upper-case letters $(A, B, \dots)$ as names for sets themselves. This is just a convention, not an ironclad rule, and of course it breaks down when one set is a member of another set.

An important special case is the empty set $\{\,\}$. Often the special symbol $\varnothing$ is used to denote the empty set.

Suppose we are given some value x and a set S, and we want to know whether x is a member of S. There is a notation for this question: the expression $x \in S$ is true if x is a member of S and otherwise false. The expression $x \in S$ is pronounced as 'x is a member of S', or 'x is an element of S', or simply as 'x is in S'. For example:

$$
\begin{aligned}
\text{dog} \in A &= \text{True} \\
\text{bat} \in A &= \text{False}
\end{aligned}
$$

Another useful notation is $x \notin S$, which is True if and only if $x \in S$ is False:

$$
\begin{aligned}
\text{dog} \notin A &= \text{False} \\
\text{bat} \notin A &= \text{True}
\end{aligned}
$$

When a set has a few elements, you can just write them out inside braces. This is how the sets A, B, and C were defined above. However, when a set has many elements this becomes tedious, and if the set has an infinite number of elements it is impossible. The set D has 101 elements, but it is more readable to use the $\dots$ notation and omit most of them. Set N, the natural numbers, has an infinite number of elements, and we cannot write them all inside braces.

There is an interesting point about the ... notation. One of the reasons we use mathematics in computing is to be precise and formal, to be absolutely sure there is no ambiguity in what we are defining. The ... notation is informal, and it relies on the intuition of the reader to understand and fill in the dots. It is easy to construct cases where different people might interpret a set defined with ... differently. For example, does $\{2, 3, \ldots, 7\}$ mean the set of numbers from 2 through 7, $\{2, 3, 4, 5, 6, 7\}$, or does it mean the set of prime numbers $\{2, 3, 5, 7\}$? If you are aware of such a problem when describing a set, you can overcome it, but how can you ever be sure that your set is really well-defined—that there is one way, and only one way, to interpret it? The problem is not so serious for finite sets, because you could—in principle—write out all the elements, and for large sets you could provide an algorithm that produces all of them. The problem is more fundamental for infinite sets.

Another standard way to define sets is the *set comprehension*. In its simplest form, a set comprehension is written as

$$\{x \mid p \; x\},$$

where $p \; x$ is simply an expression containing x which is either true or false; such an expression is called a *predicate*. The expression is pronounced 'the set of x such that $p \; x$', and it means that the set consists of exactly those objects x of which $p \; x$ is true. For example, we could define the set of even numbers as

$$\{x \mid x \in N \land \text{even } x\}.$$

The predicate here is

$$p \; x \; = \; x \in N \; \land \; \text{even } x,$$

and it is true if and only if x is a natural number that is even. In English, we would call this 'the set of all x such that x is a natural number and x is even', or simply 'the set of even natural numbers'.

A more general form of the set comprehension is

$$\{f \; x \mid p \; x\}.$$

In this case, the set consists not of the values x that satisfy the predicate, but of the results of applying the function f to those values. This form of the set comprehension is sometimes easier to use than the simpler form. For example

$$\{\sqrt{x} \mid x \in \{1, 2, 3, 4\}\}$$

defines the set $\{1.0, 1.41, 1.73, 2.0\}$.

It is important to state the set from which a variable derives its value. If this set is not stated explicitly, then we assume that it is U, the universe of discourse.

8.2   Basic Operations on Sets

There is a large number of operations that can be performed on sets, in order to compare them, define new sets, and so on. This section defines the basic set operations. In the next section, we will see how to implement these operations on a computer.

8.2.1   Subsets and Set Equality

There are several important relationships between two sets that are determined by the elements they share. The first of these is the *subset* relation. The expression $A \subseteq B$, pronounced 'A is a subset of B', is true if each element of A also appears in B. This idea is expressed formally by the following definition.

Definition 18 (Subset). Let A and B be sets. Then $A \subseteq B$ if and only if

$$\forall x. x \in A \rightarrow x \in B.$$

Two sets are equal if they contain exactly the same elements. We can define this formally using the subset relation, because if A and B contain exactly the same elements, then everything in A is also in B and vice versa. This leads to the definition of set equality.

Definition 19 (Set equality). Let A and B be sets. Then $A = B$ if and only if $A \subseteq B$ and $B \subseteq A$.

If A is a subset of B but $A \neq B$, then all the elements of A are in B but there must be some element of B that is not in A. In this case, we say that A is a *proper subset* of B, which is written as $A \subset B$. The notations are designed to help you remember them. Think of $A \subset B$ as saying that the set A is contained within B, and it is smaller than B, while $A \subseteq B$ means that possibly $A = B$; the symbols $\subset$ and $\subseteq$ are reminiscent of $<$ and $\leq$.

Definition 20 (Proper subset). Let A and B be sets. Then $A \subset B$ if and only if $A \subseteq B$ and $A \neq B$.

8.2.2   Union, Intersection, and Difference

There are several operators that take two sets and return a set as a result; the most important of these are union, intersection, and difference.

- The *union* of two sets A and B, written $A \cup B$, is the set that contains all the elements that are in either A or B (or both). Every element of $A \cup B$ must be in A or B (or both).

- The *intersection* of A and B, written $A \cap B$, is the set consisting of all the elements that are in *both* A and B.

- The *difference* of A and B, written A − B, is the set of all the elements that are in A but not in B.

Definition 21. Let A and B be sets. Then

$$A \cup B = \{x \mid x \in A \lor x \in B\}, \tag{8.1}$$
$$A \cap B = \{x \mid x \in A \land x \in B\}, \tag{8.2}$$
$$A - B = \{x \mid x \in A \land x \notin B\}. \tag{8.3}$$

Example 11. Let $A = \{1, 2, 3\}$, $B = \{3, 4, 5\}$ and $C = \{4, 5, 6\}$. Then

$$A \cup B = \{1, 2, 3, 4, 5\},$$
$$A \cap B = \{3\},$$
$$A - B = \{1, 2\}.$$

$$A \cup C = \{1, 2, 3, 4, 5, 6\},$$
$$A \cap C = \emptyset,$$
$$A - C = \{1, 2, 3\}.$$

Example 12. Let

$$I = \{\ldots, -2, -1, 0, 1, 2, \ldots\},$$
$$N = \{0, 1, 2, \ldots\},$$
$$H = \{-2^{15}, \ldots, -2, -1, 0, 1, 2, \ldots, 2^{15} - 1\},$$
$$W = \{-2^{31}, \ldots, -2, -1, 0, 1, 2, \ldots, 2^{31} - 1\}.$$

Thus I is the set of integers, N is the set of natural numbers, H is the set of integers that are representable on a computer with a 16-bit word using 2's complement number representation, and W is the set of integers that are representable in a 32-bit word. We can use these definitions to create new sets. For example, $I - W$ is the set of integers that are not representable in a word.

We can calculate the union of several sets using the $\cup$ operator. For example, the union of three sets A, B, and C can be written as

$$A \cup B \cup C,$$

and it contains all the elements that appear in one or more of the sets A, B and C. This expression is unambiguous because the $\cup$ operator is associative (see Section 8.4), which means that it makes no difference whether you interpret $A \cup B \cup C$ as $(A \cup B) \cup C$ or as $A \cup (B \cup C)$. In the same way, we can calculate the intersection of four sets with the expression

$$A \cap B \cap C \cap D.$$

Sometimes it is necessary to compute the union (or intersection) of several sets in a more general way, using operators that give the union (or intersection) of an arbitrary number of sets, rather than just two of them. These operations are often called *big union* and *big intersection*, because their operators, $\bigcup$ and $\bigcap$, are larger versions of the ordinary $\cup$ and $\cap$ operators.

Definition 22. Let C be a non-empty collection (set) of subsets of the universe U. Let I be a non-empty set, and for each $i \in I$ let $A_i \subseteq C$. Then

$$\bigcup_{i \in I} A_i = \{x \mid \exists i \in I . x \in A_i\},$$

$$\bigcap_{i \in I} A_i = \{x \mid \forall i \in I . x \in A_i\}.$$

Another way to say this is that if C is a set containing some sets, then the set of all elements of the sets in C is $\bigcup_{A \in C} A$ and the set of elements that these sets in C have in common is $\bigcap_{A \in C} A$.

$$\bigcup_{A \in C} A = \{x \mid \exists A \in C . x \in A\}$$

$$\bigcap_{A \in C} A = \{x \mid \forall A \in C . x \in A\}$$

Two sets are *disjoint* if they have no elements in common.

Definition 23. For any two sets A and B, if $A \cap B = \varnothing$ then A and B are *disjoint* sets.

Exercise 1. Given the sets $A = \{1, 2, 3, 4, 5\}$ and $B = \{2, 4, 6\}$, calculate the following sets:

  (a) $A \cup (B \cap A)$

  (b) $(A \cap B) \cup B$

  (c) $A - B$

  (d) $(B - A) \cap B$

  (e) $A \cup (B - A)$

8.2.3   Complement and Power

In many applications, there is a *universe* of all the objects that might possibly appear in any of our sets. For example, we might be working with various sets of numbers, but none of the sets will contain anything that is *not* a number. In this kind of situation, it is often convenient to define the universe explicitly as a set U, which can then be used in set expressions.

The universe is needed to define the *complement* of a set. The intuitive idea is that the complement of a set A is the set of everything that is *not* in A. However, what does 'everything' mean? There are both practical and theoretical problems if we don't define 'everything'. A practical problem is that we could get nonsensical results; for example, if we are talking about sets of people, then the complement of the set of tall people should be the set that includes short people—but not numbers, cars, and toasters, all of which appear in the set of 'everything'. The solution to these problems is to define the universe U to consist of the elements we are interested in, and then to define the complement of A to consist of the elements of U that are not in A.

Definition 24. Let U be the universe of discourse and A be a set. The *complement* of A, written A', is the set $U - A$.

If you see a set complement defined in a book or paper, look back several pages and you should find the definition of U. When you're using sets, be sure to define the universe explicitly if you are going to use complements.

Example 13. Given the universe of alphanumeric characters, the complement of the set of digits is the set of letters.

Example 14. If the universe is $\{1, 2, 3, 4, 5\}$, then $\{1, 2\}' = \{3, 4, 5\}$.

A set that contains lots of elements will have an even larger number of subsets. These subsets are themselves objects, and it is often useful to define a new set containing all of them. The set of all subsets of A is called the *powerset* of A. (In contrast, the set of all *elements* of A is just A itself.)

Definition 25. Let A be a set. The *powerset* of A, written $P(A)$, is the set of all subsets of A:

$$P(A) = \{S \mid S \subseteq A\}$$

Example 15. (Powersets)

- $P(\{\,\}) = \{\varnothing\} = \{\{\,\}\}$

- $P(\{a\}) = \{\varnothing, \{a\}\}$

- $P(\{a, b\}) = \{\varnothing, \{a\}, \{b\}, \{a, b\}\}$

- $P(\{a, b, c\}) = \{\varnothing, \{a\}, \{b\}, \{c\}, \{a, b\}, \{a, c\}, \{b, c\}, \{a, b, c\}\}$

Notice that if A contains n elements, then its powerset $P(A)$ contains 2^n elements. If A is the empty set $\varnothing$, then there is one element of $P(A)$—namely $\varnothing$—and the number of elements of $P(A)$ is indeed $1 = 2^0$. The example above shows that sets with 0, 1, 2, and 3 elements have powersets respectively containing 1, 2, 4, and 8 elements.

8.3   Finite Sets with Equality

A class of sets that is particularly important for computing are the finite sets with equality—that is, sets with a finite number of elements, and where we have an equality function that can be used to determine whether any two elements of the universe are the same.

It is possible to perform some computations with infinite sets, but there is ample opportunity for a program to go into an infinite loop. Usually we want to ensure that our programs will terminate, and this is more straightforward with finite sets.

We will use lists to represent sets; thus a set whose elements are of type a will be represented by a list of type $[a]$. A finite set will be represented by a finite list. There are several differences between lists and sets that will require care: (1) lists may have duplicate elements, (2) there is a fixed order of the elements of a list, and (3) all the elements of a list must have the same type.

In order to do any practical computing with sets, we need to be able to determine whether a given value is an element of a set. This requires, in turn, the ability to determine whether two values x and y are the same. Therefore we need an operator $==$ so that $x == y$ will give either True or False, for any values of x and y. This is a strong requirement, because there are some values where equality is not computable. For all the elementary data types, such as integers, Booleans, characters, and so on, there is no problem with determining equality. However, functions are also values that can be included in sets, and it is in general impossible to compare two functions to determine whether they are equal.

There is a special notation in Haskell to express the fact that it must be possible to compare two set elements for equality. If we say simply that a set is represented by a list of type $[a]$, it might turn out that a is one of those types whose values cannot be compared. What is needed is a way to restrict the element type a. This is done by saying that a list has type

$$Eq \ a \Rightarrow [a].$$

This denotes the type 'For every type a which can be compared for Equality, the type $[a]$'. In practice, we will also usually want the element type a to be printable. To ensure this as well, we will often use the type $(Eq, Show) \Rightarrow [a]$. For example, this would allow Int to be used as the set element type, because integers can be compared for equality and they can be printed, but it would not allow functions to be used as the set element type, as functions cannot be compared or printed.

A set A will be represented by a list containing all the elements of A. We need to be careful in computing with such lists because of two factors: the possibility of duplicate elements and the ordering of the elements.

A set contains just one instance of each of its elements; thus the sets $\{1, 2, 3\}$ and $\{1, 2, 1, 3\}$ are identical (and the second way of writing it is bad notation because 1 appears only once in the set; it should appear only once in the notation). However, the lists $[1, 2, 3]$ and $[1, 2, 1, 3]$ are different. The first one has three elements and the second has four elements. We will assume that a list represents the set of distinct values that appear in the list: for example, the list $[1, 2, 1, 3]$ represents the set $\{1, 2, 3\}$. However, the *normal form* for a list representing a set will contain each element only once. The operations on set-representing-lists will all produce their result in normal form. This makes some of the operations easier to implement, and also makes them more efficient.

There is no concept of ordering of the elements of a set. Thus $\{1, 2, 3\}$ and $\{3, 2, 1\}$ are the same set. If you print a set, there is no particular order in which the elements will appear. Lists, however, have a specific ordering of their elements; thus $[1, 2, 3]$ and $[3, 2, 1]$ are different lists and their elements will be printed in different orders. This does not cause any difficulties for implementing the set operations. In a computer program, however, it may be a good idea to define an ordering on the elements of a set in order to make the output more readable. This can be achieved by sorting the elements of the list that represents the set. However, sorting a list requires more than the ability to compare two elements for equality: we must also be able to compare them for ordering $(<, =, >)$. If a program uses sorted lists to represent sets, then a stronger type constraint is needed:

$$Ord \ a \Rightarrow [a].$$

This says that there must be an ordering on the element type a, which can be used to determine the relations $<, \leq, =, \neq, >, \geq$.

The methods for defining lists can also be used to define sets. Three methods are especially common: enumerated sets, sequences, and comprehensions. We will look at each of these methods in turn, and the following section defines functions corresponding to the set operators we have already covered.

An enumerated set (or list) is defined just by giving its elements, like $\{1, 2, 3\}$ and $[1, 2, 3]$.

A *sequence* is used when it would be too tedious to write an enumerated set or list. For example, the set of natural numbers up to 5 is $\{0, 1, 2, 3, 4, 5\}$, but it would be painful to write the set of natural numbers up to 1,000. Instead, the '...' notation is used in mathematics: $\{0, 1, 2, \ldots, 1000\}$. We can use a similar notation to define the corresponding list: $[0; 1..1000]$.

Sets are often described in mathematics using *set comprehensions*, expressions like

$$\{x^2 \mid x \in \{0, 1, \ldots, n\}\},$$

which is the set of numbers of the form x^2 where x is a natural number between 0 and n. Haskell provides an almost identical notation, the *list comprehension*. A list comprehension that expresses the set just given is

```
[x^2 | x <- [0..n]]
```

Here are some other examples:

```
[ x | x <- [0..n], x 'mod' 2 == 0]
[ x + 3 | x <- [0..n], x < 10]
```

8.3.1   Computing with Sets

This section describes a collection of functions and operators that you can use
on finite sets with equality. Each of these functions will accept a representation
with duplicated elements, but they always return results in normal form; that
is, they remove any duplicated elements from their result.

We begin by defining the type of set representations.

```
type Set a = [a]
```

The function `normalForm` determines whether there are any duplicate elements within its argument, while `normalizeSet` takes a list and removes any
duplicate elements.

```
normalForm :: (Eq a, Show a) => [a] -> Bool
normalizeSet :: Eq a => Set a -> Set a
```

Example 16. `normalForm [1,2,3]` is True, but `normalForm [1,2,1,3]` is
False because there is a repeated element, and `normalizeSet [1,2,1,3]` removes the repeated element, returning [1,2,3].

We define symbolic operators for the set operations that take two arguments:

$$
\begin{array}{lll}
\text{A+++B} & = & A \cup B \quad & (union) \\
\text{A***B} & = & A \cap B \quad & (intersection) \\
\text{A}\sim\sim\sim\text{B} & = & A - B \quad & (difference)
\end{array}
$$

The types of these operators are

```
(+++) :: (Eq a, Show a) => Set a -> Set a -> Set a
(***) :: (Eq a, Show a) => Set a -> Set a -> Set a
(~~~) :: (Eq a, Show a) => Set a -> Set a -> Set a
```

The function `subset` takes two sets A and B, and returns True if $A \subseteq B$,
while `properSubset` A B returns True if A is a proper subset of B; that is, it
returns the value of $A \subset B$. Their type signatures are:

```
subset, properSubset ::
    (Eq a, Show a) => Set a -> Set a -> Bool
```

A special function `setEq` is needed to determine whether two sets are the
same. The built-in Haskell operator `==` should not be used for this purpose, as
two lists may be different even though they represent the same set.

```
setEq :: (Eq a, Show a) => Set a -> Set a -> Bool
```

The complement of a set is defined with respect to the universe of discourse. It is convenient to define a global variable `universe` to be the universe of discourse; then we can define the set `complement` function as follows:

```
complement s = universe ~~~ s
```

Exercise 2. Work out the values of the following set expressions, and then check your answer using the Haskell expression that follows.

   (a) `[1,2,3] +++ [3]`

   (b) `[4,2] +++ [2,4]`

   (c) `[1,2,3] *** [3]`

   (d) `[] *** [1,3,5]`

   (e) `[1,2,3] ~~~ [3]`

   (f) `[2,3] ~~~ [1,2,3]`

   (g) `[1,2,3] *** [1,2]`

   (h) `[1,2,3] +++ [4,5,6]`

   (i) `([4,3] ~~~ [5,4]) *** [1,2]`

   (j) `([3,2,4] +++ [4,2]) ~~~ [2,3]`

   (k) `subset [3,4] [4,5,6]`

   (l) `subset [1,3] [4,1,3,6]`

  (m) `subset [] [1,2,3]`

   (n) `setEq [1,2] [2,1]`

   (o) `setEq [3,4,6] [2,3,5]`

   (p) `[1,2,3] ~~~ [1]`

   (q) `[] ~~~ [1,2]`

Exercise 3. The function

```
        powerset :: (Eq a, Show a) => Set a -> Set (Set a)
```

takes a set and returns its power set. Work out the values of the following expressions:

```
        powerset [3,2,4]
        powerset [2]
```

Exercise 4. The *cross product* of two sets A and B is defined as
$A \times B = \{(a,b) \mid a \in A, b \in B\}$
The function

```
crossproduct :: (Eq a, Show a, Eq b, Show b) =>
                Set a -> Set b -> Set (a,b)
```

takes two sets and returns their cross product. Evaluate these expressions:

```
crossproduct [1,2,3] ['a','b']
crossproduct [1] ['a','b']
```

Exercise 5. In the following exercise, let u be `[1,2,3,4,5,6,7,8,9,10]`, a be `[2,3,4]`, b be `[5,6,7]` and c be `[1,2]`. Give the elements of each set:

```
a +++ b
u~~~a *** (b +++ c)
c ~~~ b
(a +++ b) +++ c
u~~~a
u~~~(b *** c)
```

Exercise 6. What are the elements of the set $\{x+y \mid x \in \{1,2,3\} \land y \in \{4,5\}\}$?

Exercise 7. Write and evaluate a list comprehension that expresses the set $\{x \mid x \in \{1,2,3,4,5\} \land x < 0\}$

Exercise 8. Write and evaluate a list comprehension that expresses the set $\{x + y \mid x \in \{1,2,3\} \land y \in \{4,5\}\}$

Exercise 9. Write and evaluate a list comprehension that expresses the set $\{x \mid x \in \{1,2,3,4,5,6,7,8,9,10\} \land even\ x\}$

Exercise 10. What is the value of each of the following expressions?

```
subset [1,3,4] [4,3]
subset [] [2,3,4]

setEq [2,3] [4,5,6]
setEq [1,2] [1,2,3]
```

8.4   Set Laws

The operations on sets that we have been covering are often used for describing properties of algorithms, so we often need to be able to understand and calculate with expressions that contain several such operations. Fortunately, the set operations satisfy a number of simple laws that greatly simplify their use, just as the basic properties of ordinary addition, multiplication, etc., are helpful in ordinary algebra. In this section, we state and prove some of the most useful laws about set operations. The proofs have a standard form in which an assertion appears, followed by its justification, which may depend on previous lines in the proof. As justifications take many forms in practice, these will be more terse than those in previous proofs.

The following law says that the $\subseteq$ operation is transitive:

Theorem 68. Let A, B, and C be sets. If $A \subseteq B$ and $B \subseteq C$, then $A \subseteq C$.

Proof. Let x be any element of the universe of discourse.

1.	$A \subseteq B$	{ Premise }
2.	$x \in A \to x \in B$	{ Def. $\subseteq$ }
3.	$B \subseteq C$	{ Premise }
4.	$x \in B \to x \in C$	{ Def. $\subseteq$ }
5.	$x \in A \to x \in C$	{ Hypothetical syllogism (chain rule), (2), (4) }
6.	$\forall x.\, (x \in A \to x \in C)$	{ $\forall$ introduction }
7.	$A \subseteq C$	{ Def. $\subseteq$ }

$\square$

Exercise 11. Let A, B, and C be sets. Prove that if $A \subset B$ and $B \subset C$, then $A \subset C$.

Exercise 12. Consider the following two claims. For each one, if it is true give a proof, but if it is false give a counterexample.

(a) If $A \subseteq B$ and $B \subseteq C$, then $A \subset C$.

(b) If $A \subset B$ and $B \subset C$, then $A \subseteq C$.

8.4.1 Associative and Commutative Set Operations

The set union and intersection operators are commutative and associative.

Theorem 69. For all sets A, B, and C,

1. $A \cup B = B \cup A$

2. $A \cap B = B \cap A$

3. $A \cup (B \cup C) = (A \cup B) \cup C$

4. $A \cap (B \cap C) = (A \cap B) \cap C$

5. $A - B = A \cap B'$

Proof. We prove the second equation. Let x be any element of U. Then each line in the proof below is logically equivalent ($\leftrightarrow$) to the following line.

1. $x \in A \cap B$ { Premise }
2. $x \in A \wedge x \in B$ { Def. $\cap$ }
3. $x \in B \wedge x \in A$ { Comm. $\wedge$ }
4. $x \in B \cap A$ { Def. $\cap$ }
5. $\forall x \in U.$
        $x \in A \cap B \leftrightarrow x \in B \cap A$ { $\{\forall I\}$ }
6. $A \cap B = B \cap A.$ { Def. set equality }

$\square$

The proofs of the other equations are similar.

8.4.2   Distributive Laws

The following theorem states that the union and intersection operators distribute over each other.

Theorem 70. $A \cap (B \cup C) = (A \cap B) \cup (A \cap C)$

Proof. Let x be an arbitrary element of the universe U. Then the following expressions are equivalent ($\leftrightarrow$):

1. $x \in A \cap (B \cup C)$ { Premise }
2. $x \in A \wedge (x \in B \cup C)$ { Def. $\cap$ }
3. $x \in A \wedge (x \in B \vee x \in C)$ { Def. $\cup$ }
4. $(x \in A \wedge x \in B) \vee (x \in A \wedge x \in C)$ { Distr. $\wedge$ over $\vee$ }
5. $(x \in A \cap B) \vee (x \in A \cap C)$ { Def. $\cap$ }
6. $x \in (A \cap B) \cup (A \cap C)$ { Def. $\cup$ }
7. $\forall x \in U.$
        $x \in A \cap (B \cup C) \leftrightarrow x \in (A \cap B) \cup (A \cap C)$ { $\{\forall I\}$ }
8. $A \cap (B \cup C) = (A \cap B) \cup (A \cap C).$ { Def. set equality }

$\square$

Theorem 71. $A \cup (B \cap C) = (A \cup B) \cap (A \cup C)$

Proof. Exercise for the reader. $\square$

8.4.3   DeMorgan's Laws for Sets

Theorem 72. Let A and B be arbitrary sets. Then

$$(A \cup B)' = A' \cap B'$$

and

$$(A \cap B)' = A' \cup B'.$$

Proof. We prove that $(A \cup B)' = A' \cap B'$. Let x be any element of U. Then the following lines are equivalent:

Table 8.1: Summary of Set Notation

Elements	$a, b, c, \ldots$
Sets	$A, B, C, \ldots$
Empty set	$\{\ \}, \phi$
Enumerated set	$\{e_1, e_2, \ldots\}$
Set comprehension	$\{x \mid \cdots\}$
Cardinality	$\mid A \mid$
Member	$x \in A$
Not member	$x \notin A$
Subset	$A \subseteq B$
Not subset	$A \nsubseteq B$
Proper subset	$A \subset B$
Not proper subset	$A \not\subset B$
Union	$A \cup B$
Intersection	$A \cap B$
Set difference	$A - B$
Cross product	$A \times B$

1. $x \in (A \cup B)'$ { Premise }
2. $x \in U \wedge \neg(x \in A \cup B)$ { Def. comp }
3. $x \in U \wedge \neg(x \in A \vee x \in B)$ { Def. $\cup$ }
4. $x \in U \wedge (\neg(x \in A) \wedge \neg(x \in B))$ { DeMorgan }
5. $x \in U \wedge x \in U \wedge (\neg(x \in A) \wedge \neg(x \in B))$ { Idemp. of $\wedge$ }
6. $(x \in U \wedge \neg(x \in A)) \wedge (x \in U \wedge \neg(x \in B))$ { Comm. of $\wedge$ }
7. $x \in U - A \wedge x \in U - B$ { Def. of diff. }
8. $x \in (U - A) \cap (U - B)$ { Def. $\cup$ }
9. $(x \in A' \cap B')$ { Def. of comp. }
10. $\forall x.\ x \in (A \cup B)' \leftrightarrow x \in (A' \cap B')$ { $\forall$ introduction }
11. $(A \cup B)' = A' \cap B'$

$\square$

8.5   Summary

The notations used in set theory are listed in Table 8.1, and the laws for reasoning about sets are given in Table 8.2.

Table 8.2: Set Laws

Idempotent
$$A = A \cup A$$
$$A = A \cap A$$

Domination
$$A \cup U = U$$
$$A \cap \varnothing = \varnothing$$

Identity
$$A \cup \varnothing = A$$
$$A \cap U = A$$

Double complement
$$A = A''$$

DeMorgan's laws
$$(A \cup B)' = A' \cap B'$$
$$(A \cap B)' = A' \cup B'$$

Commutative laws
$$A \cup B = B \cup A$$
$$A \cap B = B \cap A$$

Associative laws
$$(A \cup B) \cup C = A \cup (B \cup C)$$
$$(A \cap B) \cap C = A \cap (B \cap C)$$

Distributive laws
$$A \cap (B \cup C) = (A \cap B) \cup (A \cap C)$$
$$A \cup (B \cap C) = (A \cup B) \cap (A \cup C)$$

Absorption laws
$$A \cup (A \cap B) = A$$
$$A \cap (A \cup B) = A$$

8.6 Suggestions for Further Reading

Mathematics from the Birth of Numbers, by Gullberg [16], is an interesting general survey of mathematics. It covers many of the topics in this book, including an excellent survey of elementary set theory.

Classic Set Theory [14], by Derek Goldrei, is a self-study textbook telling the full story of set theory, including construction of the real numbers, the Axiom of Choice, cardinal and ordinal numbers, and more. This book is challenging, but it conveys the sense of excitement that surrounded set theory as it was developed.

8.7 Review Exercises

Exercise 13. For the following questions, give a proof using set laws, or find a counterexample.

    (a) $(A' \cup B)' \cap C' = A \cap (B \cup C)'$

    (b) $A - (B \cup C)' = A \cap (B \cup C)$

    (c) $(A \cap B) \cup (A \cap B') = A$

    (d) $A \cup (B - A) = A \cup B$

    (e) $A - B = B' - A'$

    (f) $A \cap (B - C) = (A \cap B) - (A \cap C)$

    (g) $A - (B \cup C) = (A - B) \cap (A - C)$

    (h) $A \cap (A' \cup B) = A \cap B$

    (i) $(A - B') \cup (A - C') = A \cap (B \cap C)$

Exercise 14. The function

```
smaller :: Ord a => a -> [a] -> Bool
```

takes a value and a list of values and returns True if the value is smaller than the first element in the list. Using this function, write a function that takes a set and returns its powerset. Use `foldr`.

Exercise 15. Prove that $(A \cup B)' = ((A \cup A') \cap A') \cap ((B \cup B') \cap B')$.

Exercise 16. Using a list comprehension, write a function that takes two sets and returns True if the first is a subset of the other.

Exercise 17. What is wrong with this definition of `diff`, a function that takes two sets and returns their difference?

```
diff :: Eq a => [a] -> [a] -> [a]
diff set1 set2 = [e | e <- set2, not (elem e set1)]
```

Exercise 18. What is wrong with this definition of `intersection`, a function that takes two sets and returns their intersection?

```
intersection :: [a] -> [a] -> [a]
intersection set1 set2 = [e | e <- set1, e <- set2]
```

Exercise 19. Write a function using a list comprehension that takes two sets and returns their union.

Exercise 20. Is it ever the case that $A \cup (B - C) = B$?

Exercise 21. Give an example in which $(A \cup C) \cap (B \cup C) = \varnothing$.

Exercise 22. Prove the commutative law of set-intersection, $A \cap B = B \cap A$.

Exercise 23. Express the commutative law of set-intersection in terms of the set operations and Boolean operations defined in the `Stdm` module.

Exercise 24. Prove the associative law of set-union, $(A \cup B) \cup C = A \cup (B \cup C)$.

Exercise 25. Prove that the difference between two sets is the intersection of one with the complement of the other, which can be written as $A - B = A \cap B'$.

Exercise 26. Prove that union distributes over intersection,

$$A \cup (B \cap C) = (A \cup B) \cap (A \cup C).$$

Exercise 27. Prove DeMorgan's law for set intersection, $(A \cap B)' = A' \cup B'$.

Chapter 9

Inductively Defined Sets

In this chapter, we explore the construction of sets using induction. To understand why induction is useful, consider the problem of defining a set. The simplest method is to define a set by naming each of its elements, one by one. This is called *enumeration*. It works only for finite sets and is impractical for large sets. Another approach is to use ellipses ('...') to indicate that the set continues, but this is imprecise, and so can be ambiguous. For example, what is meant by $\{1, 2, 3, \ldots\}$? Is the next element 4 or 5? Even if you think the answer is obvious, how do you know everyone else will consider the *same* thing to be obvious? If we are enumerating the positive integers, the next element is 4, but if we are adding the two previous numbers in the series, it is 5.

9.1   The Idea Behind Induction

Induction is rather like a mathematical 'program' that calculates a proof when needed. The proof asserts that an element is a member of the set defined by induction. For example, here are two propositions:

$$0 \in S$$
$$n \in S \to n + 1 \in S$$

Together, they let us show that any natural number is in set S. To see how they do this, consider an example: we show that 2 is an element of set S. Using the propositions above, we can construct a *chain* that looks like this:

$$0 \in S$$
$$0 \in S \to 1 \in S$$
$$1 \in S \to 2 \in S$$

We then use the first assertion and *Modus Ponens* to deduce that 1 is in set S, and from that, using *Modus Ponens*, we deduce that 2 is in set S. In fact,

207

we can use the two propositions to build a chain that is as long as needed to reach *any* natural number.

Of course, if any of the links in the chain were missing, for example the proposition $0 \in S \to 1 \in S$, then we could not reach the number required. This is because we could not use *Modus Ponens* to get to the next link.

When we use an inductive definition to show that a set contains a given value v, we enumerate, or count, the values that must first be shown to be in the set before v. These values form a *sequence*, which is a set with an ordering. Computers can enumerate elements of sets in the order in which they are generated from a description of the set. We can use a computer to calculate a sequence that represents an infinite set, although we will only see a finite prefix of the entire sequence.

Let's implement what we have seen using Haskell. A set of numbers can be represented by a list; for example, the set with the numbers 1,2, and 3 is [1,2,3], and the empty set is [].

How do we implement the implications? An implication of the form $1 \in s \to 2 \in s$ can be implemented as a function that takes 1 and returns the next element, 2. If it is applied to anything other than 1, then an error message is returned:

```
imp1 :: Integer -> Integer
imp1 1 = 2
imp1 other = error "premise does not match"
```

We can implement a chain using function application. The argument of the function imp1 is an element of s. If that element matches the pattern of imp1, then imp1 can be applied to it and produce a new element of s. This is just like what we do when deciding whether we can use *Modus Ponens*: we match the premise of the implication with elements of the set; if there is a match, then we can use *Modus Ponens*, otherwise the match fails and we cannot. For example, consider the following assertions:

$$1 \in S$$
$$1 \in S \to 2 \in S$$
$$2 \in S \to 3 \in S$$

This can be implemented by the following Haskell definitions:

```
imp1 :: Integer -> Integer
imp1 1 = 2
imp1 x = error  "premise does not match"

imp2 :: Integer -> Integer
imp2 2 = 3
imp2 x = error  "premise does not match"
```

```
s :: [Integer]
s = [1, imp1 (s !! 0), imp2 (s !! 1)]
```

The function application s!!0 returns the first element of s, indexing from 0; the result is 1. The function imp11 is applied to this. Since the argument matches the pattern, the application succeeds, adding 2 to s. Then the function imp12 is applied to 2. The argument matches the pattern, so the application succeeds, adding 3 to s. The value of s is [1,2,3].

There is a difference between matching the premise of the implication with all of the elements of the set and applying imp1 to the most recently added member of s. However, in the form of induction that we are studying, it doesn't matter. The only value that could possibly match the premise is the one generated by the previous implication. But consider now this set of assertions:

$$1 \in S$$
$$2 \in S \to 3 \in S$$

This is implemented as:

```
imp1 :: Integer -> Integer
imp1 2 = 3
imp1 x = error  "premise does not match"

s :: [Integer]
s = [1, imp1 (s !! 0)]
```

In this case, we cannot use *Modus Ponens* to conclude that 3 is in the list, because nothing states that 2 is in it.

Exercise 1. Is the following a chain? You can test your conclusions by evaluating s in each case.

```
imp1 :: Integer -> Integer
imp1 1 = 2
imp1 x = error "imp1: premise does not apply"

imp2 :: Integer -> Integer
imp2 2 = 3
imp2 x = error "imp2: premise does not apply"

imp3 :: Integer -> Integer
imp3 3 = 4
imp3 x = error "imp3: premise does not apply"
s :: [Integer]
s = [1, imp1 (s !! 0), imp2 (s !! 1), imp3 (s !! 2)]
```

Exercise 2. Is the following a chain?

```
imp1 :: Integer -> Integer
imp1 1 = 2
imp1 x = error "imp1: premise does not apply"

imp2 :: Integer -> Integer
imp2 3 = 4
imp2 x = error "imp2: premise does not apply"

s :: [Integer]
s = [0, imp1 (s !! 0), imp2 (s !! 1)]
```

Exercise 3. Is the following a chain?

```
imp1 :: Integer -> Integer
imp1 0 = 1
imp1 x = error "imp1: premise does not apply"

imp2 :: Integer -> Integer
imp2 3 = 4
imp2 x = error "imp2: premise does not apply"

s :: [Integer]
s = [0, imp1 (s !! 0), imp2 (s !! 1)]
```

Exercise 4. Is the following a chain?

```
imp1 :: Integer -> Integer
imp1 0 = 1
imp1 x = error "imp1: premise does not apply"

imp2 :: Integer -> Integer
imp2 1 = 2
imp2 x = error "imp2: premise does not apply"

s :: [Integer]
s = [0, imp1 (s !! 1), imp2 (s !! 0)]
```

9.1.1 The Induction Rule

Recall the two propositions we used in the first section:

$$0 \in S$$
$$n \in S \rightarrow n+1 \in S$$

The first one is called the *base case*, and the second is called the *induction case*, or the *induction rule*. It is the induction case that generates the links of the chain which will allow us to reach any number in the set being defined.

So far, the induction has had a fixed form because we were defining a particular set. However, we could have a rule

$$n \in S \rightarrow n + 2 \in S.$$

Together with the base case $1 \in S$, this would define the odd natural numbers. Alternatively, our induction rule might be

$$n \in S \rightarrow n * 5 \in S.$$

Together with the base case, this would define the set of powers of 5.

As we will see, it can sometimes be hard to construct the correct rule, and it is necessary to debug rules to get them right.

Suppose we have a set defined by the following assertions:

$$0 \in s$$
$$x \in s \rightarrow x + 1 \in s$$

First, we want to find out whether 2 is in the set, and will use the computer to help. (Of course, we could solve this by hand, but if the induction rule is complicated, or if we want to find out whether 1,000,000 is in the set, software tools are invaluable.) We can implement the induction rule as the `increment` function:

```
increment :: Integer -> Integer
increment x = x + 1

s :: [Integer]
s = [0, increment (s !! 0), increment (s !! 1)]
```

We can load this definition and evaluate s; the last element of s is 2.

Now suppose that we want to know whether 50 is in the set. It would be very tedious to write out each element as we have been doing. The same function is applied to each element of s, so we can have the following definition of s instead:

```
s :: [Integer]
s = 0 : map increment s
```

This style of programming is known as *data recursion*. The function `map` proceeds down s, creating each value it needs and then using it. We can then get at the fiftieth element by typing s!!50.

Now we have a new format for implementing inductive definitions. We first specify the induction rule, then recursively define a list in which the base case appears first, and then the rule is mapped down the list.

Exercise 5. Given the base case $0 \in n$ and the induction rule $x \in n \rightarrow x + 1 \in n$, fix the following calculation so that 3 is in set n:

```
fun :: Integer -> Integer
fun x = x - 1

n :: [Integer]
n = 0 : map fun n
```

Exercise 6. Use the following definitions, determine whether 4 is in set s, given $1 \in s$ and the induction rule $x \in s \rightarrow x + 2 \in s$.

```
fun :: Integer -> Integer
fun x = x + 2

s :: [Integer]
s = 1 : map fun s
```

Exercise 7. Fix this calculation of the positive integers:

```
fun :: Integer -> Integer
fun x = 0

p :: [Integer]
p = 0 : map fun p
```

Exercise 8. Fix this calculation of the positive multiples of 3:

```
fun :: Integer -> Integer
fun x = x * 3

p :: [Integer]
p = map fun p
```

9.2 How to Define a Set Using Induction

We have seen that an inductive set definition has a base case and an induction rule (or induction case). There is one more clause that needs to be specified in an inductive set definition. Suppose that we have defined a set S by saying that the numbers 1, 2, and 3 are in S. How do we know that *something else* isn't also in S? If we don't say explicitly that nothing else is in S, then the specification could be satisfied by lots of different sets. It could be the set $\{1, 2, 4.5, -78, 3\}$, for example.

We want to exclude all elements that aren't introduced by the base case, or instantiations of the induction case, so we include a clause (called the extremal clause) in a set definition that states *Nothing is an element of the set unless it can be constructed by a finite number of uses of the first two clauses.*

To summarise, an inductive definition of a set consists of three parts: a base case, an induction case, and an extremal clause:

- The base case is a simple statement of some mathematical fact, such as $1 \in S$;

- The induction case is an implication in a general form, such as the proposition that

$$\forall x \in U, \ x \in S \to x + 1 \in S.$$

- The extremal clause says that nothing is in the set being defined unless it got there by a finite number of uses of the first two cases.

9.2.1 Inductive Definition of the Set of Natural Numbers

We will illustrate the method by writing an inductive definition of the natural numbers.

Definition 26. The set N of natural numbers is defined as follows:

- Base case: $0 \in N$

- Induction case: $x \in N \to x + 1 \in N$

- Extremal clause: nothing is an element of the set N unless it can be constructed with a finite number of uses of the base and induction cases.

Now we can use the base and induction cases to show formally that an arbitrary number above and including 0 is a natural number. Let's choose the number 2.

1.	$0 \in N$	Base case
2.	$0 \in N \to 1 \in N$	instantiation rule, induction case
3.	$1 \in N$	1, 2, *Modus Ponens*
4.	$1 \in N \to 2 \in N$	instantiation rule, induction case
5.	$2 \in N$	3, 4, *Modus Ponens*

Exercise 9. Here is a Haskell equation that defines the set s inductively. Is 82 an element of s?

```
s :: [Integer]
s = 0 : map ((+) 2) s
```

Exercise 10. What set is defined by the following?

```
s :: [Integer]
s = 1 : map ((*) 3) s
```

9.2.2   The Set of Binary Machine Words

Now we define a set `BinWords`, each of which is a machine word represented in binary notation. In general, a machine word can be of any length.

The base case says that the elements of another set (the set of binary digits) are also elements of `BinWords`. The induction case uses *concatenation* to create a new value from one already in the set. We represent the concatenation of a character to a string by placing them one after the other: e.g., '1' '01' is the string '101'.

Definition 27. Let `BinDigit` be the set $\{0, 1\}$. The set `BinWords` of machine words in binary is defined as follows:

- Base case:
$$x \in \texttt{BinDigit} \ \rightarrow x \ \in \texttt{BinWords}$$

- Induction case: if x is a binary digit and y is a binary word, then their concatenation xy is also a binary word:
$$(x \in \texttt{BinDigit} \ \wedge \ y \in \texttt{BinWords}) \ \rightarrow xy \in \texttt{BinWords}$$

- Extremal clause: nothing is an element of `BinWords` unless it can be constructed with a finite number of uses of the base and induction cases.

A set based on another set S in this way is given the name S^+, indicating that it is the set of all possible non-empty strings over S. The expression S^* is the same as S^+ except that S^* also includes the empty string. Thus our set `BinWords` could be written `BinDigit`$^+$.

We can write a Haskell function to calculate the set of binary words, using the inductive definition just presented. The induction function takes a binary word. It creates a new one from that number and each binary digit in turn. For example, if it is given `[1,0]`, it returns `[0,1,0]` and `[1,1,0]`.

The induction rule takes a binary word and creates two new ones, so we define the function `newBinaryWords` to do just that:

```
newBinaryWords :: [Integer] -> [[Integer]]
newBinaryWords ys = [0 : ys, 1 : ys]
```

Finally, we define the set of binary words as follows:

```
mappend :: (a -> [b]) -> [a] -> [b]
mappend f []     = []
mappend f (x:xs) = f x ++ mappend f xs

binWords = [0] : [1] :
                (mappend newBinaryWords binWords)
```

Exercise 11. Alter the definition of `newBinaryWords` and `binWords` so that they produce all of the *octal* numbers. An octal number is one that contains only the digits 0 through 7.

9.3 Defining the Set of Integers

Now we come to a more subtle problem: defining the set I of integers. Instead of just giving the final answer, we will think through the problem the way you might in real life. Each time we have a trial solution, we will examine it to see if it works and whether it can be improved.

The sets we have been defining are *well-founded*; that is, they are infinite only in one direction, and they have a least element. The set N of natural numbers is a good example of a well-founded set.

A *countable* set is one that can be counted using the natural numbers (see Chapter 11). Are the integers countable? They don't have a least element, and they are infinite in two directions, so they aren't well-founded. But we could use a trick: start from 0 and count first n then $-n$. If we think of the integers as a measuring tape that is infinitely long in two directions, we could still count the inches (or centimetres) on the tape by thinking of the tape as being folded at 0. Now the positive numbers touch the negative numbers, and each element i of the naturals counts both the positive and the negative numbers; that is, i counts $(i, -i)$.

We have just devised a way of enumerating the integers so that every element is eventually counted. This forms an excellent basis for an inductive definition of the integers. We will now work through several attempts to define the integers using induction. Some of these have problems, and we discuss how to debug them.

9.3.1 First Attempt

Attempt 1. The set I is defined as follows:

- Base case: $0 \in I$

- Induction case: $x \in I \rightarrow -x \in I$

- Extremal clause: nothing is in I unless its presence is justified by a finite number of uses of the base and induction cases.

We will now define some functions that will make our inductive definitions easier to understand and also make it possible to use the computer to help carry out experiments with the definition. Each of these is designed to take the base and induction cases as arguments and construct the data recursion automatically. The first uses `map` and the second uses `mappend`.

```
build :: a -> (a -> a) -> Set a
build a f = set
          where set = a : map f set

builds :: a -> (a -> [a]) -> Set a
builds a f = set
```

```
where set = a : mappend f set
```

Here is an implementation of the first definition:

```
nextInteger1 :: Integer -> Integer
nextInteger1 x = -x

integers1 :: [Integer]
integers1 = build 0 nextInteger1
```

Exercise 12. Use `take 10 integers1` to evaluate the first 10 integers according to this definition. Describe the set that is actually defined by Attempt 1.

9.3.2   Second Attempt

Attempt 1 doesn't work. It was based on our intuitive method making the integers countable. The problem is that according to this definition, only 0 is a member of I. When the natural numbers were defined, there was a mechanism for including new numbers by adding 1 to the integer in the premise. Something similar is needed here to include the other integers.

Attempt 2. The set I is defined as follows:

- Base case: $0 \in I$

- Induction case: $x \in I \to (x + 1 \in I \wedge x - 1 \in I)$

- Extremal clause: Nothing is in I unless its presence is justified by a finite number of uses of the base and induction cases.

Here is an implementation of Attempt 2:

```
nextIntegers2 :: Integer -> [Integer]
nextIntegers2 x = [x + 1, x - 1]

integers2 :: [Integer]
integers2 = builds 0 nextIntegers2
```

Exercise 13. Use `take 20 integers2` to evaluate the first 20 integers according to this definition. Describe the set that is actually defined by Attempt 2.

9.3.3   Third Attempt

The previous attempt gave a correct inductive definition of the integers, but there is still a problem with it, as can be seen by a simple example. Consider proving that -2 is in I:

1. $0 \in I$ base case
2. $0 \in I \rightarrow (1 \in I \wedge -1 \in I)$ instantiation, induction case
3. $1 \in I \wedge -1 \in I$ 1,2, *Modus Ponens*
4. $-1 \in I \rightarrow (0 \in I \wedge -2 \in I)$ instantiation, induction case
5. $0 \in I \wedge -2 \in I$ 3,4, *Modus Ponens*

We can be sure that this definition is correct because the induction step brings both the positive and negative numbers into I. Each element of I is incremented, guaranteeing that the naturals will be included as 1 and its successors are added. Each element of I is also decremented, ensuring that all of the negative numbers will be included as -1 and its predecessors are added.

A drawback, however, is that there are *two* ways for 0 to become a member of I. The base case puts it there, and step 5 does it again. In fact, the real defects of this definition are graphically illustrated by the fact that

```
take 20 integers2
= [0,1,-1,2,0,0,-2,3,1,1,-1,1,-1,-1,-3,4,2,2,0,2]
```

It is more elegant if the definition introduces each element only once. Furthermore, practical software based in an inductively defined set would be inefficient (and possibly even incorrect) if the elements were introduced more than once. Here is one way to try to fix the problem:

Attempt 3. The set I is defined as follows:

- Base case: $0 \in I$

- Induction case: $x \in I \rightarrow (x + 1 \in I \wedge -(x + 1) \in I)$

- Extremal clause: nothing is in I unless its presence is justified by a finite number of uses of the base and induction cases.

Here is an implementation of the third definition:

```
nextIntegers3 :: Integer -> [Integer]
nextIntegers3 x = [x + 1, -(x + 1)]

integers3 :: [Integer]
integers3 = builds 0 nextIntegers3
```

Exercise 14. Use the computer to examine the first 10 integers generated by this definition, and describe the set that is defined.

9.3.4   Fourth Attempt

This third attempt is a correct definition of the set of integers, and it is much closer to our intuition. What we want to do is say that if x is in I, then $-x$ is also in I. However, we also have to increment and decrement x somehow, so that all the integers can be included. Consider what happens when we attempt to show that -2 is in I:

1. $0 \in I$ base case
2. $0 \in I \to (1 \in I \wedge -1 \in I)$ instantiation, induction case
3. $1 \in I \wedge -1 \in I$ 1,2, *Modus Ponens*
4. $1 \in I \to (2 \in I \wedge -2 \in I)$ instantiation, induction case
5. $2 \in I \wedge -2 \in I$ 3,4, and, *Modus Ponens*

This is almost what we want, except that when the induction rule is instantiated with -1 it places 0 in I again. And,

```
take 10 integers3 =
  [0,1,-1,2,-2,0,0,3,-3,-1].
```

Therefore Attempt 3 doesn't introduce each element precisely once.

Attempt 4. The set I of integers is defined as follows:

* Base case: $0 \in I$

* Induction case:

    1. $(x \in I \wedge x \geq 0) \to x + 1 \in I$

    2. $(x \in I \wedge x < 0) \to x - 1 \in I$

* Extremal clause: nothing is in I unless its presence is justified by a finite number of uses of the base and induction cases.

Here is an implementation of the fourth definition:

```
nextInteger4 :: Integer -> Integer
nextInteger4 x = if x < 0 then x - 1 else x + 1

integers4 :: [Integer]
integers4 = build 0 nextInteger4
```

Exercise 15. Use the computer to generate some elements of the set defined by Attempt 4, and describe the result.

9.3.5 Fifth Attempt

Attempt 4 does not work, because the numbers generated from 0 must always be positive. It is necessary to go back to Attempt 4 and improve that. Furthermore, it would be better to have only one inductive case, if possible.

Attempt 5. The set I of integers is defined as follows:

- Base case: $0 \in I$

- Induction case: $(x \in I \wedge x \geq 0) \to x + 1 \in I \wedge -(x+1) \in I$

- Extremal clause: nothing is in I unless its presence is justified by a finite number of uses of the base and induction cases.

This definition is exactly what the original method of counting the integers suggested. It is implemented by the following Haskell program:

```
nextIntegers5 :: Integer -> [Integer]
nextIntegers5 x
  = if x > 0 \/ x == 0
      then [x + 1, -(x + 1)]
      else []

integers5 :: [Integer]
integers5 = builds 0 nextIntegers5
```

Exercise 16. Use the computer to evaluate the first 10 elements of the set, and describe the result.

9.4 Suggestions for Further Reading

Many of the references on set theory cited in Chapter 8 also deal with inductive definitions of sets. *Elements of Set Theory*, by Enderton [11], gives some good examples of inductively defined sets. A more advanced treatment appears in *Axiomatic Set Theory*, by Suppes [30].

9.5 Review Exercises

Exercise 17. Does ints, using the following definition, enumerate the integers? If it does, then you should be able to pick any integer and see it eventually in the output produced by ints. Will you ever see the value -1?

```
nats :: [Integer]
nats = build 0 (1 +)
```

```
negs :: [Integer]
negs = build (-1) (1 -)

ints :: [Integer]
ints = nats ++ negs
```

Exercise 18. Does `twos` enumerate the set of even natural numbers?

```
twos :: [Integer]
twos = build 0 (2 *)
```

Exercise 19. What is wrong with the following definition of the stream of natural numbers?

```
nats = map (+ 1) nats ++ [0]
```

Exercise 20. What is the problem with the following definition of the naturals?

```
naturals :: [Integer] -> [Integer]
naturals (i:acc) = naturals (i + 1:i:acc)

nats :: [Integer]
nats = naturals [0]
```

Exercise 21. Can we write a function that will take a stream of the naturals (appearing in any order) and give the index of a particular number?

Exercise 22. Using induction, define the set of roots of a given number n.

Exercise 23. Given the following definition, prove that n^3 is in set P of powers of n.

Definition 28. Given a number n, the set P of powers of n is defined as follows:

- $n^0 \in P$
- $n^m \in P \rightarrow n^{m+1} \in P$
- Nothing else is in P unless it can be shown to be in P by a finite number of uses of the base and induction rules.

Exercise 24. When is 0 in the set defined below?

Definition 29. Given a number n, the set N is defined as follows:

- $n \in N$
- $m \in N \rightarrow m - 2 \in N$
- Nothing is in N unless it can be shown to be in N by a finite number of uses of the previous rules.

Exercise 25. What set is defined by the following definition?

Definition 30. The set S is defined as follows:

- $1 \in S$
- $n \in S \wedge n \bmod 2 = 0 \to n + 1 \in S$
- $n \in S \wedge n \bmod 2 = 1 \to n + 2 \in S$
- Nothing else is in S unless it can be shown to be in S by a finite number of uses of the previous rules.

Exercise 26. Prove that 4 is in the set defined as follows:

Definition 31. The set S is defined as follows:

1. $0 \in S$
2. $n \in S \wedge n \bmod 2 = 0 \to n + 2 \in S$
3. $n \in S \wedge n \bmod 2 = 1 \to n + 1 \in S$
4. Nothing is in S unless it can be shown to be in S by a finite number of uses of the previous rules.

Exercise 27. Given the following definition, prove that the string 'yyyy' is in *YYS*.

Definition 32. The set YYS of strings containing pairs of the letter 'y' is defined as follows:

1. ""$\in YYS$
2. $s \in YYS \to$ "yy" $+\!\!+ s \in YYS$
3. Nothing else is in YYS unless it can be shown to be in YYS by a finite number of uses of rules (1) and (2).

Exercise 28. Using data recursion, define the set of strings containing the letter 'z'.

Exercise 29. Using induction, define the set of strings of spaces of length less than or equal to some positive integer n.

Exercise 30. Using recursion, define the set of strings of spaces of length less than or equal to length n, where n is a positive integer.

Exercise 31. We could have a set that consists of all the natural numbers except for 2; you can write this as $N - \{2\}$. Similarly, for every natural number x, there is a set that contains all the natural numbers except for x. Now, we could make a set *SSN* of all of these results. Write an inductive definition of *SSN*.

Exercise 32. Given the following definition, show that the set $I - \{-3\} \in SSI-$.

The set of sets of integers SSI, each of which is missing a distinct negative integer, is defined inductively as follows:

1. $I - \{-1\} \in SSI-$

2. $I - \{n\} \to I - \{n-1\} \in SSI-$

3. Nothing else is in SSI- unless it can be shown to be in SSI- by a finite number of uses of rules (1) and (2).

Exercise 33. Given the following definition, prove that -7 is in ONI. The set ONI of odd negative integers is defined as follows:

1. $-1 \in ONI$

2. $n \in ONI \to n-2 \in ONI$

3. Nothing is in ONI unless it can be shown to be in ONI by a finite number of uses of the previous rules.

Exercise 34. Using data recursion, define the set ni of negative integers.

Exercise 35. If you print the elements of

```
[(a,b) | a <- [0..], b <- [0..]]
```

will you ever see the element (1,2)?

Exercise 36. What set is given by the following definition?

Definition 33. The set S is defined as follows:

1. $1 \in S$

2. $n \in S \to n-n \in S$

3. Nothing is in S unless it can be shown to be in S by a finite number of uses of the previous rules.

Chapter 10

Relations

There are many kinds of relationship that occur in everyday life. Some of these describe how the members of a family are related to each other: parent, child, brother, sister, sibling. We could also have a relation called *is in* for cities and countries: for example, London is in Great Britain, and Paris is in France. Or we could have a relation that describes which make of car is produced by which manufacturer. Relations are used in mathematics to describe how two numbers are related to each other; for example expressions like $x < y$ and $p \geq q$ use the relations $<$ and $\geq$.

Similar examples abound in computing, and many branches of computer science use the terminology of relations to describe concepts precisely. Relations are naturally at the heart of relational databases; they are used heavily in the description of programming language syntax; they provide a good notation for representing the internal information required for web search engines, and so on.

Since relations are ubiquitous and important, it is useful to define them as mathematical objects and to describe their properties. In this chapter, we will see how to define relations using set theory and how to perform various calculations with them.

10.1 Binary Relations

A *binary relation* is used to describe the relationship between two objects. The word *binary* here means simply that there are *two* objects involved; it has nothing at all to do with binary number representations or binary files. General relationships among any number n of objects are called *n-ary relations*, but binary relations are the most important in computing, and we will restrict ourselves to those for the time being.

Definition 34. A *binary relation* R with type $R :: A \times B$ is a subset of $A \times B$, where A is the *domain* and B is the *codomain* of R. For $x \in A$ and $y \in B$, the

223

notation $x \, R \, y$ means $(x, y) \in R$.

Example 17. Let P be the set of people $\{Bill, Sue, Pat\}$, and let A be the set of animals $\{dog, cat\}$. The relation $has :: P \times A$ describes which person has which kind of animal. Suppose that Bill and Sue both have a dog, and Sue and Pat have a cat. We would represent this information by writing the following relational expressions:

Bill has dog   Sue has dog

Sue has cat   Pat has cat

but the statement '*Bill has cat*' is false. Written out in full, the relation is

$$has = \big\{ (Bill, dog), (Sue, dog), (Sue, cat), (Pat, cat) \big\}.$$

Example 18. Let R be the set of real numbers. Then $(<) \subset (R \times R)$ is the 'less than' relation and consists of the set of all ordered pairs (x, y) such that $x < y$. Because $(<)$ has an infinite number of elements, we can write it out only partially; for example,

$$(<) = \big\{ \ldots, (-35.2, -12.1), (-1, 2.7), \ldots \big\}.$$

Example 19. Many databases use relations to represent the data, as this is a good way to associate different pieces of information with each other. For example, a relation can be used to specify that a person's name is related to that person's address. These are called *relational databases*.

Suppose that you are building a relational database that maintains genealogical data about the families in a town. One of the relations in your database might be the *IsFatherOf* relation. If *John* is the father of *Mary* and *Peter*, then two pairs in the relation have *John* as their first component: $\{(John, Mary), (John, Peter)\}$. This relation states two facts: *John IsFatherOf Mary* and *John IsFatherOf Peter*.

We could also represent the relationship between *John* and his children using a single tuple: $\{(John, Mary, Peter)\}$. This is not a binary relation; it is a more general form of relation called an *n*-ary relation (where $n = 3$ in this case).

Example 20. What would have happened if the pairs had been written as $(Mary, John)$ and $(Peter, John)$? Then they would have had an entirely different meaning, asserting *Mary IsFatherOf John* and *Peter IsFatherOf John*, which is not what was intended. It is important to remember that the pairs in a relation are *ordered* pairs. The pairs (1,2) and (2,1) are not equal to each other, nor are the pairs (1,2) and (1,3).

Example 21. Let's consider two sets, *Children* and *Adults*. The set *Children* includes *Joe*, *Anne*, and *Susan*, and the set *Adults* includes *Ray*, *John*, and *Dinah*. We would like to create a relation *SmallFamilies* that pairs each child

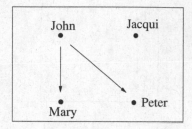

Figure 10.1: The Digraph of the *IsFatherOf* Relation

with every possible adult. This is done by taking each child in the first set and pairing it in turn with each adult in the second set; that is, we are taking the cross product of the two sets, denoted *Children* × *Adults*:

$$SmallFamilies =$$
$$\{(Joe, Ray), (Joe, John), (Joe, Dinah),$$
$$(Anne, Ray), (Anne, John), (Anne, Dinah),$$
$$(Susan, Ray), (Susan, John), (Susan, Dinah)\}$$

10.2 Representing Relations with Digraphs

Sometimes a diagram provides a good way to visualise a relation. There is a representation of binary relations called a *digraph* which is convenient for computing, and which also is well suited for diagrams. Every element of the domain and codomain in a digraph diagram is represented by a labelled dot (called an *element* or *node*) in the graph, and every pair (x, y) in the relation is represented by an arrow going from x to y (which may be called an *arrow* or *arc*).

Example 22. Figure 10.1 shows a graph illustrating the `IsFatherOf` relation. There is a node for each of *John, Mary, Peter*, and *Jacqui*. There are two arrows, from *John* to *Mary* and *John* to *Peter*.

Many graphs contain some nodes that have no arrows: for example, the node *Jacqui* in Figure 10.1. This means that we need to specify the graph by giving both the set of nodes and the set of arcs.

Some relations have the same set as their domain and codomain, so the relation has a type of the form $R :: A \times A$. In such cases, you draw the graph by writing (and labelling) a dot for every element of A and then draw in the arrows. Sometimes the domain and codomain are *disjoint*, which means that no element appears in both the domain and codomain. When this happens, it is helpful to keep the dots representing the domain together in the graph diagram, and separate from the dots representing the codomain. Figure 10.2 shows the graph for the relation $R :: A \times A = \{(1, 4), (2, 6)\}$ where $A = \{1, 2, 4, 6\}$.

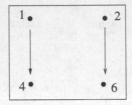

Figure 10.2: The Digraph $(\{1,2,4,6\},\{(1,4),(2,6)\})$

It is important to remember that a relation is more than just a set of ordered pairs; the domain and codomain must be specified, too. In drawing the graph of a relation, you should either draw a dot (or node) for every element of the domain and codomain, and use the layout to indicate exactly what these sets are, or you should specify them explicitly.

Definition 35. Let A be a set, and let R be a binary relation $R :: A \times A$. The *digraph D* of R is the ordered pair $D = (A, R)$.

Example 23. The digraph of the relation $R :: A \times A$, where $R = \{(1,2),(2,3)\}$ and $A = \{1,2,3\}$, is $(\{1,2,3\},\ \{(1,2),(2,3)\})$.

Example 24. The digraph of the relation $R :: A \times A$, where $R = \{(1,2),(2,3)\}$ and $A = \{1,2,3,4,5,6\}$, is $(\{1,2,3,4,5,6\},\ \{(1,2),(2,3)\})$. Note that this is a different relation than in the previous example, although the set of ordered pairs is identical. The digraph representation records the domain and codomain, giving a precise and complete description of the relation.

This has some interesting implications. For example, two graphs may show an empty relation that contains no arrows (no ordered pairs), but the relations are not equivalent unless their domains and codomains are equal.

Many relations have arcs that are connected to each other in a special way. For example, a set of arcs connected in a sequence is called a (directed) *path*.

Definition 36. A *directed path* is a set of arcs that can be arranged in a sequence, so that the end point of one arc in the sequence is the start point of the next.

Example 25. The sets $\{(1,2),(2,3),(3,4)\}$ and $\{(1,3),(3,1)\}$ are both paths, but the set $\{(1,2),(5,6)\}$ is not.

10.3   Computing with Binary Relations

It is common to compute directly with relations. Throughout this chapter, we will use the computer as a calculator for expressions on relations; this is a good way to become accustomed to all the operations on relations since they are used to specify computations formally in many specialised areas of computer science.

A relation $R :: A \times B$, with domain A and codomain B, can be represented as a list of type [(A,B)]. Each element of the list has type (A,B), and is a pair of the form (x,y) where x::A is in the domain and y::B is in the codomain.

Often we impose two restrictions on a relation in order to make it easier to compute with it: (1) there is a finite number of elements in the relation, and (2) the types of the domain and codomain must be in the classes Eq and Show, so that we can compare and print elements of the relation.

Example 26. The relation of colour complements can be represented as follows:

```
data Colour = Red | Blue | Green | Orange | Yellow | Violet
              deriving (Eq, Show)
colourComplement :: Digraph Colour
colourComplement =
 ([Red,Blue,Green,Orange,Yellow,Violet],
  [(Red,Green), (Green,Red),
   (Blue,Orange), (Orange,Blue),
   (Yellow,Violet), (Violet,Yellow)])
```

To say 'the colour complement of red is green', we would write either of the following:

$$Red \; colourComplement \; Green$$
$$(Red, Green) \in colourComplement$$

In the example above, we must include both (*Red, Green*) and (*Green, Red*) in *colourComplement*. If we omitted either one of these, we would have a different relation.

The function domain takes a relation and returns its domain:

```
domain ::
  (Eq a, Show a, Eq b, Show b) => Set (a,b) -> Set a
```

For example, domain colourComplement returns the set of colours in the domain of the relation, which is

$$\{Red, Green, Blue, Orange, Yellow, Violet\}.$$

The set is represented as a list, but there is no significance to the order of elements in the list.

The *codomain* function is similar:

```
codomain ::
  (Eq a, Show a, Eq b, Show b) => Set (a,b) -> Set b
```

Many of the operations and functions on sets that we defined in Chapter 8 are useful for working on relations, including `crossproduct` and `setEq`.

Exercise 1. Work out the values of the following expressions, and then check your answer by evaluating the expressions with the computer.

```
domain [(1,100),(2,200),(3,300)]
codomain [(1,100),(2,200),(3,300)]
crossproduct [1,2,3] [4]
```

Exercise 2. The following list comprehensions define a list of ordered pairs. What relations are represented by these lists? Give the domain and the codomain, as well as the pairs themselves.

(a)   `[(a,b) | a <- [1,2],`
              `b <- [3,4]]`

(b)   `[ (a,b) | a <- [1,2,3],`
                `b <- [1,2,3],`
                `a == b]`

(c)   `[ (a,b) | a <- [1,2,3],`
                `b <- [1,2,3],`
                `a < b]`

10.4   Properties of Relations

Many relations share interesting and useful properties. For example, we know that if person a is a sibling of person b and b is a sibling of c, then a is also a sibling of c. In a similar way, if x, y and z are numbers, and we know that $x < y$ and $y < z$, then it must also be the case that $x < z$. These two examples show that the *sibling* relation and the $(<)$ relation have essentially the same property (which is called 'transitivity'). In this section we define a variety of such relational properties.

10.4.1   Reflexive Relations

In a *reflexive relation*, every element of the domain is related to itself.

Definition 37. A binary relation R over A is *reflexive* if xRx for every element x of the domain A.

When a reflexive relation is shown in a graph diagram, there must be an arrow from every dot in the domain back to itself.

Example 27. The relation $R :: A \times A$, where $A = \{1, 2\}$ and

$$R = \{(1, 1), (1, 2), (2, 2)\},$$

is reflexive.

Example 28. Let $R :: A \times A$ be a relation, where $A = \{1, 2, 3\}$ and $R = \{(1, 1), (2, 2)\}$. R is not reflexive, but if we added $(3, 3)$ to R then it would be reflexive.

Example 29. A relation *SameFamily*, such that x *SameFamily* y for any two people x and y who are in the same family, is reflexive (because you are in the same family as yourself).

Example 30. The following relations on numbers are reflexive: equality ($=$), greater than or equal ($\geq$), and less than or equal ($\leq$).

Example 31. The following relations on numbers are not reflexive: inequality ($\neq$), less than ($<$), and greater than ($>$).

The Haskell software tools include many functions that test relations for their properties. The function `isReflexive` returns a Boolean that is `True` if the relation is reflexive:

```
isReflexive ::
  (Eq a, Show a) => Digraph a -> Bool
```

Example 32. Consider the following digraphs:

```
a = [1,2,3]
digraph1 = (a,[(1,1), (1,2), (2,2), (2,3), (3,3)])
digraph2 = (a,[(1,2), (2,3), (3,1)])
digraph3 = (a,[(1,1), (1,2), (2,2), (2,3)])
```

The first one, in `digraph1`, is reflexive, because a contains 1, 2, and 3, and all the pairs $(1, 1)$, $(2, 2)$, and $(3, 3)$ appear in `digraph1`. However, `digraph2` is not reflexive because $(1, 1)$ doesn't appear in its set of ordered pairs. (An equally good argument is that $(2, 2)$ doesn't appear, or $(3, 3)$ doesn't appear— all you have to do to show that a relation is not reflexive is to show that there is *some* element x of the domain where (x, x) doesn't appear in the set of ordered pairs.) Finally, `digraph3` is also not reflexive, because $(3, 3)$ is not in the set of ordered pairs.

10.4.2 Irreflexive Relations

A relation is *irreflexive* if no element of its domain is related to itself.

Definition 38. A binary relation R over A is *irreflexive* if, for every $x \in A$, it is *not* the case that xRx.

Example 33. The greater than ($<$) and less than ($>$) relations over numbers are irreflexive because $x < x$ and $x > x$ are always false.

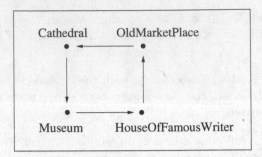

Figure 10.3: The *ByBus* Relation

As long as the domain A of a relation $R :: A \times A$ is non-empty, then it is impossible for R to be both reflexive and irreflexive. To see this, consider some element x of the domain (such an x must exist, because the domain is not empty). If R is reflexive then $(x,x) \in R$, but if R is irreflexive then $(x,x) \notin R$, and both cannot be true.

Example 34. The empty relation $R :: \varnothing \times \varnothing$ is reflexive and also irreflexive. In both cases the conditions are met vacuously.

Example 35. Many relations among people are irreflexive. For example, the relations *IsMarriedTo* and *IsChildOf* are irreflexive relations, because no one can marry themself or be their own child.

It often happens that a relation is not reflexive and it is also not irreflexive. For example, let $A = \{1, 2, 3, 4, 5\}$ be the domain and codomain of the relation $R = \{(1,3), (2,4), (3,3), (3,5)\}$. Then R is not reflexive (for example, $(1,1)$ is not in R) but it is also not irreflexive (because $(3,3)$ *is* in R).

Suppose that we are visiting a city in France and want to see several buildings by bus. We can get a bus schedule and look at it, note down the buildings and draw an arrow between each pair of nodes that the bus will visit (Figure 10.3).

```
{(Cathedral, Museum),(Museum, HouseOfFamousWriter),
  (HouseOfFamousWriter, OldMarketPlace),
  (OldMarketPlace, Cathedral)}
```

The bus is travelling in a *cycle*, a path that starts and stops at the same node. However, the ByBus relation is not reflexive: the bus isn't going to waste time cycling around one place and returning to it without going anywhere else. The ByBus relation is *irreflexive*.

The following Haskell function determines whether a binary relation is irreflexive:

```
isIrreflexive ::
  (Eq a, Show a) => Digraph a -> Bool
```

Exercise 3. For each of the following `Digraph` representations of a relation, draw a graph of the relation, work out whether it is reflexive and whether it is irreflexive, and then check your conclusion using the `isReflexive` and `isIrreflexive` functions:

```
([1,2,3],[(1,2)])
([1,2],[(1,2),(2,2),(1,1)])
([1,2],[(2,1)])
([1,2,3],[(1,2),(1,1)])
```

Exercise 4. Determine whether each of the following relations on real numbers is reflexive and whether it is irreflexive. Justify your conclusions.

(a) less than ($<$)

(b) less than or equal to ($\leq$)

(c) greater than ($>$)

(d) greater than or equal to ($\geq$)

(e) equal ($=$)

(f) not equal ($\neq$)

We can't use the `isReflexive` and `isIrreflexive` functions on relations with an infinite domain. However, if we restrict the domain to the natural numbers from 0 through 100, it's possible to represent the relations completely and check them with the software tools. The following binary relation representations, with domain $N100 = \{0, \ldots, 100\}$, are defined in the software tools:

```
lessThan_N100,     lessThanOrEq_N100,
greaterThan_N100, greaterThanOrEq_N100,
equals_N100,       notEq_N100
  :: Digraph Int
```

Using these finite relations, use the computer to check your results. Note that a partial check like this does *not* prove anything about the infinite relations, but it is guaranteed to give the correct result for a finite relation on the first hundred natural numbers.

10.4.3 Symmetric Relations

Some relations have the property that the order of two related objects does not matter; that is, if xRy it must also be true that yRx. Such a relation is called a *symmetric* relation.

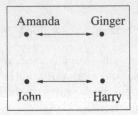

Figure 10.4: The *IsSiblingOf* Relation

Definition 39. Let $R :: A \times A$ be a binary relation. Then R is *symmetric* if $\forall x, y \in A.\ xRy \rightarrow yRx$.

Example 36. Equality on real numbers ($=$) is symmetric, because if $x = y$ then also $y = x$. The equality relation is commonly defined for sets, and it is always symmetric; in fact, one of the essential properties of an abstract equality relation is that it must be symmetric.

Example 37. The family relation *IsSiblingOf* is symmetric.

Example 38. The family relations *IsBrotherOf* and *IsSisterOf* are not symmetric: for example, the term 'Robert *IsBrotherOf* Mary' is true, but 'Mary *IsBrotherOf* Robert' is false.

When you draw the graph diagram for a symmetric relation, every arc from a to b will have a matching arc from b back to a. The notation can be simplified by putting an arrowhead on both sides of every arc.

Example 39. Here is a possible definition of the `IsSiblingOf` relation, shown in Figure 10.4:

```
{(John, Harry), (Harry, John),
 (Amanda, Ginger), (Ginger, Amanda)}
```

Example 40. The relation $R = \{(1,2),(2,1),(2,3),(3,2)\}$ is symmetric.

Example 41. The relation $R = \{(1,2),(1,3),(3,1)\}$ is not symmetric, because $(1,2) \in R$ but $(2,1) \notin R$.

Exercise 5. Is the family relation *IsChildOf* symmetric?

Exercise 6. Suppose we have a relation $R :: A \times A$, where A is non-empty and reflexive, but it has *only* the arcs required in order to be reflexive. Is R symmetric?

Exercise 7. In the definition of a symmetric relation, can the variables x and y can be instantiated by a single node?

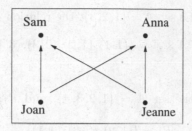

Figure 10.5: The *IsChildOf* Relation

10.4.4 Antisymmetric Relations

An *antisymmetric* relation is one where for all distinct values a and b, it is *never* the case that both aRb and bRa.

Example 42. The less-than relation ($<$) is antisymmetric, because it cannot be true that $x < y$ and also $y < x$.

Example 43. The family relation *IsChildOf* is antisymmetric; if x is a child of y, than y must be the *parent*—not the child—of x. For example, suppose the *IsChildOf* relation contains the following ordered pairs (Figure 10.5):

$$\{(Joan, Sam), (Jeanne, Sam), (Joan, Anna), (Jeanne, Anna)\}$$

Notice that this relation never has both a pair (x, y) and also a pair (y, x).

The antisymmetric property is defined formally as follows:

Definition 40. A binary relation $R :: A \times A$ is *antisymmetric* if

$$\forall x, y \in A.\ xRy \wedge yRx \to x = y.$$

The graph of an antisymmetric relation may contain some cycles; for example the relation $R = \{(1, 2), (2, 3), (3, 1)\}$ has a cycle from 1 to 2 to 3 and back to 1, and the relation $R_2 = \{(1, 1)\}$ has a trivial cycle containing just 1. However, if an antisymmetric relation does have a cycle, then the length of the cycle cannot be 2, although it may be 1, or greater than 2. In other words, this graph will have no cycles of length 2, but it can have cycles of any other length.

Example 44. Given the set $A = \{1, 2, 3\}$, the relation

$$R :: A \times A = \{(1, 2), (2, 1), (2, 3), (3, 1)\}$$

is not anti-symmetric because both $(1, 2)$ and $(2, 1)$ appear in the set of ordered pairs.

Example 45. Given the set $A = \{1, 2, 3\}$, the relation

$$R :: A \times A = \{(1,1), (1,2), (2,3), (3,1)\}$$

is anti-symmetric.

Example 46. Given the set $A = \{1, 2, 3, 4\}$, and $R_1, R_2, R_3 :: A \times A$, the relations

$$R_1 = \{(1,2), (2,3), (4,1)\}$$

and

$$R_2 = \{(1,1), (2,2)\}$$

are both antisymmetric, but

$$R_3 = \{(1,3), (3,1), (2,3), (3,2)\}$$

is not antisymmetric.

If a relation $R :: A \times A$ is antisymmetric, both of the following statements must be true:

$$\forall x, y \in A. \ x \neq y \ \rightarrow \ \neg(xRy \wedge yRx)$$
$$\forall x, y \in A. \ x \neq y \ \rightarrow \ \neg xRy \vee \neg yRx$$

Both propositions say that for two distinct elements of the domain, the graph diagram of R contains at most one arrow connecting them.

Example 47. Suppose that we were misanthropic and thought people didn't treat each other well in general. When told that a *Helps* b and b *Helps* a, we might retort that a and b must therefore be the same person! We could express this gloomy view of the world as

$$\forall x, y \in WorldPopulation. \ x \ Helps \ y \ \wedge \ y \ Helps \ x \ \rightarrow \ x = y.$$

The software tools define the following functions, which determine whether a finite binary relation is symmetric or antisymmetric:

```
isSymmetric, isAntisymmetric ::
(Eq a, Show a) => Digraph a -> Bool
```

Exercise 8. First work out whether the relations in the following expressions are symmetric and whether they are antisymmetric, and then check your conclusions by evaluating the expressions with Haskell:

```
isSymmetric ([1,2,3],[(1,2),(2,3)])
isSymmetric ([1,2],[(2,2),(1,1)])
isAntisymmetric ([1,2,3],[(2,1),(1,2)])
isAntisymmetric ([1,2,3],[(1,2),(2,3),(3,1)])
```

Exercise 9. Which of the following relations are symmetric? Antisymmetric?

    (a) The empty binary relation over a set with four nodes;

    (b) The = relation;

    (c) The $\leq$ relation;

    (d) The < relation.

10.4.5 Transitive Relations

If x, y, and z are three people, and you know that x is a sister of y and y is a sister of z, then x must also be a sister of z. Similarly, if you know that $x < y$ and also that $y < z$, it must also be the case that $x < z$. Relations that have this property are called *transitive* relations.

Definition 41. A binary relation $R :: A \times B$ is *transitive* if

$$\forall x, y, z \in A. \ xRy \wedge yRz \rightarrow xRz.$$

Example 48. The relation $R = \{(1,2),(2,3),(1,3)\}$ is transitive because it contains $(1,3)$, which is required by the presence of $(1,2)$ and $(2,3)$.

Example 49. The relation $R = \{(1,2),(2,3)\}$ is not transitive because there are pairs $(1,2)$ and $(2,3)$ but there is no pair $(1,3)$.

The $(=)$ relation is transitive, as is the *IsAncestorOf* relation.

Example 50. The *IsMarriedTo* relation is not transitive. It is certainly symmetric, because if x *IsMarriedTo* y then it must also be the case that y *IsMarriedTo* x. Suppose, however, that x and y are two married people. Then (x, y) and (y, x) are both in the relation, so, if it were transitive, then (x, x) would also need to be in the relation. Nobody is married to themself, so this cannot be, and the relation is not transitive.

Example 51. Suppose we are flying from one city to another. The relation *FlightTo* describes the point-to-point flights that are available: for example, $(London, Paris) \in FlightTo$ because there is a direct flight from London to Paris. This relation is not transitive, because there are flights from many small cities to London, but those small cities don't have direct flights to Paris. However, the *ReachableByAir* relation *is* transitive. In effect, the airlines define the *FlightTo* relation, and the travel agents extend this to the more general *ReachableByAir* relation, which may involve several connecting flights.

As the previous example suggests, a binary relation R can be extended to make a new binary relation R^T, such that $R \subseteq R^T$ and R^T is transitive. This often entails adding several new ordered pairs. For example, suppose we have a relation *CityMap* that defines direct street connections, so that $(x, y) \in$

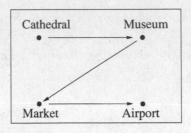

Figure 10.6: The *CityMap* Relation

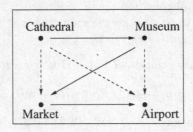

Figure 10.7: The Transitive *CityMap* Relation

CityMap if there is a street connecting x directly with y (Figure 10.6). The relation could be defined (for a small city) as

$$\{(Cathedral, Museum), (Museum, Market), (Market, Airport)\}.$$

The *CityMap* relation is not transitive, because there is a street path from *Cathedral* to *Market*, but no street connects them directly. Just adding the pair (*Cathedral, Market*) is not enough to make the relation transitive; a total of three ordered pairs must be added. These are shown as dashed arrows in Figure 10.7. The new pairs that we added to the relation are

$$\{(Cathedral, Market), (Cathedral, Airport), (Museum, Airport)\}.$$

A transitive relation provides a short cut for every path of length 2 or more. To make a relation transitive, we must continue adding new pairs until the new relation is transitive. This process is called taking the *transitive closure* of the relation.

The software tools contain a definition of the following function, which determines whether a finite binary relation is transitive:

```
isTransitive ::
  (Eq a, Show a) => Digraph a -> Bool
```

Exercise 10. Determine by hand whether the following relations are transitive, and then check your conclusion using the computer:

```
isTransitive ([1,2],[(1,2),(2,1),(2,2)])
isTransitive ([1,2,3],[(1,2)])
```

Exercise 11. Determine which of the following relations on real numbers are transitive: $(=)$, $(\neq)$, $(<)$, $(\leq)$, $(>)$, $(\geq)$.

Exercise 12. Which of the following relations are transitive?

(a) The empty relation;

(b) The *IsSiblingOf* relation;

(c) An irreflexive relation;

(d) The *IsAncestorOf* relation.

10.5 Relational Composition

We can think of a relation $R :: A \times B$ as taking us from a point $x \in A$ to a point $y \in B$, assuming that $(x, y) \in R$. Now suppose there is another relation $S :: B \times C$, and suppose that $(y, z) \in S$, where $z \in C$. Using first R and then S, we get from x to z, via the intermediate point y.

We could define a new relation that describes the effect of doing first R and then S. This is called the *composition* of R and S, and the notation for it is $R; S$.

Example 52. Suppose that we have a relation *Flight* over the set *City*, where $(a, b) \in Flight$ if there is an airline flight from a to b. There is also a relation *BusTrip* over *City*, and $(c, d) \in BusTrip$ if there is a bus connection from c to d. Now, we are interested in a relation that describes where we can go, starting from a city with an airport. The relation *Flight*; *BusTrip* consists of the set of pairs (x, y) such that you can get from x to y by flying first to some intermediate city y, and then taking the bus from y on to z.

The use of a semicolon (;) as the operator for relational composition is common, but not completely standard. Many older mathematics books omit the relational composition operator, using RS to mean the relational composition of R and S. Computer scientists often prefer to make all operators explicit. The use of a semicolon is intended to suggest sequencing: just as *statement1 ; statement2* in an imperative programming language means 'First execute *statement1* and then execute *statement2*', the relational composition $R; S$ means 'First apply the relation R and then apply the relation S'.

Relational composition is defined formally as follows.

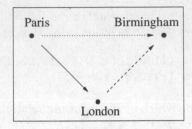

Figure 10.8: The *Route1; Route1* Relation

Definition 42. Let $R_1 :: A \times B$ be a relation from set A to set B, and $R_2 :: B \times C$ be a relation from set B to set C. Their *relational composition* is defined as follows:

$$R_1; R_2 \quad :: \quad A \times C$$
$$R_1; R_2 \quad = \quad \{(a,c) \mid a \in A \wedge c \in C \wedge (\exists b \in B. \ (a,b) \in R_1 \wedge (b,c) \in R_2)\}$$

The definition just says formally that $R_1; R_2$ consists of all the pairs (a,c), such that there is an intermediate connecting point b. This means that $(a,b) \in R_1$ and $(b,c) \in R_2$.

Example 53. When we compose two relations, any two links between a and b in the first relation and b and c in the second produce a new link between a and c. Suppose we have a relation `Route1` linking Paris and London and `Route2` linking London and Birmingham. The composition of `Route1` and `Route2` yields a new route relation which shows that it is possible to travel taking first `Route1` and then `Route2`, starting from Paris and ending at Birmingham (Figure 10.8). In our diagram, the arcs are of three different patterns because they belong to three separate relations.

Sometimes it is useful to compose a relation with itself. A common situation is to start with a relation like *Flight*, which represents trips consisting of just one flight, starting from one city and ending in another one. The composition *Flight; Flight* describes all the trips that can be made using exactly two connecting flights.

Another example arises in databases, where queries often cause the database to derive new information from the facts already available. If the system can predict the requirements of some common queries, then some of this new information can be represented as facts, represented as a new relation, speeding up the execution of future queries.

Suppose that we need to know whether a, who died of a hereditary disease, was a blood relative of b. This could mean calculating all of the descendants of a, then checking to see whether b is among them. It might be better to save some of the work done (space permitting) in calculating the descendants of a,

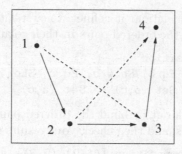

Figure 10.9: The $R_1; R_1$ Relation

so that when we need to know whether a was a blood relative of c, some of the work need not be repeated.

As an example, when determining whether **Joseph** and **Jane** are blood relations, we discover that *Joseph IsBloodRelationOf Sarah, Sarah IsBloodRelationOf Jane*, and *Jane IsBloodRelationOf Joel*. During this process, we add the newly discovered fact to the database: *Joseph IsBloodRelationOf Jane*. Now, when we have a query asking whether Joseph is a blood relation of Joel, the new link represents the two links between Joseph and Jane. This reduces the number of links to be traversed.

In creating the composition of two relations, we look for arcs in the first relation that have terminal nodes matching the starting nodes of arcs in the second relation. This operation requires that we systematically check all arcs in R_1 against all arcs in R_2.

Example 54. Let's calculate a relational composition by hand. Let

$$R_1 = \{(1,2), (2,3), (3,4)\}.$$

The composition $R_1; R_1$ is worked out by deducing all the ordered pairs that correspond to an application of R_1 followed by an application of R_1.

First we find all the ordered pairs of the form $(1, x)$. R_1 has only one ordered pair starting with 1; this is $(1, 2)$. This means the first application of R_1 goes from 1 to 2, and the $(2, 3)$ pair means that the second application goes to 3. Therefore the composition $R_1; R_1$ should contain a pair $(1, 3)$. Next, consider what happens starting with 2: the $(2, 3)$ pair goes from 2 to 3, and looking at all the available pairs $\{(1, 2), (2, 3), (3, 4)\}$ shows that 3 then goes to 4. Finally, we see what happens when we start with 3: the first application of R_1 goes from 3 to 4, but there is no pair of the form $(4, x)$. This means that there cannot be any pair of the form $(3, x)$ in the composition $R_1; R_1$. The result of all these comparisons is $R_1; R_1 = \{(1, 3), (2, 4)\}$ (Figure 10.9). In our diagram, the new relation is indicated by arrows with dashes.

The calculation in Example 54 is straightforward and tedious—well suited for computers. The software tools define a function `relationalComposition` that implements this calculation: it defines a new relation giving two existing ones, by working out all the ordered pairs in their relational composition.

```
relationalComposition ::
  (Eq a, Show a, Eq b, Show b, Eq c, Show c) =>
    Set (a,b) -> Set (b,c) -> Set (a,c)
```

Exercise 13. First work out by hand the ordered pairs in the following relational compositions, and then check your results using the computer:

```
relationalComposition [(1,2),(2,3)] [(3,4)]
relationalComposition [(1,2)] [(1,3)]
```

Exercise 14. (a) Find the composition of the following relations:

$$\{(Alice,\ Bernard),\ (Carol, Daniel)\} \text{ and } \{(Bernard, Carol)\}.$$

(b) $\{(a,b),(aa,bb)\}$ and $\{(b,c),(cc,bb)\}$

(c) $R; R$, where the relation R is defined as

$$R = \{(1,2),(2,3),(3,4),(4,1)\}.$$

(d) $\{(1,2)\}$ and $\{(3,4)\}$

(e) The empty set and any other relation.

10.6   Powers of Relations

As we saw in the previous section, the composition *Flight; Flight* defines the relation describing all possible trips that consist of two connected flights. More generally, we might want to define a relation defining all possible trips that consist of n connected flights, where n is a natural number. This is called the nth power of the relation. For a relation R, the nth power is the composition $R; R; \cdots ; R$, where R appears n times, and its notation is R^n. Notice in particular that $R^2 = R; R$, and $R^1 = R$. It is also convenient to define R^0 to be the identity relation.

When a relation R is composed with itself n times, producing R^n, a path of length n in R from a to b causes there to be a single link (a, b) in the power relation R^n.

Suppose that we have to calculate the relationships between several people in our database, and that the original facts are these (Figure 10.10):

Andrew	*IsParentOf*	*Beth*
Beth	*IsParentOf*	*Ian*
Beth	*IsParentOf*	*Joanna*
Ian	*IsParentOf*	*William*
William	*IsParentOf*	*Tina*

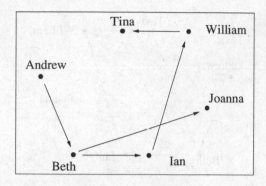

Figure 10.10: The *IsParentOf* Relation

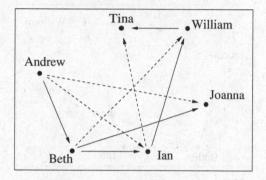

Figure 10.11: The Relation *IsParentOf*^{2}

Now, we will calculate the powers of this relation. The 0th power is just the identity relation, and the first power *IsParentOf*^{1} is simply the *IsParentOf* relation. The higher powers will tell us the grandparents, great grandparents, and great great grandparents. You should expect to see that each of the new relations *IsParentOf*^{2}, *IsParentOf*^{3}, and *IsParentOf*^{4} connect up the starting and ending points of a path 2, 3, and 4 arcs long within the original *IsParentOf* relation. In the following diagrams, the arrows with dashes indicate relations defined as a power, while all other arrows belong to the IsParentOf relation.

If we compose the *IsParentOf* relation with itself (i.e., *IsParentOf*^{2}), we have the grandparent relation (Figure 10.11):

Andrew	*IsGrandParentOf*	*Ian*
Andrew	*IsGrandParentOf*	*Joanna*
Beth	*IsGrandParentOf*	*William*
Ian	*IsGrandParentOf*	*Tina*

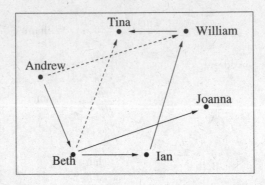

Figure 10.12: The *IsParentOf*^{3} Relation

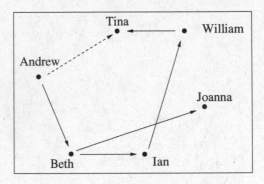

Figure 10.13: The *IsParentOf*^{4} Relation

Now if we compose the *IsGrandParentOf* relation with the original relation, we obtain the *great grand parent* relation (Figure 10.12):

$$Andrew \quad IsGreatGrandParentOf \quad William$$
$$Beth \quad IsGreatGrandParentOf \quad Tina$$

Figure 10.13 shows the composition of the *IsGreatGrandParentOf* relation with *IsParentOf*; thus we have just calculated the fourth power of the *IsParentOf* relation.

$$Andrew \quad IsGreatGreatGrandParentOf \quad Tina$$

We will now give the formal definition of relational powers. The definition is recursive, because we have to define the meaning of R^n for all n. The base case of the recursion will be R^0, which is just the identity relation (it's like taking *zero* flights from a city, which leaves you where you started). The recursive case defines R^{n+1} using the value of R^n.

Definition 43. Let A be a set and let $R :: A \times A$ be a relation defined over A. The nth power of R, denoted R^n, is defined as follows:

$$R^0 = \{(a,a) \mid a \in A\}$$
$$R^{n+1} = R^n; R$$

Example 55. Using the formal definition, we calculate R^4, where

$$R = \{(2,3), (3,2), (3,3)\}.$$

R^0 is just the identity (equality) relation, which contains a reflexive loop for every node. By the definition, $R^1 = R^0; R = R$, because the identity relation composed with R just gives R. The first nontrivial calculation is to find $R^2 = R^1; R = R; R$. We have to take each pair (a,b) in R, and see whether there is a pair (b,c); if so, we need to put the pair (a,c) into R^2. The result of this calculation is $R^2 = \{(2,2), (2,3), (3,2), (3,3)\}$.

Now we have to calculate $R^3 = R^2; R$. We compose

$$\{(2,2), (2,3), (3,2), (3,3)\}$$

with

$$\{(2,3), (3,2), (3,3)\},$$

which yields

$$\{(2,2), (2,3), (3,2), (3,3)\}.$$

At this point, it's helpful to notice that $R^3 = R^2$. In other words, composing R^2 with R just gives R^2 back, and we can do this any number of times. This means that any further powers will be the same as R^2—so we have found R^4 without needing to do lots of calculations with ordered pairs.

Relational composition is associative. This causes the powers of a relation to follow algebraic laws that are similar to the corresponding laws for powers on numbers. For example, $R^{(a+b)} = R^a; R^b$.

A relation whose domain is $\{x_0, \ldots, x_{n-1}\}$ is *cyclic* if it contains a cycle of ordered pairs of the form (x_0, x_1), (x_1, x_2), (x_2, x_3), (x_3, x_4), $\ldots$, (x_{n-1}, x_0). That is, the relation is cyclic if there is a cycle comprising all the elements of its domain.

Consider what happens to a cyclic relation as we calculate its powers. The relation is defined as

$$R = \{(a,b), (b,c), (c,a)\}.$$

The first power R^1 is just R. The second power R^2 is calculated by working out the ordered pairs in $R; R$; the result is

$$R^2 = \{(a,b), (b,c), (c,a)\}; \{(a,b), (b,c), (c,a)\}$$
$$= \{(a,c), (b,a), (c,b)\}.$$

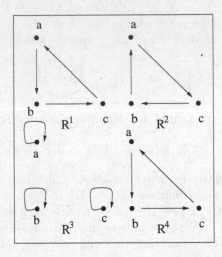

Figure 10.14: Four Powers of a Cyclic Relation

This contains only paths between the start and end points of all the paths of length two in the original relation. Now the third power is

$$R^3 \;=\; \{(a,c),(b,a),(c,b)\}; \{(a,b),(b,c),(c,a)\}$$
$$\;=\; \{(a,a),(b,b),(c,c)\}.$$

This result contains only arcs connecting the origin and destination points of paths of length *three* in the original relation. What will happen next?

$$R^4 \;=\; \{(a,a),(b,b),(c,c)\}; \{(a,b),(b,c),(c,a)\}$$
$$\;=\; \{(a,b),(b,c),(c,a)\}$$

Just what we might have expected: each of these arcs represents a path of length 4, so we have started round the cycle again. What can we now say about the powers of this relation? They repeat in a cycle. $R^4 = R^1$, $R^5 = R^2$ and in general $R^{n+3} = R^n$ (Figure 10.14).

The software tools file defines the following function, which takes a set and returns the equality relation on that set.

```
equalityRelation ::
  (Eq a, Show a) => Set a -> Relation a
```

There is also a function that calculates the power of a relation:

```
relationalPower ::
  (Eq a, Show a) => Digraph a -> Int -> Relation a
```

Exercise 15. Work out the values of these expressions, and then evaluate them using the computer:

```
equalityRelation [1,2,3]
equalityRelation ([]::[Int])
```

Exercise 16. Calculate the following relational powers by hand, and then evaluate them using the computer.

```
relationalPower ([1,2,3,4],[(1,2),(2,3),(3,4)]) 1
relationalPower ([1,2,3,4],[(1,2),(2,3),(3,4)]) 2
relationalPower ([1,2,3,4],[(1,2),(2,3),(3,4)]) 3
relationalPower ([1,2,3,4],[(1,2),(2,3),(3,4)]) 4
```

Exercise 17. Why do we not need to check the ordered pairs in R while calculating $R^0; R$?

Exercise 18. Why can we stop calculating powers after finding that two successive powers are the same relation?

Exercise 19. What is R^4 where R is $\{(2,2),(4,4)\}$?

Exercise 20. What is the relationship between adding new ordered pairs to make a relation transitive and taking the power of a relation?

Exercise 21. Suppose a set A contains n elements. How many possible relations with type $R :: A \times A$ are there?

Exercise 22. Given the relation $\{(a,b),(b,c),(c,d),(d,e)\}$, how many times would we have to compose this relation with itself before the empty relation is produced?

Exercise 23. Given the set $A = \{1,2,3\}$ and the relation $R :: A \times A$ where $R = \{(3,1),(1,2),(2,3)\}$, what is the value of R^2? R^3?

10.7 Closure Properties of Relations

In computing applications, we normally want to keep the specification of a relation as small and readable as possible, so that it can be defined and maintained accurately. On the other hand, some computations may require the relation to have some special properties (for example, symmetric or transitive). These special properties would require adding a large number of ordered pairs to the relation, making it harder to maintain. There is a standard technique used in such situations—we define *two* relations:

1. A basic relation, containing just the essential information, is specified;

2. A larger relation is derived from the basic one by adding the ordered pairs required to give it the special properties that are needed.

When circumstances change, only the basic relation is edited by hand. The derived relation is recalculated using a computer.

Example 56. An airline keeps a set of all the flights they offer. This is represented by a relation *Flight*, where $(a, b) \in Flight$ if the airline has a direct flight from a to b. However, when a customer asks a question like 'Do you fly from Glasgow to Seattle?', the airline needs a *transitive* relation: if $(Glasgow, New York) \in Flight$ and also $(New York, Seattle) \in Flight$, the answer should be *yes*. Thus the airline's flight-planning staff define the basic relation *Flight*, but the sales staff work with a derived relation that is similar to *Flight*, but which is transitive.

A relation derived in this way is called the *closure* of the basic relation:

Definition 44. The *closure* of a relation R with respect to a given property is the smallest possible relation that contains R and has that property.

Closure is suitable for adding properties that require the *presence* of certain ordered pairs. For example, you can take the symmetric closure of a relation by checking every existing pair (x, y), and adding the pair (y, x) if it isn't already present. However, closure is *not* suitable for properties that require the *absence* of certain ordered pairs. For example, the relation $R = \{(1,1), (1,2), (2,3)\}$ does not have an irreflexive closure, as that would need to contain $(1,1)$ (because the closure must contain the basic relation), yet it *must not* contain $(1,1)$ (in order to be irreflexive).

You can give a relation a property such as reflexivity, or transitivity, by creating its reflexive or transitive closure. Notice, however, that the new relation may no longer have all of the properties of the original relation. For example, suppose that a relation is irreflexive, as in $\{(1,2), (2,1)\}$. The smallest possible transitive relation containing this one also has the arcs $(1,1)$ and $(2,2)$, which means that it is no longer irreflexive.

10.7.1 Reflexive Closure

The reflexive closure of a relation contains all of the arcs in the relation together with an arc from each node to itself. For example, consider the set

$$\{Red, Orange, Yellow, Green, Blue, Violet\}.$$

The relation *ContainsPrimaryColour* is given in Figure 10.15 by the solid arcs. It is not reflexive, because the colours that are not primary colours do not contain reflexive arcs. However, the relation *ContainsColour* is reflexive (given by all the arcs in Figure 10.15), and is in fact the reflexive closure of *Contains-PrimaryColour*.

The formal definition that follows defines the reflexive closure of a relation R as the smallest reflexive relation containing the arcs of R.

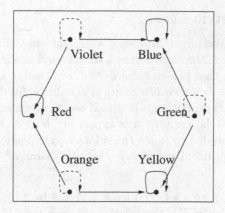

Figure 10.15: The *ContainsColour* and *ContainsPrimaryColour* Relations

Definition 45. Let A be a set, and let $R :: A \times A$ be a binary relation over A. The *reflexive closure* of R is the relation R' such that R' is reflexive, R' is a superset of R, and for any reflexive relation R'', if R'' is a superset of R, then R'' is a superset of R'. The notation $r(R)$ denotes the reflexive closure of R.

Example 57. The reflexive closure of the relation $\{(1,2),(2,3)\}$ over the set $\{1,2,3\}$ is $\{(1,1),(2,2),(3,3),(1,2),(2,3)\}$.

Example 58. The relation $\{(1,1),(1,2),(3,3),(2,3)\}$ is *not* the reflexive closure of $\{(1,2),(2,3)\}$, because it is missing the reflexive arc $(2,2)$.

The following theorem provides a straightforward method for calculating the reflexive closure of a relation:

Theorem 73. Let A be a set, let E be the equality relation on A, and let R be a binary relation defined over A. Then $r(R) = R \cup E$.

All we have to do to calculate the reflexive closure $R :: A \times A$ is to add self-loops (x,x) for all of the nodes x in the set A over which the relation R is defined.

The software tools file provides the following function, which automates the calculation of reflexive closures:

```
reflexiveClosure ::
    (Eq a, Show a) => Digraph a -> Digraph a
```

Exercise 24. Work out the following reflexive closures by hand, and then check your results using the computer:

```
reflexiveClosure ([1,2,3],[(1,2),(2,3)])
reflexiveClosure ([1,2],[(1,2),(2,1)])
```

Exercise 25. What is the reflexive closure of the relation $R; R$, where R is defined as $\{(1,2),(2,1)\}$?

10.7.2 Symmetric Closure

In maintaining a genealogical database, we might enter an ordered pair (a, b) stating that person a *IsMarriedTo* person b, but we don't want to enter another pair saying explicitly that person b *IsMarriedTo* person a, in order to save time typing. However, the *IsMarriedTo* relation should certainly be symmetric. In order to derive this information, the database must calculate the *symmetric closure* of the basic relation that was typed in. It does this by adding only those arcs that are needed to make the relation symmetric.

The formal definition of symmetric closure is similar[1] to the definition of reflexive closure.

Definition 46. Let A be a set, and let $R :: A \times A$ be a binary relation over A. The *symmetric closure* of R is the relation R' such that R' is symmetric, R' is a superset of R, and for any symmetric relation R'', if R'' is a superset of R, then R'' is a superset of R'. The notation $s(R)$ denotes the symmetric closure of R.

Sometimes it is useful to turn around a relation and use its ordered pairs in reverse. This is called the *converse* of the relation:

Definition 47. Let A and B be sets, and let $R :: A \times B$ be a binary relation from A to B. The *converse* of R, written R^c, is the binary relation from B to A defined as follows:

$$R^c = \{(b, a) | (a, b) \in R\}.$$

This definition says that if you reverse the order of the components in a relation's arcs, you create its converse. The symmetric closure of a relation is the union of the relation and its converse.

The converse operation provides an alternative way to calculate the symmetric closure of a relation. The idea is that we take the original relation R, and add to it the set of reversed ordered pairs, which is R^c. The following theorem states this formally:

Theorem 74. Let A be a set and let $R :: A \times A$ be a binary relation over A. Then the symmetric closure $s(R) = R \cup R^c$.

Example 59. Suppose that we had a relation $\{(1, 1), (2, 2)\}$ and wanted to create its symmetric closure. How many arcs would need to be added? None, because the relation is already symmetric.

Example 60. On the other hand, suppose that we had the relation $\{(1, 2)\}$. It is not symmetric, so when creating its symmetric closure, we add the arc $(2, 1)$ (Figure 10.16).

The following Haskell function calculates the symmetric closure of a binary relation:

---

[1] A useful tip for learning to read mathematics: many definitions follow standard forms, and this makes it easier to read and understand new ones.

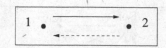

Figure 10.16: The Symmetric Closure of $\{(1,2)\}$

```
symmetricClosure ::
  (Eq a, Show a) => Digraph a -> Digraph a
```

Exercise 26. Work out the following symmetric closures by hand, and then calculate them using the computer window:

```
symmetricClosure ([1,2],[(1,1),(1,2)])
symmetricClosure ([1,2,3],[(1,2),(2,3)])
```

Exercise 27. What is the symmetric reflexive closure of the relation

$$\{(a,b),(b,c)\}?$$

Hint: take the reflexive closure first, followed by the symmetric closure of the result.

Exercise 28. Find the reflexive symmetric closure of the relation $\{(a,c)\}$.

10.7.3 Transitive Closure

The transitive closure is one of the most important operations on relations. You can define a relation that describes one step; the transitive closure of the relation then describes the effect of taking n steps, for any n.

Example 61. Suppose that *Flight* is the relation where $(a,b) \in$ *Flight* if there is a direct flight from city a to city b. Then the transitive closure of *Flight* is the relation consisting of pairs (a,b) where b is *reachable* by air from a.

Consider now how to calculate the transitive closure of a relation. As an example, suppose that you need to define the *IsDescendantOf* relation in a database of people. The database contains records for *Zoe, Bruce, Gina, Annabel, Dirk, Kay,* and *Don.* We start with the *IsChildOf* relation, defined as follows:

$$\{(Zoe, Bruce), (Gina, Zoe),$$
$$(Bruce, Annabel), (Dirk, Kay),$$
$$(Kay, Don), (Annabel, Kay)\}$$

Observe that the relation *IsDescendantOf* should have all these arcs, and many more. For example, *Gina* is a descendant of *Bruce.*

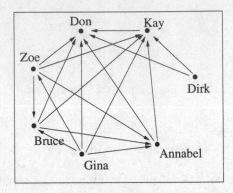

Figure 10.17: The *IsDescendantOf* Relation

How many more arcs need to be added? There should be an arc between the origin and destination points of each path of length 2 or more in the *IsChildOf* relation. In other words, we need the transitive closure of *IsChildOf*.

The powers of the *IsChildOf* relation are as follows:

$$
\begin{aligned}
Is\_Child\_Of^2 &= \{(Zoe, Annabel), (Gina, Bruce), \\
&\quad (Bruce, Kay), (Dirk, Don), (Annabel, Don)\} \\
Is\_Child\_Of^3 &= \{(Zoe, Kay), (Gina, Annabel), (Bruce, Don)\} \\
Is\_Child\_Of^4 &= \{(Zoe, Don), (Gina, Kay)\} \\
Is\_Child\_Of^5 &= \{(Gina, Don)\}
\end{aligned}
$$

The *IsDescendantOf* relation contains the union of all of these powers (Figure 10.17):

$$
\begin{aligned}
IsDescendantOf = \\
\{(Zoe, Bruce), (Gina, Zoe), (Bruce, Annabel), (Dirk, Kay), \\
(Kay, Don), (Annabel, Kay), (Zoe, Annabel), (Gina, Bruce), \\
(Bruce, Kay), (Dirk, Don), (Annabel, Don), (Zoe, Kay), \\
(Gina, Annabel), (Bruce, Don), (Zoe, Don), (Gina, Kay), \\
(Gina, Don)\}
\end{aligned}
$$

The calculations we have just gone through suggest an algorithm for calculating the transitive closure of any relation: calculate all the powers R^1, R^2, and so on, up to R^n, and take their union. There are n nodes in the digraph, so the longest possible path (ignoring cycles) must be no more than $n-1$ elements long. The transitive closure must provide a short cut for each path, which is why we must include a power of the relation for each possible path length. This leads also to a way to define the transitive closure formally:

Definition 48. Let A be a set of n elements, and let $R :: A \times A$ be a binary relation over A. The *transitive closure* of R is defined as

$$t(R) = \bigcup_{i=1}^{n} R^i.$$

For example, if a set A has four elements, then the transitive closure of a relation $R :: A \times A$ would be

$$R^1 \cup R^2 \cup R^3 \cup R^4.$$

The *IsDescendantOf* relation is the union of as many powers of the *IsChildOf* relation as there are people in the *IsChildOf* relation's domain.

Example 62. Using the definition, we calculate the transitive closure of $R = \{(1,2),(2,3),(3,2),(3,4),(4,4)\}$. There are four elements in the set over which the relation is defined (we haven't specified otherwise), so we shall need to calculate the union of the relations R, R^2, R^3, and R^4 (Figure 10.18). First, we calculate:

$$
\begin{aligned}
R^2 &= \{(1,3),(2,2),(2,4),(3,3),(3,4),(4,4)\} \\
R^3 &= \{(1,2),(1,4),(2,3),(2,4),(3,2),(3,4),(4,4)\} \\
R^4 &= \{(1,3),(1,4),(2,2),(2,4),(3,3),(3,4),(4,4)\}
\end{aligned}
$$

The union of all of these relations is the transitive closure of R:

$$\{(1,2),(1,3),(1,4),(2,2),(2,3),(2,4),(3,2),(3,3),(3,4),(4,4)\}$$

The following function calculates the transitive closure of a relation, using the definition:

```
transitiveClosure ::
    (Eq a, Show a) => Digraph a -> Digraph a
```

Exercise 29. Work out the following transitive closures by hand, and then evaluate them using the computer:

```
transitiveClosure ([1,2,3],[(1,2),(2,3)])
transitiveClosure ([1,2,3],[(1,2),(2,1)])
```

Exercise 30. Given a digraph $(\{1,2,3,4\},\{(1,2)\})$, what can we do to speed up the transitive closure algorithm, which requires that we take as many powers of this relation as there are nodes in the digraph?

Exercise 31. Find the transitive symmetric closure and the symmetric transitive closure of the following relations:

(a) $\{(a,b),(a,c)\}$

(b) $\{(a,b)\}$

(c) $\{(1,1),(1,2),(1,3),(2,3)\}$

(d) $\{(1,2),(2,1),(1,3)\}$

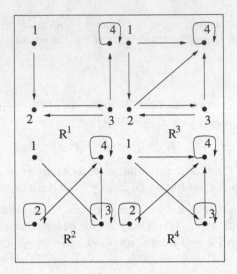

Figure 10.18: The Powers of $\{(1,2),(2,3),(3,2),(3,4),(4,4)\}$

10.8   Order Relations

An *order relation* specifies an ordering that can be used to create a sequence from the elements of its domain. Order relations are extremely important in computing, because data values often need to be placed in a well-defined sequence for processing. The standard mathematical relations less-than $(<)$ and less-than-or-equal $(\leq)$ are examples of order relations, but there are many more.

One of the most fundamental properties of an order relation is transitivity: if a precedes b in the ordering, and b precedes c, then we surely want a also to precede c. However, some of the other properties of relations may be present or absent, so there are several different kinds of order relation. We will examine these in turn, starting with partial orders.

10.8.1   Partial Order

A *partial order* puts at least some of the elements in its domain into sequence, but not necessarily all of them. There could also be several sequences within a partial order, without any ordering between elements belonging to different subsequences.

Example 63. Suppose that a database of people contains records that specify the breed of dog owned—for those people who have a dog. The records of dog owners could be ordered alphabetically using the breed name, producing a sequence of dog owners. However, this ordering would not include the people

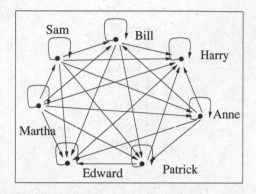

Figure 10.19: The *IsYoungerOrSameAgeAs* Partial Order

who don't have dogs, so it is only a *partial* order. Of course, it might happen that everyone (or no one) owns a dog, in which case we would still technically have a partial order. That is, it is possible that the entire partial order is sorted using some ordering; the point is just that this is not required.

Example 64. Consider the problem of ordering all the records in the database by people's names. Some names are common, so there might be more than one record per name. Therefore this is a partial order.

Example 65. We are programming with a data structure that contains ordered pairs (x, y), and we define an ordering such that the pair (x_1, y_1) precedes the pair (x_2, y_2) if $x_1 \leq x_2 \wedge y_1 \leq y_2$. This is a partial order, because it doesn't specify the ordering between $(1, 4)$ and $(2, 3)$.

The formal definition of partial orders is stated using the properties of relations that we have already defined:

Definition 49. A binary relation R over a set A is a *partial order* if it is reflexive, antisymmetric, and transitive.

Example 66. Suppose that we used age to order our database records. Our *IsYoungerOrSameAgeAs* relation is reflexive, antisymmetric, and transitive, so it is a partial order. Figure 10.19 gives its digraph.

Poset diagrams

The purpose of drawing the graph diagram for a relation is to make it easier to understand. It often defeats the purpose to include all of the relation's arcs in the diagram, as there are so many of them. For a partial order, there is no point in drawing the reflexive and transitive arcs, because we know they must be there anyway and they clutter the diagram and make it hard to see the important arcs that tell us about the ordering.

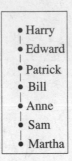

Figure 10.20: A Poset Diagram of the *IsYoungerOrSameAgeAs* Relation

A *poset* (partially ordered set) diagram is a relation diagram for partial orders, where the distracting transitive and reflexive arcs are omitted. It is important to state explicitly that the diagram shows a partial order (or a poset); without knowing this fact, a reader would not know that the relation also contains the reflexive and transitive arcs.

When drawing a poset diagram, we position it so that all of the arcs point upwards. All of the arcs of the partial order that are *not* implied by the reflexive or transitive properties must be drawn explicitly. We remove the reflexive and transitive arcs and the arrowheads of the remaining arcs.

If there is a directed path between two nodes in a poset diagram, then those nodes are comparable. An element of a partial order may be comparable to some of the elements but not to the others.

Example 67. We redraw Figure 10.19 so that it is a poset diagram (Figure 10.20). Now, it is easy to see the record ordering.

Weakest and Greatest Elements of a Poset

The following definitions give the standard terminology used to describe how two elements of a partial order are related to each other:

Definition 50. If there is a directed path from x to y in a partial order (i.e., if x precedes y in the partial order), then x *is weaker than* y. The mathematical notation for this $x \sqsubseteq y$. If $x \sqsubseteq y$ is false, then we write $x \not\sqsubseteq y$.

Definition 51. Two nodes x and y in a partial order are *incomparable* if $x \not\sqsubseteq y \wedge y \not\sqsubseteq x$. That is, x and y are incomparable if there is no directed path from x to y, and there is also no directed path from y to x.

In a finite set of numbers, there must be a unique smallest element and a unique greatest element. However, a poset might have *several* least elements. For example, if x and y are incomparable, but they are both weaker than all

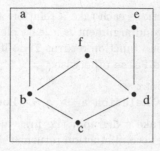

Figure 10.21: A Poset Diagram

the other elements of the poset, then both are least elements. Similarly, there may be several greatest elements. The following definitions define the sets of least and greatest elements formally:

Definition 52. The set of *least elements* of a poset P is

$$\{x \in P \mid \forall y \in P. \ (x \sqsubseteq y \lor (x \not\sqsubseteq y \land y \not\sqsubseteq x))\}.$$

That is, the least elements of P are the elements that are either incomparable to or weaker than any other element.

Definition 53. The set of *greatest elements* of a poset P is

$$\{x \in P \mid \forall y \in P. \ (y \sqsubseteq x \lor (x \not\sqsubseteq y \land y \not\sqsubseteq x))\}.$$

That is, the greatest elements of P are the elements that are either incomparable to or greater than any other element.

Example 68. Suppose that a family in our database has the following children: *Ray*, aged 17, *Tom*, aged 6, and the twins *Belle* and *Eunice*, aged 5. We can define a partial order ($\sqsubseteq$) based on age, such that $x \sqsubseteq y$ if x is younger than or the same age as y. Even though there are twins, there are no incomparable elements of this poset, since *Belle* $\sqsubseteq$ *Eunice* and also *Eunice* $\sqsubseteq$ *Belle*. However, both of the twins satisfy the requirements for the least element. The set of greatest nodes is $\{Ray\}$ and the set of weakest nodes is $\{Belle, Eunice\}$.

Example 69. Figure 10.21 shows a poset where the set of weakest elements is $\{c\}$, and the set of greatest elements is $\{a, f, e\}$.

The following Haskell function, defined in the software tools file, takes a digraph and returns **True** if the digraph represents a partial order and **False** otherwise.

```
isPartialOrder ::
  (Eq a, Show a) => Digraph a -> Bool
```

The following two functions each take a relation and an element. The first
one returns `True` if the second argument is a least element in the relation, and
`False` otherwise. The second function returns `True` if the element is a greatest
element in the relation and `False` otherwise.

```
isWeakest, isGreatest ::
  (Eq a, Show a) => Relation a -> a -> Bool
```

These functions each take a digraph; the first function returns the set of
weakest elements while the second function returns the set of greatest elements:

```
weakestSet ::   (Eq a, Show a) => Digraph a -> Set a
greatestSet :: (Eq a, Show a) => Digraph a -> Set a
```

Exercise 32. Work out by hand whether the following digraphs are partial
orders, and then check your results using the computer:

```
isPartialOrder ([1,2,3], [(1,2),(2,3)])
isPartialOrder
  ([1,2,3], [(1,2),(2,3),(1,3),(1,1),(2,2),(3,3)])
```

Exercise 33. Calculate the following by hand, and then evaluate using the
computer:

```
isWeakest [(1,2),(2,3),(1,3),(1,1),(2,2),(3,3)] 2
isWeakest [(1,2),(1,3),(1,1),(2,2),(3,3)] 3

isGreatest [(1,2),(2,3),(1,3),(1,1),(2,2),(3,3)] 3
isGreatest [(1,2),(1,3),(1,1),(2,2),(3,3)] 1
```

Exercise 34. Calculate the following by hand, and then evaluate using the
computer:

```
weakestSet ([1,2,3,4],
             [(1,4),(1,3),(1,2),(1,1),
              (2,3),(2,4),(2,2),(3,4),
              (3,3),(4,4)])
weakestSet ([1,2,3,4],
             [(1,4),(1,2),(1,1),(2,4),
              (2,2),(3,4),(3,3),(4,4)])
greatestSet ([1,2,3,4],
             [(1,2),(3,4),(1,1),(2,2),(3,3),(4,4)])
greatestSet ([1,2,3,4],
             [(2,3),(3,4),(2,4),(1,1),(2,2),(3,3),(4,4)])
```

Exercise 35. What are the greatest and weakest elements in a poset diagram
that contains the following arcs:

(a) $\{(a,b),(a,c)\}$

(b) $\{(a,b),(c,d)\}$

(c) $\{(a,b),(a,d),(b,c)\}$

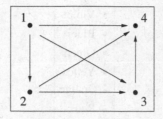

Figure 10.22: A Quasi Order

10.8.2   Quasi Order

A *quasi order* is similar to a partial order, except that it is irreflexive:

Definition 54. A binary relation R over a set A is a *quasi order* if it is irreflexive and transitive.

Example 70. The relation $(<)$ on numbers is a quasi order, but $(\le)$ is not.

Notice that the definition of a quasi order doesn't mention symmetry. Can a quasi order be symmetric? Suppose there are two elements x and y, such that $x \sqsubseteq y$. If the quasi order were symmetric, then we would also have $y \sqsubseteq x$, and since it is also transitive, we then have $x \sqsubseteq x$, which violates the requirement that a quasi order be irreflexive. This argument would not apply, of course, in a trivial quasi order where no two elements are related by $\sqsubseteq$, but non-trivial quasi orders cannot be symmetric.

We should also inquire whether a quasi order can be (or must be) antisymmetric. By definition, it is antisymmetric if $x \sqsubseteq y \wedge y \sqsubseteq x \to x = y$ for any two elements x and y. Now, if we choose x and y to be the same, then $x \sqsubseteq y \wedge y \sqsubseteq x$ is false, because the quasi order is irreflexive. This means the logical implication is vacuously true. If we choose x and y to be different, then $x \sqsubseteq y \wedge y \sqsubseteq x$ is again false (as we have just shown while discussing symmetry). In all cases, therefore, the definition of antisymmetry is satisfied, but vacuously.

The conclusion is that quasi orders may be symmetric, but only if they are trivial, and they are always antisymmetric, but only because they satisfy the definition vacuously. The properties of symmetry and antisymmetry are uninteresting for quasi orders.

Example 71. Figure 10.22 gives the graph diagram for the quasi order $(<)$ on the set $\{1, 2, 3, 4\}$.

The following function takes a digraph and returns **True** if the relation it represents is a quasi order, and **False** otherwise:

```
isQuasiOrder ::
  (Eq a, Show a) => Digraph a -> Bool
```

Figure 10.23: A Chain of the Rainbow Colours

Exercise 36. Work out the following expressions, and evaluate them with the computer:

```
isQuasiOrder ([1,2,3,4],[(1,2),(2,3),(3,4)])
isQuasiOrder ([1,2,3,4],[(1,2)])
```

10.8.3   Linear Order

A *linear order* or *total order* is like a partial order, except that it requires that *all* of the relation's elements must be related to each other.

Example 72. The ($\leq$) and ($\geq$) relations on real numbers are total orders: any two numbers x and y can be compared with each other, and it is guaranteed that either $x \leq y$ or $y \leq x$ will be true (and both are true if $x = y$).

Definition 55. A *linear order* is a partial order defined over a set A in which for each element a and b in A, either $a \sqsubseteq b$ or $b \sqsubseteq a$.

Example 73. Suppose that the database recorded the exact time at which each child was born. We could then use a form of $\leq$ to order the children within the families. This information could be useful in a study of the influence of primogeniture on the medical history of an aristocracy.

The elements of a linear order can be said to form a *chain*. When we draw the graph diagram for a chain, we omit the arcs that are implied by transitivity and reflexivity. Without these extra arcs, and because no element can be incomparable to the others, the diagram looks like a real chain. For example, the colours of the rainbow are often given as a chain starting with *Red* and ending with *Violet*. As *Red* light has the longest wavelength and *Violet* the shortest, the relation that imposes this chain ordering on the set of six colours is the $\leq$ relation on the wave frequency (Figure 10.23).

The *isLinearOrder* function takes a digraph and returns **True** if it represents a linear order, and **False** otherwise.

```
isLinearOrder :: Eq a => Digraph a -> Bool
```

Exercise 37. Evaluate the following expressions, by hand and using the computer:

```
isLinearOrder
    ([1,2,3],[(1,2),(2,3),(1,3),(1,1),(2,2),(3,3)])
isLinearOrder
    ([1,2,3],[(1,2),(1,3),(1,1),(2,2),(3,3)])
```

10.8.4 Well Order

A *well order* is a total (or linear) order that has a least element; furthermore, every subset of a well order must have a least element.

The existence of a least element is significant because it provides a base case for recursive functions and for inductive proofs. Note that any total order that has a finite number of elements must have a least element. Some total orders with an infinite number of elements have a least element, and others do not.

Example 74. The ($\leq$) relation on the set $N = \{0, 1, 2, \ldots\}$ of natural numbers is a total order. Furthermore, N has a least element, because $\forall x \in N.\ 0 \leq x$. Therefore ($\leq$) on N is a well order.

Example 75. The ($\leq$) relation on the set $Z = \{\ldots, -2, -1, 0, 1, 2, \ldots\}$ of integers is a total order. However, Z does not have a least element, because

$$\forall x \in Z. \exists y \in Z. (y \leq x \land y \neq x).$$

Therefore ($\leq$) on Z is only a total order, and not a well order.

Definition 56. Given a set S and a binary relation R over S, R is a *well order* if R is a linear order and every subset of S that is not empty contains a least element.

Well orders are important because they support induction, and they are countable. Informally, a countable set is a set in which an arbitrary item can eventually be processed by a computer. The set could be infinite: for example, the set of natural numbers is infinite, but *every* element of that set would eventually be reached if we just work on 0, 1, 2, ... in sequence. For example, if a computer started printing the natural numbers, we would eventually see the number 4058000023. However, if it started printing an uncountable set such as the irrational numbers, then it might get stuck printing an infinite number of irrationals *without reaching* the number we are interested in.

Example 76. In our database, each record is given a numeric key that is unique. As there are a finite number of keys in the database, the $\leq$ relation over these keys is a well order.

Exercise 38. We have been watching a computer terminal. Is the order in which people come and use the terminal a total order?

Exercise 39. Is it always possible to count the elements of a linear order?

Exercise 40. Can a set that is not a well order be countable?

10.8.5   Topological Sort

Computers are good at doing one task at a time, in sequence. When an algorithm is working on a data structure, it needs to know which element of the data structure to work on next. Often there is an order relation that must be followed (for example, we might want to output the items in a database in alphabetical order). If we have a total order on the elements of the data structure, the algorithm can use that to find the next piece of work. If, however, we have only a partial order on the data items, then there are several possible orders in which items could be processed while still respecting the order relation. Often we don't care *which* order is used—we just want the algorithm to find one and proceed with the work.

The process of taking a partial order and putting its elements into a total order is called *topological sorting*.

Example 77. Some compilers analyse the order in which procedures call each other. Such a compiler could construct a 'dependency graph' for the program it is translating, where each node corresponds to a procedure, and arcs in the graph correspond to procedure calls. The dependency graph is a partial order. Now, suppose the compiler generates object code for the procedures in the order of their appearance in the call graph, so that the lowest-level procedures are processed first and the highest level ones are done last, in order to make as many procedure calls as possible into forward references. The compiler uses a topological sort to produce the total order in which it prints the information. The first name in the total order will be a procedure that doesn't call any other procedure, while the last is the top-level procedure with which the program starts execution.

There is a simple and general algorithm for topological sorting. Choose x, one of the elements that is greatest in the set A, and make it the first in the sequence. Now do the same for the set $A - \{x\}$, and continue until A is empty.

Example 78. Suppose that we have a relation that expresses the call graph:

{('A','B'),('B','B'),('B','C'),('B','D'),('C','D')}

What would the topological sort of this graph be? First, the functions that call no other function would appear, followed by the functions that call them, followed by the functions calling *them*, etc. The result would be 'D', 'C', 'B', 'A' (Figure 10.24).

Example 79. What is the topological sort of $\{(1,2),(1,3)\}$ given the nodes 1,2,3? The sequence 3,2,1 or 2,3,1.

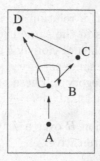

Figure 10.24: A Call Graph

The following function takes a digraph and returns a topological sort of its relation.

```
topsort ::
  (Eq a, Show a) => Digraph a -> Set a
```

Exercise 41. Check to see that the following partial orders are not, in fact, total orders. Use the computer to generate a total order, using a topological sort.

```
topsort ([1,2,3,4],[(1,2),(1,3),(2,3),(1,4),(2,4),
                    (1,1),(2,2),(3,3),(4,4)])
topsort ([1,2,3],[(1,2),(1,3),(1,4),(1,1),(2,2),(3,3)])
```

10.9 Equivalence Relations

Some relations can be used to break a set up into several categories or 'partitions', where each element of the set belongs to just one of the categories. Such a relation is called an *equivalence relation*.

Example 80. In organising a personal telephone list, it is convenient to organise the set S of people's names into 26 sets corresponding to the first letter in the name. In other words, we are making a section of the telephone list for each letter of the alphabet.

Example 81. Given a genealogy database, we would expect to see many queries about membership in families. We might assume that persons a and b are in the same family if they have the same last name, or there might be some other way to define what a family is (but it needs to have the property that *every* person belongs to *exactly one* family). One common query might be 'Is a in the same family as b?'. Once the *InSameFamilyAs* relation is defined, it provides all the information needed to organise all people into families.

These examples suggest that a relation we are using in this way needs to be reflexive (everything belongs in the same category as itself), symmetric (if a is in the same category as b, then obviously b must be in the same as a), and transitive (if a and b are in the same category, and so are b and c, then a and c must also be in the same category). These observations lead to the formal definition of an equivalence relation:

Definition 57. A binary relation R over a set A is an *equivalence relation* if it is reflexive, symmetric, and transitive.

Example 82. Suppose that everyone in the database lives in a real location somewhere in the world. We can represent the world as a map, and then partition the map into small areas by using a *LivesIn SameLocationAs* relation.

The equivalence classes of a non-empty equivalence relation can be thought of as a *partition* of the set into disjoint subsets. Now we define this term formally:

Definition 58. A *partition* P of a non-empty set A is a set of non-empty subsets of A such that

- For each subset S_1 and S_2 of P, either $S_1 = S_2$ or $S_1 \cap S_2 = \emptyset$;

- $A = \bigcup_{S \in P} S$.

Example 83. For example, let's consider the set of people's last names and the relation `HasNameStartingWithSameLetterAs`. This relation divides the set into 26 subsets. There can be no overlaps between the subsets, and the set of names is the union of the subsets.

Example 84. Computer keyboards can generate several characters from most of the keys, depending on whether the Control, Shift, or Meta keys are already down. We could define a relation *IsOnSameKeyAs*, which would partition the set of ASCII characters into equivalence classes, one for each key.

A good example of an equivalence relation, which is frequently used in computing applications, comes from the mathematical *modulus* (mod) operation on integers. The expression e mod k gives the remainder produced when dividing e by k; and the value of e mod k is a number between 0 and $k - 1$. Now, every number x which is a multiple of k will have the property that x mod $k = 0$, and we can build a set of all these numbers. Similarly, there is a set of numbers that have the same remainder when divided by 1, by 2, and so on when divided by k. To do this, we define the *congruence* relation as follows:

Definition 59. Let k be a positive integer, and let a and b be integers. If there is an integer n such that

$$(a - b) = n \times k,$$

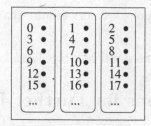

Figure 10.25: A Congruence Equivalence Relation

then a is congruent to b (modulo k). The mathematical notation for this statement is

$$a \equiv b \ (\text{mod } k).$$

It is necessary to ensure that k is a *positive* integer—positive means greater than 0—in order to avoid dividing by 0. We can define a relation, called the congruence relation, for all k:

Definition 60. The congruence relation C_k is defined for all natural k such that $k > 0$, as follows: aC_kb if and only if $a \equiv b \ (\text{mod } k)$.

The congruence relation is useful because it is an equivalence relation:

Theorem 75. For all natural $k > 0$, the congruence relation C_k is an equivalence relation.

Example 85. Consider partitioning the integers by congruence (C_3) (see Figure 10.25). This gives rise to three sets: all the integers that are of the form $n \times 3$ (this is just the set of multiples of 3), the integers of the form $n \times 3 + 1$, and the integers of the form $n \times 3 + 2$.

The following functions in the software tools module create the smallest possible equivalence relation from a digraph and determine whether a given relation is an equivalence relation. They do this by taking a digraph and calculating its transitive symmetric reflexive closure.

```
equivalenceRelation ::
  (Eq a, Show a) => Digraph a -> Digraph a
isEquivalenceRelation ::
  (Eq a, Show a) => Digraph a -> Bool
```

Exercise 42. Evaluate the following expressions using the computer:

```
equivalenceRelation ([1,2],[(1,1),(2,2),(1,2),(2,1)])
equivalenceRelation ([1,2,3],[(1,1),(2,2)])

isEquivalenceRelation ([1,2],[(1,1),(2,2),(1,2),(2,1)])
isEquivalenceRelation ([1],[])
```

Exercise 43. Does the topological sort require that the graph's relation is a partial order?

Exercise 44. Can the graph given to a topological sort have cycles?

10.10   Suggestions for Further Reading

Extensive discussions of relations can be found in *Discrete Mathematics in Computer Science*, by Stanat and McAllister [29], and *Discrete Mathematics*, by Ross and Wright [26].

Relations are a basic tool used for a wide variety of applications. Two good examples are relational data bases [7] and circuit design [19].

10.11   Review Exercises

Exercise 45. Which of the following relations is an equivalence relation?

    (a) *InTheSameRoomAs*

    (b) *IsARelativeOf*

    (c) *IsBiggerThan*

    (d) The equality relation

Exercise 46. Given a non-empty antisymmetric relation, does its transitive closure ever contain symmetric arcs?

Exercise 47. What relation is both a quasi order and an equivalence relation?

Exercise 48. Write a function that takes a relation and returns True if that relation has a power that is the given relation.

Exercise 49. A quasi order is transitive and irreflexive. Can it have any symmetric loops in it?

Exercise 50. Given an antisymmetric irreflexive relation, could its transitive closure contain reflexive arcs?

Exercise 51. Write a function that takes a relation and returns True if all of its powers have fewer arcs than it does.

Exercise 52. Write a function that takes a relation and returns True if the relation is smaller than its symmetric closure.

Exercise 53. Given the partial order

$$\{(A, B), (B, C), (A, D)\},$$

which of the following is not a topological sort?

```
[D,C,B,A]
[C,B,D,A]
[D,C,A,B]
```

Exercise 54. Is a reflexive and symmetric relation ever antisymmetric as well?

Exercise 55. Given a relation containing only a single path of length n, how many arcs can be added by its symmetric transitive closure?

Exercise 56. Given a relation containing only a cycle of length n containing all of the nodes in the domain, which power will be reflexive?

Exercise 57. Can we write a function that determines whether the equality relation over the positive integers is reflexive?

Exercise 58. Why can't partial orders have cycles of length greater than 1?

Exercise 59. Is the last power of a relation always the empty set?

Exercise 60. The following list comprehension gives the arcs of a poset diagram. What kind of order relation does the diagram represent?

```
[(a,a+1) | a <- [1..]]
```

Exercise 61. Is the composition of a relation containing only a single cycle with its converse the equality relation?

Exercise 62. Give examples of partial orders in which the set of greatest elements is the same as the set of weakest elements.

Chapter 11

Functions

A function is an abstract model of computation: you give it some input, and it produces a result. The essential aspect is that the result is completely determined by the input: if you repeatedly apply the same function to the same argument, you will always obtain the same result. Examples of functions include:

- An inquiry to a telephone directory service: you supply a person's name, and the service provides the corresponding telephone number;

- The mathematical sin function: you give it an angle, and the function returns the sin of that angle;

- An addition circuit in a computer's processor; you give it a pair of binary numbers, and it returns the binary representation of their sum.

In contrast, a weather prediction service would *not* be modelled as a function, because the answer to 'What will the weather be tomorrow?' changes from day to day.

Functions are an important tool, both in mathematics and in computing, because they provide a mechanism for abstraction. A function is a 'black box': to use it, you need to know the interface, but not the internal details about how the function is defined.

There are many ways to formalise the function concept. This chapter looks at the ones that are most important for computer science. We start with one of the most common mathematical approaches, which treats a function as a special kind of relation, and then we will consider a more algorithmic way of defining functions. It is good to remember, however, that the concept of 'function' is abstract, and there are many different formalisms that can be used to define it.

11.1   The Graph of a Function

In this section, we examine one of the most common mathematical techniques for defining functions. This approach uses a set of ordered pairs, and it brings out a close connection between functions and relations.

A function specifies, for any particular input value x, the value y of the result. This can be represented as an ordered pair (x, y), and the entire function is represented as a set of ordered pairs. This representation of a function is called a *function graph*, and it is similar to the digraph used to represent relations.

A function is a relation, but some additional properties are required. A relation digraph is a set of ordered pairs, with no restrictions: thus, a relation R might contain two ordered pairs (x, y_1) and (x, y_2), and we would write $x\,R\,y_1$ and also $x\,R\,y_2$. This would not be acceptable for a function, though, because it would mean that given the argument x there are two possible results y_1 and y_2. A function must return a unique result for any argument. Accordingly, we define a function to be a relation with an extra requirement that only one result may be specified for each argument:

Definition 61. Let A and B be sets. A function f with type $A \to B$ is a relation with domain A and codomain B, such that

$$\forall x \in A.\ \forall y_1 \in B.\ \forall y_2 \in B.\quad ((x, y_1) \in f \land (x, y_2) \in f) \to y_1 = y_2.$$

A is called the *argument type* and B is called the *result type* of the function.

The definition says that a function is a set of ordered pairs, just like a relation. The set of ordered pairs is called the *graph* of the function. If an ordered pair (x, y) is a member of the function, then $x \in A$ and $y \in B$. Furthermore, the definition states formally that if the result of applying a function to an argument x could be y_1 but it could alternatively be y_2, then it must be that $y_1 = y_2$. This is just a way of saying that there is a unique result corresponding to each argument.

Example 86. The set $\{(1, 4), (1, 5)\}$ cannot be the graph of a function, because it contains two pairs with the same first element but different second elements: $(1, 4)$ and $(1, 5)$ (Figure 11.1). A function must return just one result for any argument; it can't choose among several alternatives.

Example 87. The set of ordered pairs $\{(1, 2),\ (2, 2),\ (3, 4)\}$ is the graph of a function. It doesn't matter that several arguments, 1 and 2, produce the same result 2.

An expression denoting the result produced by a function when presented with an input x is called a *function application*. For example, $\sin(2 \times \pi)$ is an application of the sin function to the argument $2 \times \pi$. The following definition specifies the syntax, type, and value of a function application:

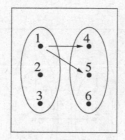

Figure 11.1: A Relation That Is Not a Function

Definition 62. An *application* of the function f to the argument x, provided that $f :: A \to B$ and $x :: A$, is written as either $f\ x$ or as $f(x)$, and its value is y if the ordered pair (x, y) is in the graph of f; otherwise the application of f to x is undefined:

$$f\ x \ = \ y \quad \leftrightarrow \quad (x, y) \in f$$

A type can be thought of as the set of possible values that a variable might have. Thus the statement '$x :: A$', which is pronounced 'x has type A', is equivalent to $x \in A$. The type of the function is written as $A \to B$, which suggests that the function takes an argument of type A and transforms it to a result of type B. This notation is a clue to an important intuition: the function is a black box (you can think of it as a machine) that turns arguments into results.

If $x \in A$ and there is a pair $(x, y) \in f$, then we say that '$f\ x$ is defined to be y'. However, if $x \in A$ but there is no pair $(x, y) \in f$, we say '$f\ x$ is undefined'. A shorthand mathematical notation for saying '$f\ x$ is undefined' is '$f\ x = \bot$', where the symbol $\bot$ denotes an undefined value.

It often turns out that some of the elements of A and B don't actually appear in any of the ordered pairs belonging to a function graph. The subset of A consisting of arguments for which the function is actually defined is called the *domain* of the function. Similarly, the subset of B consisting just of the results that actually can be returned by the function is called the *image*. These sets are defined formally as follows:

Definition 63. The domain and the image of a function f are defined as:

$$\begin{aligned} domain\ f \ &= \ \{x \mid \exists y.\ (x, y) \in f\} \\ image\ f \ &= \ \{y \mid \exists x.\ (x, y) \in f\} \end{aligned}$$

This definition says that the domain of a function is the set of all x such that (x, y) appears in its graph, while its image is the set of all y such that (x, y) appears. Thus a function must be defined for every element of its domain, and it must be able to produce every element of its image (given the right

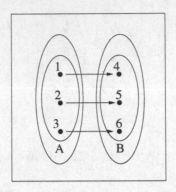

Figure 11.2: Function Type

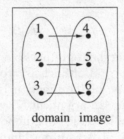

Figure 11.3: Domain and Image of a Function

argument). However, the argument type and the result type may contain extra elements that do not appear in the function graph (Figure 11.2).

Unfortunately, the terminology for functions is not entirely standard. Many authors use 'codomain' to refer to a function's argument type, but others define the codomain of a function differently. Many authors define the *range* of a function to be its image; others define it to be the result type. Whenever you are reading a document that uses any of these terms, you need to check the definitions given *in that document*.

Example 88. The set $\{(1,4),(2,5),(3,6)\}$ is the graph of a function (Figure 11.3). The domain is $\{1,2,3\}$ and the image is $\{4,5,6\}$. The function can have any type of the form $A \to B$, provided that $\{1,2,3\} \subseteq A$ and $\{4,5,6\} \subseteq B$.

Example 89. Let function $f :: Integer \to Integer$ be defined as

$$f = \{(0,1),(1,2),(2,4),(3,8)\}.$$

The argument type of f is $Integer = \{\ldots,-2,-1,0,1,2,\ldots\}$, and its domain is

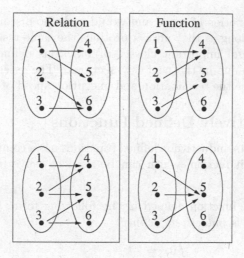

Figure 11.4: Two Relations and Two Functions

$\{0, 1, 2, 3\}$. The result type is *Integer* $= \{\dots, -2, -1, 0, 1, 2, \dots\}$, and the image is $\{1, 2, 4, 8\}$.

Example 90. Figure 11.4 shows the graph diagrams for two relations and two functions.

11.2 Functions in Programming

The 'function as a graph' idea used to model a function mathematically is not exactly the same as a function written in a programming language, although both are realisations of the same idea. The difference is that a set of ordered pairs specifies *only* what result should be produced for each input; there is no concept of an algorithm that can be used to obtain the result. The function graph approach 'pulls the result out of a hat'. In contrast, a function in a programming language is represented solely by the algorithm, and the only way to determine the value of $f\,x$ is to execute the algorithm on input x. A programming language function is a method for computing results; a mathematical function is a set of answers.

Besides providing a method for obtaining the result, a programming language function has a behaviour: it consumes memory and time in order to compute the result of an application. For example, we might write two sorting functions, one that takes very little time to run on a given test sample and one that takes a long time. We would regard them as different algorithms, and would focus on that difference as being important. The graph model of a function lacks any notion of speed and would make no distinction between the two algorithms as long as they always produce the same results.

There are several important classes of functions defined by algorithms, which we will examine in the next few sections. The essential questions we are interested in are the termination and execution speed of the function.

11.2.1   Inductively Defined Functions

As its name suggests, inductively defined functions use a computation structure that is similar to induction, which can be used to prove properties of these functions.

Definition 64. A function defined in the following form, where h is a non-recursive function, is *inductively defined*:

$$
\begin{aligned}
f\,0 &= k \\
f\,n &= h(f\,(n-1))
\end{aligned}
$$

When the argument is 0, the function returns a constant k, and when the argument n is positive, it calls itself recursively on a smaller argument $n-1$; the function can then use h to perform further calculations with the result of the recursive call.

As long as h always returns a result, an inductively defined function will always produce a result when applied to any nonnegative argument.

Example 91. The function defined below the sum $\sum_{i=0}^{n} i$, counting backwards from n down to 0. It is written in the form required for inductively defined functions, letting $k = 0$ and $h\,x = n + x$.

```
f 0 = 0
f n = n + f (n-1)
```

Example 92. The add function, defined below, is inductively defined over the second argument.

```
add :: Int -> Int -> Int
add n 0 = 0
add n k = n + add n (k - 1)
```

Example 93. The '91' function is recursive, but is not inductively defined: it calls itself recursively on a *larger* argument, and it performs yet another recursion on the result (f91 (x+11)) of the first recursion. When applied to n, this function returns 91 if $0 \le n \le 100$, and otherwise it returns $n - 10$.

```
f91 :: Int -> Int
f91 x = if x > 100
           then x - 10
           else f91 (f91 (x + 11))
```

11.2.2 Primitive Recursion

Computability theory is the branch of computer science that studies the properties of functions viewed as algorithms. One of the most important classes of algorithm is the set of *primitive recursive functions*, which are defined with a more flexible pattern than inductively defined functions. Primitive recursive functions are essentially equivalent to algorithms that can be expressed with looping structures, such as **for** loops, that are guaranteed to terminate.

Definition 65. A function f is *primitive recursive* if its definition has the following form, where g and h are primitive recursive functions.

$$
\begin{aligned}
f\,0\,x &= g\,x \\
f\,(k+1)\,x &= h\,(f\,k\,x)\,k\,x
\end{aligned}
$$

This definition specifies the standard form for a primitive recursive function. Any function that can be transformed into this form is primitive recursive, even if its definition doesn't obviously match the definition.

Example 94. The `sqr` function, which takes a natural number x and returns x^2, is primitive recursive, as shown by the following definition. This function satisfies the requirements vacuously, as it does not actually use recursion. All 'basic functions' that do not require recursion can be handled the same way, so they are all primitive recursive.

```
sqr x = f 0 x
  where f 0 x = g x
        g x = x*x
```

Example 95. The `factorial` function can be written in the standard primitive recursive form:

```
factorial k = f k undefined
  where f 0 x = 1
        f (k+1) x = (k+1) * (f k x)
```

The function `f` performs a recursion over `k`, starting from the argument to `factorial` and counting down to 0. Since factorial can be calculated simply by multiplying together all the numbers in this countdown, the `x` argument is not actually required, and `f` ignores the value of `x`. The definition of `factorial` could pick any arbitrary value for `x`, and `undefined` $= \perp$ is as good as any. Since `x` is not used, `factorial` doesn't make full use of the power of primitive recursion. The following calculation shows how the application `factorial 4` can be reduced to 24 (notice that $\perp$ is repeatedly copied but never used):

$$factorial\ 4$$
$$=\ f\ 4\ \perp$$
$$=\ 4 \times f\ 3\ \perp$$
$$=\ 4 \times (3 \times f\ 2\ \perp)$$
$$=\ 4 \times (3 \times (2 \times f\ 1\ \perp))$$
$$=\ 4 \times (3 \times (2 \times (1 \times f\ 0\ \perp)))$$
$$=\ 4 \times (3 \times (2 \times (1 \times 1)))$$
$$=\ 4 \times (3 \times (2 \times 1))$$
$$=\ 4 \times (3 \times 2)$$
$$=\ 4 \times 6$$
$$=\ 24$$

Example 96. The following function is not primitive recursive, because if the argument is odd (except for 1) it calls itself recursively with the same arguments. However, if the argument is a power of 2, then the recursion will terminate with f 1, and the function will return the logarithm (base 2) of its argument.

```
f 0 = 0
f 1 = 0
f x =
  if even x
  then 1 + f (x 'div' 2)
  else f x
```

11.2.3 Computational Complexity

The *computational complexity* of a function is a measure of how costly it is to evaluate. The memory consumption and the time required are common measures of the cost of a function.

Recursion can create some very expensive computations. A famous example is Ackermann's function:

Definition 66. Ackermann's function is

```
ack 0 y = y+1
ack x 0 = ack (x-1) 1
ack x y = ack (x-1) (ack x (y-1))
```

The ack function is easy to evaluate for small arguments, but the time it takes grows extremely quickly as x and y increase. Books on computability theory and algorithmic complexity show why this happens, but it is interesting to make a table for yourself of ack x y for small values of the arguments.

11.2.4 State

A function always returns the same result, given the same argument. This kind of repeatability is essential: if $\sqrt{4} = 2$ today, then $\sqrt{4} = 2$ also tomorrow. Some computations do not have this property. For example, many programming languages provide a 'function' that returns the current date and time of day, and the result returned from such a query will definitely be different tomorrow. The entire set of circumstances that can affect the result of a computation is called the *state*.

Example 97. The state of a computer system includes the current date and time of day, as well as the contents of the file system. Thus a 'function' that queries the date, or the amount of free space on disk, will not return the same result every time it is called. These are not true functions, although some programming languages use the keyword `function` erroneously to refer to them.

Example 98. As consumers, we expect to have to trade money for products. The interface between us and those that sell these products is a functional one. However, we also have to take into consideration things like depreciation over time, or wear and tear, or product expiration dates. These issues concern the state of the items for which we trade money.

Example 99. Some programming languages allow a 'function' to modify the value of a global variable. Even if such a 'function' always returns the same result for each argument value, its behaviour is not in principle describable by a function graph. A 'function' in a program that modifies the global state is not a mathematical function.

Notwithstanding these examples, it is possible to describe computations with state using pure mathematical functions. The idea is to include the state of the system as an extra argument to the function. For example, suppose we need a function f that takes an integer and returns an integer, but the result might also depend on the state of the computer system (perhaps the time of day, or the contents of the file system). We can handle this by defining a new type `State` that represents all the relevant aspects of the system state, and then providing the current state to f:

```
f :: State -> Int -> Int
```

Now f can return a result that depends on the time of day, even though it is a mathematical function. Given the same system state and the same argument, it will always return the same result.

Programs that need to manipulate the state can be written as pure functions, with the state made explicit and passed as an argument to each function that uses it. When a program needs to use the state frequently, however, it becomes awkward to use explicit `State` arguments; this clutters up the program, and errors in keeping track of the state can be hard to find.

Imperative languages solve this problem by making the state implicit, and allowing *side effects* to modify the state. This is a simple way to allow algorithms to use the system state. The cost of this approach is that reasoning about the program is more difficult. In mathematics, if you have an equation of the form $x = y$, you can replace x by y, or vice versa. This is called *substituting equals for equals* or *equational reasoning*. Unfortunately, equational reasoning doesn't work in general in imperative languages. If a function f depends on the system state, it is not even true that f x = f x. It is still possible to reason formally about imperative programs, for example using the *weakest precondition method*, but this is more complex than equational reasoning.

Haskell takes a different approach: it provides a mechanism allowing you to define operations that use the state implicitly. The mechanism is called a monad, and it is used with do expressions in Haskell. The technique is explained in [32].

11.3 Higher Order Functions

A distinction is often made between data (numbers, characters, etc.) and algorithms (code to be executed on a computer). For example, 23 and [1,2,3] are data values, while the `length` function is code to be executed. Many programming languages treat functions as code, and disallow their use as data. This means that the arguments and result of a function application must be data; functions themselves cannot be used as arguments or results.

In both mathematics and functional programming languages, this restriction is removed: computer code—in the form of functions—can be used as ordinary data values. This means, for example, that you can store functions in data structures. You can also pass a function like `length` to some other function, which might use it; you can also write a function that does some computation, and then produces a brand new function which it returns. Functions like this are called *higher order functions*.

Definition 67. A *first order function* has ordinary (non-function) arguments and result. A *higher order function* is one that either takes a function as an argument, or returns a function as its value, or both.

Example 100. The Haskell function `map` is a higher order function, because its first argument is a function which `map` will apply to each element of the second argument, which is a list. The type of `map` is

```
map :: (a->b) -> [a] -> [b]
```

This type reveals that the function is higher order, since the first argument has a type that contains an arrow, indicating that this argument is a function.

Example 101. The `length` function is not higher order. As its type makes plain, the argument is a list type and the result is an integer:

```
length :: [a] -> Int
```

We will now look in detail at the various kinds of higher order functions: functions that take other functions as arguments and functions that return functions as results. We will also compare two methods for allowing a function to take several arguments: building a tuple so that all the arguments are packaged in a data structure, and using higher order functions to take the arguments one at a time.

11.3.1 Functions That Take Functions as Arguments

Any function that takes another function as an argument is higher order. This kind of higher order function will have a type something like the following:

$$f :: (\cdots \to \cdots) \to \cdots$$

We have already seen many examples of such functions in Haskell; map and foldr are typical. Generally, this variety of higher order function will also take a data argument, and it will apply its function argument to its data argument in a special way.

Example 102. As we have already seen, the map function takes a data structure (which must be a list of data values) and applies its function argument to each element of the list.

```
map :: (a->b) -> [a] -> [b]
map f [] = []
map f (x:xs) = f x : map f xs
```

Example 103. The following function performs an operation twice on the data argument x. The operation to be performed is specified by the first argument f, which is a function.

```
twice :: (a -> a) -> a -> a
twice f x = f (f x)
```

Notice that the type of f is more restricted in twice than it is in map. The reason for this is that nothing in map constrains either the argument type or the result type of f, so the function can have the general function type a->b. In twice, however, the result returned by f is used as the argument to another application of f. This means that f must have the less general function type a->a.

Higher order functions provide a flexible and powerful approach to user-defined control structures. A *control structure* is a programming language construct that specifies a sequence of computations. Examples of control structures in imperative programming languages include **for** loops, **while** loops, the **if** statement, and the like.

It is often possible to define a higher order function that implements a control structure. For example, let `ys` be a list of length `n`. Then the Haskell equation

```
ys = map f xs
```

is similar to the following **for** loop (written in a common imperative style):

```
for i = 1 to n
  y[i] := f (x[i]);
```

Thus `map` describes an iteration that computes a list of values, where the ith element of the result is computed from the ith element of the argument by applying the function `f`. Similarly,

```
y = foldl f a xs
```

is similar to the following imperative loop:

```
y := a;
for i := 1 to n do
  y := f y x[i]
```

11.3.2 Functions That Return Functions

Any function that returns another function as its result is higher order, and its type will have the following form:

$$f :: \cdots \rightarrow (\cdots \rightarrow \cdots)$$

To understand this kind of higher order function, it is helpful to study the function graph in detail. First, we define some first order functions to be used in the examples:

```
ident, double, triple, quadruple :: Int -> Int

ident 1 = 1
ident 2 = 2
ident 3 = 3

double 1 = 2
double 2 = 4
double 3 = 6

triple 1 = 3
triple 2 = 6
triple 3 = 9
```

```
quadruple 1 = 4
quadruple 2 = 8
quadruple 3 = 12
```

These simple functions perform multiplication on small arguments. For example, `double` takes a number x (which must be 1, 2 or 3) and it returns $2 \times x$. The graph of a first order function is a straightforward set of ordered pairs; the functions just defined have the following graphs:

$$
\begin{aligned}
\texttt{ident} &= \{(1,1),(2,2),(3,3)\} \\
\texttt{double} &= \{(1,2),(2,4),(3,6)\} \\
\texttt{triple} &= \{(1,3),(2,6),(3,9)\} \\
\texttt{quadruple} &= \{(1,4),(2,8),(3,12)\}
\end{aligned}
$$

Now we define a higher order function, `multby`, which takes one argument of type `Int` and returns a function with type `Int->Int`:

```
multby :: Int -> (Int->Int)
multby 1 = ident
multby 2 = double
multby 3 = triple
multby 4 = quadruple
```

This function simply looks at its first argument x, an integer that must be in $\{1,2,3,4\}$, and it returns another function. Now consider the value of `multby 3 2`. This is syntactically equivalent to `(multby 3) 2`, and we can evaluate the expression using the function definitions. Notice that `multby 3` returns a function that multiplies things by 3, so `multby 3 2` evaluates to 6:

```
multby 3 2
  = (multby 3) 2      syntax rule of Haskell
  = triple 2          definition of multby (third equation)
  = 6                 definition of triple (second equation)
```

If a function returns another function as its result, then its graph will be a set of ordered pairs (x, y) where x is the argument to the function and y is another function graph. Figure 11.5 shows the graph of `multby`:

$$
\begin{aligned}
\texttt{multby} = \{ &(1, \{(1,1),(2,2),(3,3)\}) \\
&(2, \{(1,2),(2,4),(3,6)\}) \\
&(3, \{(1,3),(2,6),(3,9)\}) \\
&(4, \{(1,4),(2,8),(3,12)\})\}
\end{aligned}
$$

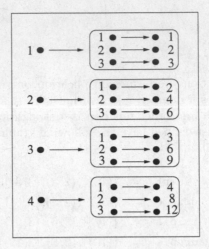

Figure 11.5: Graph of the multby Higher Order Function

11.3.3   Multiple Arguments as Tuples

Technically, a function (in either mathematics or Haskell) takes exactly one argument and returns exactly one result. There are two ways to get around this restriction. One method is to package multiple arguments (or multiple results) in a tuple. Suppose that a function needs two data values, x and y. The caller of the function can build a pair (x, y) containing these values, and that pair is now a single object which can be passed to the function.

Example 104. The following function takes two numbers and adds them together:

```
add :: (Integer,Integer) -> Integer
add (x,y) = x+y
```

The function is called using an application that builds a suitable tuple. Thus f (3,4) applies add to the pair (3,4); when (x,y) is matched with (3,4), the effect is to define x to be 3 and y to be 4 in the body of the function. The graph of the function is an infinite set, as Integer is a type with an infinite number of values. The graph has the following form:

$$
\begin{aligned}
&\text{add} = \\
&\{\cdots, \\
&\quad \cdots, \quad ((0,-2),-2), \quad ((0,-1),-1), \quad ((0,0),0), \quad ((0,1),1), \quad \cdots \\
&\quad \cdots, \quad ((1,-2),-1), \quad ((1,-1),0), \quad ((1,0),1), \quad ((1,1),2), \quad \cdots \\
&\quad \cdots, \quad ((2,-2),0), \quad ((2,-1),-1), \quad (2,0),2), \quad ((2,1),3), \quad \cdots \\
&\quad \cdots \\
&\}
\end{aligned}
$$

11.3.4 Multiple Results as a Tuple

A function must return exactly one result, but sometimes in practice we want one to return several pieces of information. In such cases, the multiple pieces can be packaged into a tuple, and the function can return that as a single result. This technique is analogous to passing several arguments as a tuple.

Example 105. The following Haskell function takes an integer x, and returns two results $x - 1$ and $x + 1$ packaged in a pair (i.e., a 2-tuple):

```
addsub1 :: Integer -> (Integer,Integer)
addsub1 x = (x-1, x+1)
```

The function graph is

```
addsub1 =
  {  ...,
     (-2,(-3,-1)),  (-1,(-2,0)),  (0,(-1,1)),  (1,(0,2)),
     (2,(1,3)),     (3,(2,4)),    (4,(3,5)),   (5,(4,6)),
     ...
  }
```

11.3.5 Multiple Arguments with Higher Order Functions

Higher order functions provide another method for passing several arguments to a function. Suppose that a function needs to receive two arguments, $x :: a$ and $y :: b$, and it will return a result of type c. The idea is to define the function with type

$$f :: a \to (b \to c).$$

Thus f takes only one argument, which has type a, and it returns a function with type $b \to c$. The result function is ready to be applied to the second argument, of type b, whereupon it will return the result with type c. This method is called *Currying*, in honour of the logician Haskell B. Curry (for whom the programming language Haskell is also named).

The graph of f is a set of ordered pairs; the first element of each pair is a data value of type a, while the second element is the function graph for the result function. That function graph contains, in effect, the information that f obtained from the first argument.

Example 106. The following function, `mult`, is similar to `add` (see example 104); apart from using * rather than +, the only difference is that `mult` is higher order, and takes its arguments one at a time:

```
mult :: Integer -> (Integer->Integer)
mult x y = x*y
```

The graph of `mult` is a set of ordered pairs of the form (k, f_k), where k is the value of the first argument to `mult`, and f_k is the graph of a function that takes a number and multiplies it by k:

```
mult =
  {···,
    (-1,  {...,  (-1, 1),   (0, 0),   (1, -1),   (2, -2),   ...}),
    (0,   {...,  (-1, 0),   (0, 0),   (1, 0),    (2, 0),    ...}),
    (1,   {...,  (-1, -1),  (0, 0),   (1, 1),    (2, 2),    ...}),
    (2,   {...,  (-1, -2),  (0, 0),   (1, 2),    (2, 4),    ...}),
    (3,   {...,  (-1, -3),  (0, 0),   (1, 3),    (2, 6),    ...}),
    ···
  }
```

11.4   Total and Partial Functions

Recall that the domain of a function $f :: A \to B$ is a subset of A consisting of all the elements of A for which f is defined. There are two sets that can be used to describe the possible arguments of f. The argument type A is generally thought of as a constraint: if you apply f to x, then it is required that $x \in A$ (alternatively, $x :: A$). If this constraint is violated then the application $f\ x$ is meaningless. The domain of f is the set of arguments for which f will produce a result. Naturally, domain f must be a subset of A. However, there is an important distinction between functions where the domain is the same as the argument type, and functions where the domain is a proper subset of the argument type.

Definition 68. Let $f :: A \to B$ be a function. If domain $f = A$ then f is a *total* function. If domain $f \subset A$ then f is a *partial* function.

If f is a partial function, and $x \in$ domain f, then we say that $f\ x$ is defined. There are several standard ways to describe an application $f\ y$ where $y :: A$ but $y \notin$ domain f. It is common, especially in mathematics, to write '$f\ y$ is undefined'. Another approach, frequently used in theoretical computer science, is to introduce a special symbol $\bot$ (pronounced bottom), which stands for an undefined value. This allows us to write $f\ y = \bot$.

Example 107. The following function has argument type `Integer`, but its domain is $\{1, 2, 3\}$:

```
f :: Integer -> Char
f 1 = 'a'
f 2 = 'b'
f 3 = 'c'
```

The expression `f 1` is defined, and its value is the character `'a'`. The expression `f 4` is undefined, and it has no value. Another way to say this is that `f 1 = 'a'` and `f 4 = ` $\bot$. The graph of `f` is $\{(1, 'a'), (2, 'b'), (3, 'c')\}$.

Partial functions are useful when describing the behaviour of programs. Generally, the type of a function can be used for compile-time analysis by the compiler, but the domain may be difficult or impossible to work out from the function's definition.

Example 108. The function `sqrt ::  Float->Float` takes the square root of its argument. The argument type is Float, and the compiler uses that constraint to detect errors in the program. Thus if a program contains an application like `sqrt "cat"` the compiler will produce an error message, because a character string is not an element of the type `Float`. However, the compiler cannot determine whether a numeric argument to `sqrt` will be positive or negative.

Example 109. The following function can be applied to any integer, but it is defined only for 1:

```
justone :: Int -> Int
justone 1 = 3
```

If this function is applied to anything that isn't an integer, the compiler will produce a type error message, and the program cannot be executed. If it is applied to 2, no error is detected at compile time but a runtime error message will be produced, for example:

```
> justone 2

Program error: {justone 2}
```

Many programming languages have the property that some type errors are not detectable by the compiler, and the application of a function to an argument of the wrong type is likely to crash the program. The Haskell type system is designed carefully so that the compiler guarantees that all type errors will be detected at compile time; it is impossible for the program to crash at runtime due to a type error. Unix programmers using C are accustomed to running a program and getting the message `segmentation fault`; this is caused by a type error (for example, if an integer value is used as an address). Such errors are rare in Haskell; even though the compiler knows nothing about the domains of functions, it is able to catch most errors just by checking their types.

If a function is applied to a value that has the right type, but that is not in the function's domain, then a runtime error occurs. Sometimes the program, or the system, is able to detect this and produce an error message. For example, if the `sqrt` function is applied to -2, an error message will be produced explaining what happened.

Some runtime error messages are generated automatically, but Haskell also allows you to implement such error messages yourself. The function `error` takes a string argument, which is an error message; if an application of `error` is evaluated, the string is printed and the program execution is terminated.

Example 110. The argument type of the following function is `Integer`, but its domain is the set of non-negative integers. Suppose that it would be a runtime error to evaluate an application of f to a negative number, but suppose also that we would like an informative error message if this happens. The `error` function can be used to terminate the execution with a tailor-made message. A common technique, illustrated here, is to construct the error message string, including pertinent information about the argument. In this case, we simply convert x to a string, with `show x`, and include that in the message.

```
f :: Integer -> Integer
f x =
  if x >= 0
    then 10*x
    else error ("f was applied to a negative number "
                ++ show x ++ ".  Don't do it again!")
```

Here are the results of evaluating two applications of f. The argument is in the domain of f in the first application, but not in the second.

```
> f (2+2)
40
> f (2-5)

Program error: f was applied to a negative
  number -3.  Don't do it again!
```

Unfortunately, it is not possible to detect all errors at runtime, and when an undetectable error occurs it is impossible to print a useful error message. Sometimes a recursive function goes into an infinite loop, and there is simply no output at all.

Example 111. The `infinite_loop` function takes an argument and calls itself with the same argument. Because it makes no progress toward a termination condition, any application of this function will run forever. The user must interrupt the execution, typically by typing CTRL-C or by clicking on Stop.

```
infinite_loop :: a -> a
infinite_loop x = infinite_loop x
```

An application of a total function will always terminate and produce a result. There are three possible outcomes from evaluating an application of a partial function to an argument: the application could terminate with a result; it could produce a runtime error message; or it could go into an infinite loop. For example, the following function will terminate if its argument is even, but it will go into an infinite loop if the argument is odd:

```
haltIfEven x =
  if even x
```

```
      then x
      else infinite_loop x
```

If we try applying `haltIfEven` to some arguments, we might quickly discover that `haltIfEven 6` $\Rightarrow$ `6`. However, when we attempt to evaluate the expression `haltIfEven 7` there will not be any output at all; the computation will just go on and on. When a computer is running a program but not producing any output, it would be useful to know whether it just needs more time, or whether it is stuck in an infinite loop. For this particular example it is obvious when the function will terminate and when it will loop forever, but what about more complicated functions where this is not obvious?

It would be extremely useful to have a function called `wouldHalt` that takes an *arbitrary* function `f` and an argument value `x`, and that returns `True` if and only if `f` would halt if we actually evaluate `f x`:

```
wouldHalt :: (Integer->Integer) -> Integer -> Bool
```

This is called the *Halting Problem*.

Obviously we cannot implement `wouldHalt` by actually evaluating `f`. Suppose we write it something like this:

```
wouldHalt :: (Integer->Integer) -> Integer -> Bool
wouldHalt f x =
   if f x == f x
      then True
      else False
```

The problem is that if `f x` goes into an infinite loop, then `wouldHalt` will never get the opportunity to return `False`. It will always return `True` if the result should be `True`, but it will go into an infinite loop if the result should be `False`.

Since `wouldHalt` cannot actually evaluate `f x`, it must instead analyse the definition of `f`, somehow figure out how it works, and then decide whether `f x` would halt. We could easily define a simplified `wouldHalt` that can handle `haltIfEven` and similar functions. Could we extend it, with enough effort, so that our function solves the Halting Problem? Unfortunately this is impossible:

Theorem 76. There does not exist a function `wouldHalt` such that for all `f` and `x`,

$$\text{wouldHalt } f\ x = \begin{cases} \text{True}, & \text{if } f\ x \text{ terminates} \\ \text{False}, & \text{if } f\ x \text{ does not terminate} \end{cases}$$

Proof. Define the function `paradox` as follows:

```
paradox :: Integer -> Integer
paradox x =
   if wouldHalt paradox x
   then paradox x
   else 1
```

Now consider the expression `paradox x`. One of the following two cases must hold:

- Suppose `paradox x` halts and produces a result. Then

$$\text{wouldHalt paradox x} \Rightarrow \text{True};$$

  therefore the definition reduces to `paradox x = paradox x`, so `paradox x` does not halt. This is a contradiction; therefore it is impossible that `paradox x` halts.

- Suppose `paradox x` does not halt. Then

$$\text{wouldHalt paradox x} \Rightarrow \text{False}.$$

  The definition then simplifies to `paradox x = 1`, and it halts.

To summarise, if `paradox x` halts then it does not halt, and if it does not halt then it halts! There is no possibility which avoids contradiction. Therefore the function `wouldHalt` does not exist.   □

The proof that the Halting Problem is unsolvable was discovered in the 1930s by Alan Turing. This is one of the earliest and most important results of computability theory. One of the commonest methods for proving that a function does *not* exist is to show how it could be used to solve the Halting Problem, which is unsolvable. This theorem also has major practical implications: it means that some software tools that would be very useful cannot actually be implemented.

A consequence of the unsolvability of the Halting Problem is that it is impossible to write a Haskell function that determines whether another Haskell function is total or partial. However, we can introduce data structures to represent the graphs of partial functions, with an explicit value that represents ⊥. The software tools file takes this approach, as described below.

The definition of a function requires that every element of the domain be mapped to some element of the result type, which can be the undefined value. This means that we need to use a new type to represent the result returned by a function, called `FunVals`. This type has two kinds of element. The first is `Undefined`, which means that the result value is ⊥. The other is called `Value`, and takes an argument that is the actual value returned by the function.

Now we can define a predicate that returns `True` if its argument is a partial function (in other words if some member of its result type is undefined), and `False` otherwise:

```
isPartialFunction
  :: (Eq a, Show a, Eq b, Show b)
  => Set a -> Set b -> Set (a,FunVals b) -> Bool
```

There is also a function `isFun` that takes a relation, and determines whether the relation is also a function:

```
isFun :: (Eq a, Show a, Eq b, Show b) =>
    Set a -> Set b -> Set (a,FunVals b) -> Bool
```

Exercise 1. Decide whether the following functions are partial or total, and then run the tests on the computer:

(a)
```
isPartialFunction
    [1,2,3] [2,3]
    [(1,Value 2),(2,Value 3),(3,Undefined)]
```

(b)
```
isPartialFunction
    [1,2] [2,3]
    [(1,Value 2),(2,Value 3)]
```

Exercise 2. Work out the following expressions, by hand and using the computer:

```
isFun [1,2,3] [1,2] [(1,Value 2),(2,Value 2)]
isFun [1,2,3] [1,2] [(1,Value 2),(2,Value 2),
                     (3,Value 2),(3,Value 1)]
isFun [1,2,3] [1,2] [(1,Value 2),
                     (2,Value 2),(3,Value 2)]
```

Exercise 3. What is the value of `mystery` x where `mystery` is defined as:

```
mystery :: Int -> Int
mystery x = if mystery x == 2 then 1 else 3
```

Exercise 4. What is the value of `mystery2` x where `mystery2` is defined as:

```
mystery2 :: Int -> Int
mystery2 x = if x == 20 then 2 + mystery2 x else 3
```

11.5 Function Composition

It is often possible to structure a computation as a sequence of function applications organised as a pipeline: the output from one function becomes the input to the next, and so on. In the simplest case, there are just two functions in the pipeline: the input x goes into the first function, g, whose output goes into the second function f (Figure 11.6).

Definition 69. Let $g :: a \to b$ and $f :: b \to c$ be functions. Then the *composition* of f with g, written $f \circ g$, is a function such that:

$$
\begin{aligned}
(f \circ g) &:: a \to c \\
(f \circ g)\, x &= f\,(g\,x)
\end{aligned}
$$

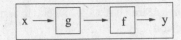

Figure 11.6: Functional Composition $(f \circ g)\, x \;=\; f\,(g\,x)$

When you think of composition as a pipeline, the input x goes into the first function g; this produces an intermediate result $g\,x$, which is the input to the second function f, and the final result is $f\,(g\,x)$. Notice, however, that in the notation $f \circ g$, the functions are written in backwards order. This may be unfortunate, but this is the standard definition of function composition, and you need to be familiar with it. Just remember that $f \circ g$ means *first* apply g, *then* f.

The $\circ$ symbol is an operator that takes two functions and produces a new function, just as $+$ is an operator that takes two numbers and produces a new number. The first argument to $\circ$ is $f :: b \to c$, and the second argument is $g :: a \to b$, and the entire composition takes an input $x :: a$ and returns a result $(f\,(g\,x)) :: c$. Therefore the $\circ$ operator has the following type:

$$(\circ) \;::\; (b \to c) \to (a \to b) \to (a \to c)$$

Haskell has a built-in operator for function composition. Since $\circ$ is unfortunately not an ASCII character, the full stop character '.' is used instead to denote function composition.

Definition 70. The Haskell function composition operator is

```
(.) :: (b -> c) -> (a -> b) -> (a -> c)
(f.g) x = f (g x)
```

Example 112. Suppose that you wish to increment the second elements of a list of pairs called lstpairs. You could write this as

```
map increment (map snd lstpairs)
```

but it is often clearer to write it as

```
map (increment.snd) lstpairs
```

This helps the reader to see that several operations are going to be applied in turn to each element of the list.

Composition is often used to define a processing pipeline, as shown in Figure 11.6, but we often need more than two functions in the pipeline. Figure 11.7 shows a typical example, where four functions are connected together using $\circ$. The following theorem states an extremely important property of the $\circ$ operator, which makes it easier to define such pipelines:

Theorem 77. Functional composition ($\circ$) is associative. That is, for all functions $h :: a \to b, g :: b \to c, f :: c \to d$,

$$f \circ (g \circ h) = (f \circ g) \circ h,$$

and both the left- and right-hand sides of the equation have type $a \to d$.

Proof. We need to prove that the two functions $f \circ (g \circ h)$ and $(f \circ g) \circ h$ are extensionally equal. To do this we need to choose an arbitrary $x :: a$, and show that both functions give the same result when applied to x. That is, we need to prove that the following equation holds:

$$(f \circ (g \circ h))\; x = ((f \circ g) \circ h)\; x.$$

This is done by straightforward equational reasoning, repeatedly using the definition of $\circ$:

$$
\begin{aligned}
&((f \circ g) \circ h)\; x \\
&= (f \circ g)\; (h\; x) \\
&= f\; (g\; (h\; x)) \\
&= f\; ((g \circ h)\; x) \\
&= (f \circ (g \circ h))\; x
\end{aligned}
$$

$\square$

The significance of this theorem is that we can consider a complete pipeline as a single black box; there is no need to structure it two functions at a time. Similarly, you can omit the redundant parentheses in the mathematical notation. The following compositions are all identical:

$$
\begin{aligned}
&(f_1 \circ f_2) \circ (f_3 \circ f_4) \\
&f_1 \circ (f_2 \circ (f_3 \circ f_4)) \\
&((f_1 \circ f_2) \circ f_3) \circ f_4 \\
&f_1 \circ ((f_2 \circ f_3) \circ f_4) \\
&(f_1 \circ (f_2 \circ f_3)) \circ f_4
\end{aligned}
$$

The parentheses in all of these expressions have no effect on the meaning of the expression, and they make the notation harder to read, so it is customary to omit them and write simply

$$f_1 \circ f_2 \circ f_3 \circ f_4.$$

Notice, however, that you must put parentheses around an expression denoting a function when you apply it to an argument. For example,

$$(f_1 \circ f_2 \circ f_3 \circ f_4)\; x$$

is the correct way to apply the pipeline to the argument x. It would be incorrect to omit the outer parentheses, as in

$$f_1 \circ f_2 \circ f_3 \circ f_4\; x,$$

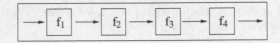

Figure 11.7: A Pipeline Defined via Functional Composition $f_4 \circ f_3 \circ f_2 \circ f_1$

because function application takes precedence over all other operations, and the meaning would be equivalent to

$$f_1 \circ f_2 \circ f_3 \circ (f_4\ x),$$

which is not what was intended. In functional programming it is common to see a pipeline of functions connected with the $\circ$ operator (written as '.' in Haskell); if this expression is applied to an argument, then there is just one pair of parentheses around the whole pipeline, but otherwise no parentheses are needed. For example, here are two Haskell equations, defining a function g and a data value y:

```
g = f1 . f2 . f3 . f4
y = (f1 . f2 . f3 . f4) x
```

The software tools contain an implementation of the following function; its two arguments are functions represented as graphs, and it returns the composition, also represented as a graph.

```
functionalComposition
  :: (Eq a, Show a, Eq b, Show b, Eq c, Show c)
  => Set (a,FunVals b) -> Set (b,FunVals c)
     -> Set (a,FunVals c)
```

Exercise 5. Work out the values of the following expressions, and then check your result by evaluating them with the computer:

```
map (increment.increment.increment) [1,2,3]
map ((+ 2).(* 2)) [1,2,3]
```

Exercise 6. Using the definitions below, work out the type and graph of f.g, and check using the computer.

```
f : Int -> String
f 1 = "cat"
f 2 = "dog"
f 3 = "mouse"

g : Char -> Int
g 'a' = 1
```

```
g 'b' = 2
g 'c' = 2
g 'd' = 3
```

Exercise 7. Functions are often composed with each other in order to form a pipeline that processes some data. What does the following expression do?

```
((map (+ 1)).(map snd)) xs
```

Exercise 8. Sometimes access to deeply nested constructor expressions is performed by function composition. What is the value of this expression?

```
(fst.snd.fst) ((1,(2,3)),4).
```

11.6 Properties of Functions

We have used four sets to characterise a function: the argument type, the domain, the result type, and the image. There are several useful properties of functions that concern these four sets, and we will examine them in this section.

11.6.1 Surjective Functions

A surjective function has an image that is the same as its result type (sometimes called the range). Thus the function can, given suitable input, produce any of the elements of the result type.

Definition 71. A function $f :: A \to B$ is *surjective* if

$$\forall b \in B. \, (\exists a \in A. \, f \, a \, = \, b).$$

Example 113. The function even :: Integer -> Bool has a result type with only two elements, True and False. Both of these values are in the function's image, as demonstrated by the applications even 2 = True and even 3 = False. Therefore every element of the result type is also an element of the image, and even is surjective.

Example 114. The times_two function takes an integer and doubles it:

```
times_two :: Int -> Int
times_two n = 2 * n
```

The image of the function is the set of even integers; this is a proper subset of the result type, so times_two is not surjective.

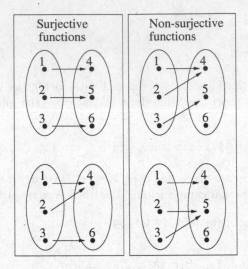

Figure 11.8: Two Surjective Functions and Two That Are Not Surjective

Example 115. The `increment` function takes an integer argument and adds 1 to it. The result type is `Integer`, and the image is also `Integer`, because every integer is 1 greater than its predecessor. Thus the image set is the same as the result type, and the function is surjective.

Example 116. Let $A = \{1, 2, 3\}$ and $B = \{4, 5\}$. Define $f :: A \to B$ as $\{(1, 4), (2, 5), (3, 4)\}$. Then f is surjective, because there is an ordered pair whose second component is 4, and the same is also true of 5.

Example 117. Figure 11.8 shows two surjective functions and two that are not surjective. The domain and image of each function is circled.

If the result type of a function is larger than its domain, then the function cannot be surjective.

Example 118. Let $A = \{2, 3\}$ and $B = \{4, 5, 6\}$. If $f :: A \to B$ then it is not surjective, since it is not possible for it to contain three ordered pairs $(x_1, 4)$, $(x_2, 5)$ and $(x_3, 6)$ such that x_1, x_2 and x_3 are all elements of A and are all distinct.

Example 119. The following Haskell functions are surjective:

```
not :: Bool -> Bool
member v :: [Int] -> Bool
increment :: Int -> Int
id :: a -> a
```

All of these functions have result types that are no larger than their domain types.

Example 120. The following Haskell functions are not surjective. The `length` function returns only zero or a positive integer, and `abs` returns the absolute value of its argument, so both functions can never return a negative number. The `times_two` function may be applied to an odd or an even number, but it always returns an even result.

```
length :: [a] -> Int
abs :: Int -> Int
times_two :: Int -> Int
```

The software tools file defines the following function, which takes a graph representation of a function and determines whether it is surjective:

```
isSurjective
  :: (Eq a, Show a, Eq b, Show b)
  => Set a -> Set b -> Set (a,FunVals b) -> Bool
```

Exercise 9. Decide whether the functions represented by the graphs in the following examples are surjective, and then check using the computer:

```
isSurjective [1,2,3] [4,5]
             [(1, Value 4), (2, Value 5), (3, Value 4)]

isSurjective [1,2,3] [4,5]
             [(1, Value 4), (2, Value 4), (3, Value 4)]
```

Exercise 10. Which of the following functions are surjective?

(a) $f :: A \to B$, where $A = \{1, 2\}$, $B = \{2, 3, 4\}$ and $f = \{(1, 2), (2, 3)\}$.

(b) $g :: C \to D$, where $C = \{1, 2, 3\}$, $D = \{1, 2\}$ and
$g = \{(1, 1), (2, 1), (3, 2)\}$.

Exercise 11. Which of the following functions are *not* surjective, and why?

(a) `map increment :: [Int] -> [Int]`

(b) `take 0 :: [a] -> [a]`

(c) `drop 0 :: [a] -> [a]`

(d) `(++) xs :: [a] -> [a]`

11.6.2 Injective Functions

The essential requirement that a relation must satisfy in order to be a function is that it maps each element of its domain to *exactly one* element of the image. An injective function has a similar property: each element of the image is the result of applying the function to *exactly one* element of the domain.

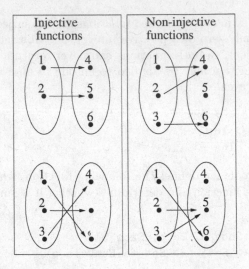

Figure 11.9: Two Injective Functions and Two That Are Not Injective

Definition 72. The function $f :: A \to B$ is *injective* if

$$\forall a, a' \in A.\ a \neq a' \to f\,a \neq f\,a'$$

Example 121. Let $A = \{1, 2\}$ and $B = \{3, 4, 5\}$. Define $f :: A \to B$ as $\{(1,3), (2,5)\}$. Then f is injective, because 3 appears in only one ordered pair, $(1,3)$, and 5 appears in only one ordered pair, $(2,5)$.

Example 122. Let $A = \{1, 2, 3\}$ and $B = \{4, 5\}$. Define $g :: A \to B$ as $\{(1,4), (2,5), (3,5)\}$. Then g is not injective because it contains the ordered pairs $(2,5)$ and $(3,5)$, which have different argument values but the same result value.

Example 123. Figure 11.9 shows four functions, two of which are injective.

Example 124. The following Haskell functions are injective:

```
(/\) True :: Bool -> Bool
increment :: Int -> Int
id :: a -> a
times_two :: Int -> Int
```

Example 125. The following Haskell functions are not injective. The `length` function is not injective, because it will map both `[1,2,3]` and `[3,2,1]` to the same number, 3. There are infinitely many other examples that would suffice to show that `length` is not injective, but you only have to give one. The function `take n` is not injective because it will map both $[a_1, a_2, \ldots, a_n, x]$ and $[a_1, a_2, \ldots, a_n, y]$ to the same result $[a_1, a_2, \ldots, a_n]$, even if $x \neq y$.

```
length :: [a] -> Int
take n :: [a] -> [a]
```

The software tools file defines the following function, which determines whether a function (specified by its graph) is injective. The first argument is the function's domain, represented as a set; the second argument is its result type, also represented as a set; and the third argument is its graph.

```
isInjective
  :: (Eq a, Show a, Eq b, Show b)
  => Set a -> Set b -> Set (a, FunVals b) -> Bool
```

Example 126. Let $A = \{1, 2, 3\}$ and $B = \{4, 5, 6\}$. Define three functions as follows:

$$
\begin{aligned}
f_1, f_2, f_3 \;\; &:: \;\; A \to B \\
f_1 \;\; &= \;\; \{(1,4), (2,6), (3,5)\} \\
f_2 \;\; &= \;\; \{(1,4), (2,4), (3,5)\} \\
f_3 \;\; &= \;\; \{(1,4), (3,5)\}
\end{aligned}
$$

We can use the software tools to explore the properties of various compositions of these functions. First we need to represent the function graphs in Haskell:

```
fun_domain = [1,2,3]
fun_codomain = [4,5,6]

fun1 = [(1, Value 4), (2, Value 6), (3, Value 5)]
fun2 = [(1, Value 4), (2, Value 4), (3, Value 5)]
fun3 = [(1, Value 4), (2, Undefined), (3, Value 5)]
```

Now we can try various experiments:

```
isInjective  fun_domain fun_codomain
   (functionalComposition fun1 fun2)
isInjective  fun_domain fun_codomain
   (functionalComposition fun1 fun3)
isInjective  fun_domain fun_codomain
   (functionalComposition fun2 fun3)
```

Exercise 12. Determine whether the functions in these examples are injective, and check your conclusions using the computer:

(a)   `isInjective [1,2,3] [2,4]`
                 `[(1,Value 2),(2,Value 4),(3,Value 2)]`

(b)   `isInjective [1,2,3] [2,3,4]`
                 `[(1,Value 2),(2,Value 4),(3,Undefined)]`

Exercise 13. Which of the following functions are injective?

    (a) $f :: A \to B$, where $A = \{1, 2\}$, $B = \{1, 2, 3\}$ and $f = \{(1, 2), (2, 3)\}$.

    (b) $g :: C \to D$, where $C = \{1, 2, 3\}$, $D = \{1, 2\}$ and $g = \{(1, 1), (2, 2)\}$.

Exercise 14. Suppose that $f :: A \to B$ and A has more elements than B. Can f be injective?

11.6.3   The Pigeonhole Principle

The Pigeonhole Principle encapsulates a common sense form of reasoning about the relationship between two finite sets. It says that if A and B are finite sets, where $|A| > |B|$ then no injection exists from A to B. In other words, since each element of the domain must be assigned a pigeonhole in the image, and the domain is bigger than the image, then there must be an element left over, *because at most one pigeon fits in a pigeonhole*. This principle is frequently used in set theory proofs, especially proofs of theorems about functions. There are a variety of ways in which to state the pigeonhole principle formally.

Theorem 78 (Pigeonhole Principle). Let A and B be finite sets, such that $|A| > |B|$ and $|A| > 1$. Let $f :: A \to B$. Then

$$\exists a_1, a_2 \in A. \ (a_1 \neq a_2) \wedge (f \, a_1 = f \, a_2).$$

11.7   Bijective Functions

Definition 73. A function is *bijective* if it is both surjective and injective. An alternative name for 'bijective' is *one-to-one and onto*. A bijective function is sometimes called a *one-to-one correspondence*.

Example 127. Figure 11.10 shows some bijective functions and some that are not bijective.

    The domain and image of a bijective function must have the same number of elements. This is stated formally in the following theorem:

Theorem 79. Let $f :: A \to B$ be a bijective function. Then $|\text{domain } f| = |\text{image } f|$.

Proof. Suppose that the domain A is larger than the image B. Then f cannot be injective, by the Pigeonhole Principle. Now suppose that B is larger than A. Then not every element of B can be paired with an element of A: there are too many of them, so f cannot be surjective. Thus a function is bijective only when its domain and image are the same size.   □

    A bijective function must have a domain and image that are the same size, and it must also be surjective and injective. As before, we assume that

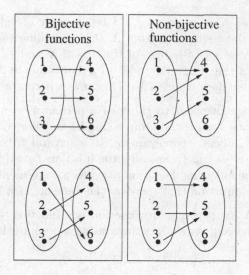

Figure 11.10: Two Bijective Functions and Two That Are Not Bijective

these functions are finite. The following function, defined in the software tools file, takes a domain, codomain, and a function, and it determines whether the function is bijective.

```
isBijective
  :: (Eq a, Show a, Eq b, Show b)
  => Set a -> Set b
  -> Set (a,FunVals b) -> Bool
```

Exercise 15. Determine whether the following functions are bijective, and check your conclusions using the computer:

```
isBijective [1,2] [3,4] [(1,Value 3),(2,Value 4)]
isBijective [1,2] [3,4] [(1,Value 3),(2,Value 3)]
```

11.7.1 Permutations

Definition 74. A *permutation* is a bijective function $f :: A \to A$; i.e., it must have the same domain and image.

Example 128. The identity function is a permutation.

The only thing a permutation function can do is to shuffle its input; it cannot produce any results that do not appear in its input.

Example 129. Let $A = \{1, 2, 3\}$ and let $f :: A \to A$ be defined by the graph $\{(1, 2), (2, 3), (3, 1)\}$. Then f is a permutation.

Example 130. Let $X = [x_1, x_2, \ldots, x_n]$ be an array of values, and let $Y = [y_1, y_2, \ldots, y_n]$ be the result of sorting X into ascending order. Define $A = \{1, 2, \ldots, n\}$ to be the set of indices of the arrays X and Y. We can define a function $f :: A \to A$ that takes the index of a data value in X and returns the location of that same data value in Y. Then f is a permutation.

Sometimes it is convenient to think of a permutation as a function that reorders the elements of a list; this is often simpler and more direct than defining a function on the indices. For example, it is natural to say that a sorting function has type $[a] \to [a]$. Technically, the function f used in Example 130 is a permutation. The following definition provides a convenient way to represent a permutation as a function that reorders the elements of a list:

Definition 75. A *list permutation function* is a function $f :: [a] \to [a]$ that takes a list of values and rearranges them using a permutation g, such that

$$xs \mathbin{!!} i = (f \; xs) \mathbin{!!} (g \; i).$$

Example 131. The functions `sort` and `reverse` are list permutation functions.

If you rearrange a list of values and then rearrange them again, you simply end up with a new rearrangement of the original list, and that could be described directly as a permutation. This idea is stated formally as follows:

Theorem 80. Let $f, g :: A \to A$ be permutations. Then their composition $f \circ g$ is also a permutation.

The following function, defined in the software tools file, determines whether a function is a permutation. The first two arguments are the domain and result type, which must be finite sets with equality, and the third argument is the function graph.

```
isPermutation
  :: (Eq a, Show a)
  => Set a -> Set a
     -> Set (a,FunVals a) -> Bool
```

Exercise 16. Let $A = \{1, 2, 3\}$ and $f :: A \to A$, where $f = \{(1,3),\ (2,1),\ (3,2)\}$. Is f bijective? Is it a permutation?

Exercise 17. Determine whether the following functions are permutations, and check using the computer:

```
isPermutation
  [1,2,3] [1,2,3]
  [(1,Value 2),(2, Value 3),(3, Undefined)]
isPermutation
  [1,2,3] [1,2,3]
  [(1,Value 2),(2, Value 3),(3, Value 1)]
```

Exercise 18. Is f, defined below, a permutation?

```
f :: Integer -> Integer
f x = x+1
```

Exercise 19. Suppose we know that the composition $f \circ g$ of the functions f and g is surjective. Show that f is surjective.

11.7.2 Inverse Functions

A function $f :: A \to B$ takes an argument $x :: A$ and gives the result $(f\ x) :: B$. The inverse of the function goes the opposite direction: given a result $y :: B$, it produces the argument $x :: A$, which would cause f to yield the result y.

Not all functions have an inverse. For example, if both $(1,5)$ and $(2,5)$ are in the graph of a function, then there is no unique argument that yields 5. Therefore the definition of inverse requires the function to be a bijection.

Definition 76. Let $f :: A \to B$ be a bijection. Then the *inverse* of f, denoted f^{-1}, has type $f^{-1} :: B \to A$, and its graph is

$$\{(y,x) \mid \exists x, y.\ (x,y) \in f\}.$$

Example 132. Let $A = \{1,2,3\}$, $B = \{4,5,6\}$ and let $f :: A \to A$ have the graph
$\{(1,4),(2,5),(3,6)\}$. Then its inverse is $f^{-1} = \{(4,1),(5,2),(6,3)\}$.

Example 133. The Haskell function `decrement` is the inverse of `increment`. Similarly, `increment` is the inverse of `decrement`.

```
increment, decrement :: Integer -> Integer
increment x = x+1
decrement x = x-1
```

Exercise 20. Suppose that $f :: A \to A$ is a permutation. What can you say about f^{-1}?

11.8 Cardinality of Sets

One of the most fundamental properties of a set is its size, and this must be defined carefully because of the subtleties of infinite sets. Bijections are a crucial tool for reasoning about the sizes of sets.

A bijection is often called a one-to-one correspondence, which is a good description: if there is a bijection $f :: A \to B$, then it is possible to associate each element of A with exactly one element of B, and vice versa. This is really what you are doing when you count a set of objects: you associate 1 with one of them, 2 with the second, and so on. When you have associated all the objects with a number, then the number n associated with the last one is the number of objects. Thus the number of objects in S is n, if there is a bijection $f :: \{1, 2, \ldots, n\} \rightarrow S$. This idea is used formally to define the size of a set, which is called its *cardinality*:

Definition 77. A set S is *finite* if and only if there is a natural number n such that there is a bijection mapping the natural numbers $\{0, 1, \ldots, n - 1\}$ to S. The *cardinality* of S is n, and it is written as $|S|$.

In other words, if S is finite then it can be counted, and the result of the count is its cardinality (i.e., the number of elements it contains).

We would also like to define what it means to say that a set is infinite. It would be meaningless to say 'a set is infinite if the number of elements is infinity', because infinity is not a natural number. We need to find a more fundamental property that distinguishes infinite sets from finite ones.

A relevant observation is that we can make a one-to-one correspondence between the set N of natural numbers and the set E of even numbers, and yet there are natural numbers that are not even.

$$0 \quad 1 \quad 2 \quad 3 \quad 4 \quad \cdots$$
$$0 \quad 2 \quad 4 \quad 6 \quad 8 \quad \cdots$$

Now, we can calculate the ith element of the second row by applying f to the ith element of the first row, where $f :: N \rightarrow E$ is defined by $f\, x = 2 \times x$. Furthermore, f is an injective function, and E is a *proper* subset of N. It would certainly be impossible to find an f with these properties for a finite set. This suggests a method for defining infinite sets:

Definition 78. A set A is *infinite* if there exists an injective function $f :: A \rightarrow B$ such that B is a proper subset of A.

We can use the properties of a function over a finite domain A and result type B to determine their relative cardinalities:

- If f is a surjection then $|A| \geq |B|$.

- If f is an injection then $|A| \leq |B|$.

- If f is a bijection then $|A| = |B|$.

Earlier we discussed counting the elements of a finite set, placing its elements in a one-to-one correspondence with the elements of $\{1, 2, \ldots, n\}$. Even though there is no natural number n which is the size of an infinite set, we can use a similar idea to define what it means to say that two sets have the same size, *even if they are infinite*:

Definition 79. Two sets A and B have the same cardinality if there is a bijection $f :: A \rightarrow B$.

Example 134. Let $A = \{1, 2, 3\}$ and $B = \{\text{cat}, \text{mouse}, \text{rabbit}\}$. Define $f ::$ $A \to B$ as

$$f = \{(1, \text{cat}), (2, \text{mouse}), (3, \text{rabbit})\}.$$

Now f is surjective and injective (you should check this), so it is a bijection. Hence the cardinality of B is

$$|B| = 3.$$

The previous example may look unduly complicated, but the point is that exactly the same technique can be used to investigate the sizes of infinite sets.

Example 135. We can place the set I of integers into one-to-one correspondence with the set N of natural numbers:

$$
\begin{array}{ccccccccc}
N = & 0 & 1 & 2 & 3 & 4 & 5 & 6 & \ldots \\
I = & 0 & -1 & 1 & -2 & 2 & -3 & 3 & \ldots
\end{array}
$$

This is done with the function $f :: I \to N$, defined as:

$$
fx = \begin{cases} 2 \times x, & \text{if } x \geq 0 \\ -2 \times x - 1, & \text{if } x < 0 \end{cases}
$$

Now f is a bijection (you should check that it is), so I has the same cardinality as N.

We have already established that the cardinality of the set of even numbers is the same as the cardinality of N, so this is also the same as the cardinality of I. The size of the set of integers is the same as the size of the set of integers that are non-negative and even!

Definition 80. A set S is *countable* if and only if there is a bijection $f :: N \to S$.

A set is countable if it has the same cardinality as the set of natural numbers. In daily life, we use the word *counting* to describe the process of enumerating a set with $1, 2, 3, \ldots$, so it is natural to call a set countable if it can be enumerated—even if the set is infinite.

Exercise 21. Explain why there cannot be a finite set that satisfies Definition 78.

Exercise 22. Suppose that your manager gave you the task of writing a program that determined whether an arbitrary set was finite or infinite. Would you accept it? Explain why or why not.

Exercise 23. Suppose that your manager asked you to write a program that decided whether a function was a bijection. How would you respond?

11.8.1   The Rational Numbers Are Countable

It turns out that some infinite sets are countable and others are not, as we will see shortly. In computer science applications, it is particularly important to know whether an infinite set is countable, since a computer performs a sequence of operations that is countable. It is often possible for a computer to print out the elements of a countable set; it will never finish printing the entire set, but any specific element will eventually be printed. However, if a set is not countable, then the computer will not even be able to ensure that a given element will eventually be printed.

A rational number is a fraction of the form x/y, where x and y are integers. We can represent a ratio as a pair of integers, the first being the numerator and the second the denominator.

Now suppose that we want to enumerate all of these ratios: we need to put them into one-to-one correspondence with N. Our goal is to create a series of columns, each of which has an index n indicating its place in the series. A column gives all possible fractions with n as the numerator.

$$
\begin{array}{lllll}
(1,1) & & & & \\
(1,2) & (2,1) & & & \\
(1,3) & (2,2) & (3,1) & & \\
(1,4) & (2,3) & (3,2) & (4,1) & \\
(1,5) & (2,4) & (3,3) & (4,2) & (5,1) \\
\ \ \vdots & \ \ \vdots & \ \ \vdots & \ \ \vdots & \ \ \vdots
\end{array}
$$

Every line in this sequence is finite, so it can be printed completely before the next line is started. Each time a line is printed, progress is made on all of the columns and a new one is added. Every ratio will eventually appear in the enumeration. Thus the set Q of rational numbers can be placed in one-to-one correspondence with N, and Q is countable.

Exercise 24. The software tools file contains a definition of a list named `rationals`, which uses the enumeration illustrated above. Try evaluating the following expressions with the computer:

```
take 3 rationals
take 15 rationals
```

11.8.2   The Real Numbers Are Uncountable

Obviously, if $A \subseteq B$ we cannot say that set B is larger than set A, since they might be equal. Surprisingly, even if we know that A is a *proper* subset of B, $A \subset B$, we *still* cannot say that B has more elements: as the previous section demonstrated, it is possible that both are infinite but countable. For example, the set of even numbers is a proper subset of the set of natural numbers, yet both sets have the same cardinality!

It turns out that some infinite sets are not countable. Such a set is so much bigger than the set of natural numbers that there is no possible way to make a one-to-one correspondence between it and the naturals. The set of real numbers has this property. In this section we will explain why, and introduce a technique called *diagonalisation* which is useful for showing that one infinite set has a larger cardinality than another.

We will not give a formal definition of the set of real numbers, but will simply consider a real number to be a string of digits. Consider just the real numbers x such that $0 \leq x < 1$; these can be written in the form $.d_0 d_1 d_2 d_3 \ldots$. There is no limit to the length of this string of digits.

Now, suppose that there is some clever way to place the set of real numbers into a one-to-one correspondence with the set of natural numbers (that is, suppose the set of reals is countable). Then we can make a table, where the ith row contains the ith real number x_i, and it contains the list of digits comprising x_i. Let us name the digits in that list $d_{i,0} d_{i,1} d_{i,2} \ldots$. Thus $d_{i,j}$ means the jth digit in the decimal representation of the ith real number x_i. Here, then, is the table which—it is alleged—contains a complete enumeration of the set of real numbers:

$$.d_{00} \quad d_{01} \quad d_{02} \quad d_{03} \quad \ldots$$
$$.d_{10} \quad d_{11} \quad d_{12} \quad d_{13} \quad \ldots$$
$$.d_{20} \quad d_{21} \quad d_{22} \quad d_{23} \quad \ldots$$
$$\ldots \quad \ldots \quad \ldots \quad \ldots \quad \ldots$$

Now we are going to show that this list is incomplete by constructing a new real number y which is definitely not in the list. This number y also has a decimal representation, which we will call $.d_{y0} d_{y1} d_{y2} \ldots$. Now we have to ensure that y is different from x_0, and it is sufficient to make the 0th digit of y (i.e., d_{y0}) different from the corresponding digit of x_0 (i.e., d_{00}). We can do this by defining a function *different* :: *Digit* $\rightarrow$ *Digit*. There are many ways to define this function; here is one:

$$\textit{different } x = \begin{cases} 0, & \text{if } x \neq 0 \\ 1, & \text{if } x = 0 \end{cases}$$

It doesn't matter exactly how *different* is defined, as long as it returns a digit that is different from its argument.

We need to ensure that y is different from x_i for *every* $i \in N$, not just for x_0. This is straightforward: just make the ith digit of y different from the ith digit of x_i:

$$d_{yi} = \textit{different } x_{ii}.$$

Now we have defined a new number y; it is real, since it is defined by a sequence of digits, and it is different from x_i for any i. Furthermore, our construction of y did not depend on knowing how the enumeration of x_i worked: for *any alleged enumeration whatsoever* of the real numbers, our construction will give a new real number that is not in that list. The conclusion, therefore, is that it is impossible to set up a one-to-one correspondence between the set R of reals and the set N of naturals. R is infinite and uncountable.

11.9   Suggestions for Further Reading

The books on set theory cited in Chapter 8 also explain the basic properties of functions.

In Section 11.4, we proved that the Halting Problem is unsolvable. This result has connections to undecidability (see the readings suggested for Chapter 6) and it is fundamental to computability theory, which is covered in many standard textbooks.

An interesting class of function is cryptography algorithms. These range from simple ciphers, which provide interesting applications of the basic properties of functions, all the way to modern public key systems. A history of the subject is given by Singh [27].

11.10   Review Exercises

Exercise 25. A program contains the expression (f.g) x.

(a) Suppose that when this is evaluated, the g function goes into an infinite loop. Does this mean that the entire expression is $\perp$?

(b) Now, suppose that the application of f goes into an infinite loop. Does this mean that the entire expression is $\perp$?

Exercise 26. Each part of this exercise is a statement that might be *correct or incorrect*. Write Haskell programs to help you experiment, so that you can find the answer.

(a) Let $f \circ g$ be a function. If f and g are surjective then $f \circ g$ is surjective.

(b) Let $f \circ g$ be a function. If f and g are injective then $f \circ g$ is injective.

(c) If $f \circ g$ is bijective then f is surjective and g is injective.

(d) If f and g are bijective then $f \circ g$ is bijective.

Exercise 27. The argument and result types given here are sets, not expressions or types in Haskell. Given the functions

```
f : {1,2,3} -> {4,5,6}
f 1 = 4
f 2 = 6
f 3 = 5


g : {4,5,6} -> {1,2,3}
g 4 = 1
```

```
g 5 = 1
g 6 = 2
```

what is

```
(g o f) 1
(g o f) 3
(f o g) 4
(f o g) 5
```

Exercise 28. State the properties of the following functions:

```
f : {3,4,5} -> {3,4,5}
f 3 = 4
f 4 = 5
f 5 = 3


g : {0,1,2} -> {0,1,2}
g 0 = 0
g 1 = 1
g 2 = 2


h : {3,4,5} -> {3,4,5}
h 4 = 3
h 5 = 4
h 3 = 5
```

Exercise 29. Given the functions

```
f : {x,y,z} -> {7,8,9,10}
f x = 8
f y = 10
f z = 7


g : {7,8,9,10} -> {x,y,z}
g 7 = x
g 8 = x
g 9 = x
g 10 = x


h : {7,8,9,10} -> {7,8,9,10}
h 7 = 10
h 8 = 7
h 9 = 8
h 10 = 9
```

describe the following functions:

```
g o f
h o f
g o h
```

Exercise 30. Given the domain and codomain $\{1,2,3,4,5\}$, which of the following are functions?

```
f 1 = 2
f 2 = 3
f 3 = 3
f 3 = 4
f 4 = 4
f 5 = 5


g 1 = 2
g 2 = 1
g 3 = 4
g 4 = 4
g 5 = 3


h 1 = 2
h 2 = 3
h 3 = 4
h 4 = 1
```

Exercise 31. Determine which of the following definitions are partial functions over the set $\{1,2,3\}$.

```
f 1 = undefined
f 2 = 1
f 3 = 2

g 1 = 3
g 2 = 2
g 3 = 1

h 1 = undefined
h 2 = undefined
h 3 = undefined
```

Exercise 32. The following functions are defined over the sets $\{1,2,3\}$ and $\{7,8,9,10\}$.

```
f 1 = 7
f 2 = 8
f 3 = 9
```

```
g 7 = 1
g 8 = 2
g 9 = 3
g 10 = 1


h 1 = 3
h 2 = 2
h 3 = 1
```

Which of the following are surjections?

```
h o h
f o g
g o f
h o f
g o h
```

Exercise 33. The functions f, g, and h are defined over the sets $\{1, 2, 3\}$ and $\{4, 5, 6\}$; which of them are injections?

```
f 1 = 4
f 2 = 5
f 3 = 5


g 4 = 1
g 5 = 2
g 6 = 3


h 4 = 1
h 5 = 1
h 6 = 1
```

Exercise 34. Consider the following functions defined over the sets $\{1, 2, 3\}$ and $\{6, 7, 8, 9\}$; which of them are bijections?

```
f 6 = 1
f 7 = 2
f 8 = 3
f 9 = 3


g 1 = 3
g 2 = 2
g 3 = 1


h 1 = 6
h 2 = 7
h 3 = 8
```

```
g o g
h o f
f o h
```

Exercise 35. Which of these functions is a partial function?

```
function1 True = False
function1 False = function1 False

function2 True = True
function2 False = True
```

Exercise 36. Using `normalForm` and `map`, write a function that takes a list of pairs and determines whether the list represents a function. You can assume in this and the following questions that the domain is the set of first elements of the pairs and the image is the set of second pair elements.

Exercise 37. Using `normalForm` and `map`, define a function `isInjection` so that it returns True if the argument represents an injective function and False otherwise.

Exercise 38. Is it possible to write a function that determines whether a list of pairs represents a surjective function *without* passing in the codomain of the function?

Exercise 39. How much information would you need to know about a Haskell function in order to be able to tell that it is not the identity function?

Exercise 40. Write a function with type

```
compare
  :: (Eq a, Eq b, Eq c, Show a, Show b, Show c)
  => (a -> b)
     -> (b -> c) -> (a -> c) -> a -> Bool
```

that takes three functions f, g, and h and determines whether `f o g = h` for some value of type a.

Exercise 41. Is this definition of `isEven` inductive?

```
isEven :: Int -> Bool
isEven 0 = True
isEven 1 = False
isEven n = isEven (n-2)
```

Exercise 42. Is this definition of `isOdd` inductive?

```
isOdd 0 = False
isOdd 1 = True
isOdd n =
   if (n < 0) then isOdd (n+2) else isOdd (n-2)
```

Exercise 43.

$A = \{1, 2, 3, 4, 5\}$
$B = \{6, 7, 8, 9, 10\}$
$D = \{7, 8, 9, 10\}$
$C = \{a, b, c, d, e\}$
$f :: A \to B, f = \{(1, 7), (2, 6), (3, 9), (4, 7), (5, 10)\}$
$g = \{(6, b), (7, a), (6, d), (8, c), (10, b)\}$

1. Is f a function? Why or why not?

2. Is f injective (that is, one-to-one)? Why or why not?

3. Is f surjective (that is, onto)? Why or why not

4. Is g a function? Why or why not?

Exercise 44. Let $A = 1, 2, \ldots, n$. Suppose $f :: A \to P(A)$, where $P(A)$ is the power set of A.

1. Prove that f is not surjective.

2. Suppose $g :: X \to Y$ and the set $(g.f)(A)$ is a subset of A (same A and same f as before), where the dot operator (.) stands for function composition. State the relationships among A, X, and Y.

3. Can g be injective? Why or why not?

4. Define f and g with the above domains and ranges such that $(f.g)$ is bijective (again, the dot operator (.) stands for function composition).

Part IV

Applications

Chapter 12

The AVL Tree Miracle

12.1 How to Find a Folder

Suppose you have a bunch of file folders (physical ones, those manila-coloured things, not folders on the computer system). Each folder contains some documents and is identified by a number on its filing tab. Every now and then someone gives you an identifying number and asks you to retrieve the corresponding folder. How much work is that going to be?

It depends on the number of file folders and the care you take in organizing them. If you have only a few folders, it doesn't matter much. You can just scatter them around on the desktop, and when you're asked to get one, look around until you find it. It might be in the first folder you pick up, or the second, and the worst that can happen is that you will have to look at all of the folders before you find the one you want. Since there aren't very many, it won't take long.

That's fine when there are only a few folders, but what if there were hundreds? Then, the frustration in finding the right folder might motivate you to rethink the arrangement. Scattering the folders around the desktop would be an unattractive way to keep track of them. You might go out and buy a four-drawer filing cabinet and arrange the folders in the cabinet in order by their identifying numbers. When somebody asks you for a particular folder, you quickly locate the folder by relying on the numbering. Much better, eh?

How much better? Say you have 100 folders and you were using the old desktop method. Sometimes you would find it on the first go, sometimes the second, sometimes the 100th. Because the folders are scattered on the desktop at random, no folder is more likely to be the one you're looking for than any other folder, so the average search time would be the average of the numbers $1, 2, \ldots 100$, which is about 50.

Now, how about the filing-cabinet method? The time it takes you here depends on how much you take advantage of the numbering, but it turns out that, no matter how you do it (unless you're extremely stupid about it), the typical search time will be much improved over the desktop method.

To get down to specifics, let's suppose you proceed by looking at the identifying number and comparing it to the number on the middle file in your four-drawer cabinet. That would be the first file in the third drawer, assuming all the drawers contain the same number of files. If the number you're looking for is bigger than the one on the middle file, then you know to look in the bottom half of the folders. If it's smaller, you look in the top half. That is, you've narrowed down the search to half as many candidate folders as you started with.

As you are no fool, you do the same thing again. That is, you compare the number on the middle folder among the remaining candidates and, for the next go, stick with the top half or the bottom half of the candidates, depending on whether the number you're looking for is smaller than or bigger than the number on the middle candidate. At each stage, you eliminate half of the folders. When you started, the folder you're looking for might have been any of the 100 folders. After the first step, you can limit your search to 50 folders. After the second step, 25 folders. Then 12, 6, 3, and 1. Of course, you might have found it somewhere along the way, but the worst that can happen is that you don't find it until you narrow it down to one folder. That takes six steps, worst case, so the average case must be at least that good, which is more than eight times better than the 50-step average required in the desktop method.

Good as that sounds, it gets even better. If you have 1,000 folders, instead of just 100, the first go knocks it down to 500 folders, then 250, 125, 62, 31, 15, 7, 3, and 1. That is, nine steps, max, for the filing cabinet method, over fifty times better than the 500-step average for the desktop method. If you have a million folders, the improvement is even more dramatic. With a million folders, the filing cabinet method has a twenty-five-thousand-fold advantage over the desktop method. It gets better and better, the more folders you have.

12.2   The Filing Cabinet Advantage

Let's work out a formula for the filing-cabinet advantage. If you have n folders, the most steps you have to make in your search is the number of times you have to successively halve n to get down to one. That number is the base-two logarithm of n. Actually, it's the integer portion of that number. You can throw the fraction away. The notation $\lfloor x \rfloor$ means the integer part of x; for example $\lfloor 3.5 \rfloor = 3$.

Here are the formulas for the typical number of steps in the two methods: desktop versus filing cabinet, as functions of the number of folders.

$$D(n) = \lfloor n/2 \rfloor \quad \textit{desktop method steps}$$
$$C(n) = \lfloor log_2 n \rfloor \quad \textit{cabinet method steps}$$

The ratio $D(n)/C(n)$ expresses the advantage of the filing-cabinet method over the desktop method. As the examples indicated, it gets better and better as n, the number of folders, gets larger. For a billion folders, there is more than a ten-million-fold advantage, not to mention the fact that the desktop would be long buried by that time, and the folders would be scattered all over the office building.

12.3 The New-File Problem

The filing-cabinet method performs an algorithm known as a **binary search**. It's a simple idea, and probably one that almost anyone would think of, faced with the problem of organizing a large number of folders for retrieval purposes. It works amazingly well, as long as the number of folders remains fixed.

However, if, on occasion, a new folder needs to be added to the collection, it can be a problem. The problem usually doesn't come up on the first few new folders. For those, there will be room to squeeze them into the drawers where they belong. The problem comes up when the number of folders keeps growing. Eventually, there is no room left in the drawer where a new folder belongs, so some of the folders in that drawer have to be moved down to another drawer to make room. What's worse, the next drawer may also be full. The problem can cascade further and further along. With a whole office building filled with cabinets for billions of folders, putting in one new folder can take a lot of heavy lifting.

In fact, with an office building filled with cabinets, it will be advantageous to keep the folders in as few cabinets as possible. If they were scattered out, and many cabinets were nearly empty, there could be a lot of walking involved to find the appropriate cabinet.

In other words, to keep down the walking, folders must be kept in the smallest number of drawers possible. This means that the drawers involved are almost always full, and whenever a new folder comes in, a bunch of folders have to be moved around to accommodate the new one.

How much moving around? As in the desktop method, it depends on where the new folder goes. If it goes at the end, no problem. Just put it at the end of the last non-empty drawer. But, if it goes at the beginning, all of the folders must be moved down one slot to accommodate the new one. Bummer!

On the average, just as with finding folders using the desktop method, about half of the folders must be moved to accommodate a new folder. Something similar happens if a folder needs to be discarded. Again, on the average, about half of the folders must be moved to keep the folders compressed into the smallest number of drawers.

12.4   The AVL Miracle

In a computer, the filing-cabinet method is normally implemented using an array to store the folders. The identifying numbers on the folders are arranged in order in the array, and to find a particular identifying number, the software looks first at the middle element of the array, then at the middle element of the narrowed ranged resulting from the first comparison of identifying numbers, and so on – that is, a binary search is performed in the array.

When a new folder comes in for the array, it must be inserted in the place required by its identifying number (so that the numbers remain in order), and that means that all of the elements beyond that point must be moved down to make room. This can be time-consuming if there are, say, billions of folders in the array.

The same thing happens in deleting a folder from the array. All the folders beyond the point of deletion must be moved up to eliminate the gap. In either case, the number of array elements to be moved is, on the average, about half of the number of folders stored in the array. That's not good, but what can be done about it? This is not an easy problem to fix. In fact, it can't be fixed with array-based methods.

It turns out that in the 1960s (ancient history in the computing business), two Russian mathematicians, Adelson-Velski and Landis, figured out a way to store folders in a tree structure so that finding a folder, adding a new folder, and removing a folder all require only about $log(n)$ steps, at most, where n is the number of folders stored in the tree. The structure these mathematicians invented, together with their insertion and deletion methods, is known as the AVL tree.

The simple part of the Adelson-Velski/Landis solution is to store the folders in a tree that facilitates binary search. Except for leaf-nodes, each node in the tree stores a folder and has a left subtree and a right subtree. (Leaf nodes just mark the boundaries of the tree. They don't store folders, and they don't have subtrees.) All of the folders in the left subtree have identifying numbers that are smaller than that of the folder in the node, itself, and all of the folders in the right subtree have identifying numbers that are larger. To find a folder, look at the root. If the folder stored at the root has the sought-for identifying number, deliver that folder. If not, look at the left subtree if the sought-for identifying number is smaller than the one in the root, and look in the right subtree if it is larger.

If a folder with the sought-for identifying number is in the tree, it will be found. If not, a leaf-node will be encountered, in which case deliver a signal indicating that the required folder is missing.

That's the simple part. The hard part has to do with insertion and deletion. The number of steps in the search depends on where the folder occurs in the tree, but at worst, the number of steps cannot exceed the number of levels in the tree. Nodes in a tree have subtrees, and the roots of subtrees have subtrees, and so on. Eventually, the desired folder is encountered, or a leaf is encountered, and the search cannot continue. The maximum number of stages in this search, for all possible search paths in the tree, is known as the *height* of the tree. AVL trees maintain a balance among the heights of subtrees, at all levels, as nodes are inserted and deleted. By maintaining balance, the AVL tree preserves a high ratio between number of folders it contains and the height of the tree.

In fact, the height of an AVL tree is approximately the base-2 logarithm of the number of folders it contains. That means every search will terminate within $log_2\ n$ steps, where n is the number of folders stored in the tree.

If you want to get a feeling for just how ingenious the AVL solution is, try to find a way to insert and delete folders into a tree that maintains order (left subtrees have folders with smaller identifying numbers, right subtrees larger) and balance (left subtrees have about the same number of nodes as right subtrees, at all levels). To match the effectiveness of AVL trees, your method will have to be able to insert or delete a folder in about $log(n)$ steps, where n is the number of folders stored in the tree. After a few hours work, you'll see why we call this method the "AVL miracle".

12.5 Search Trees and Occurrence of Keys

It's a long road from here to the complete AVL solution. As usual, the road starts with formulas and equations that make the ideas amenable to mathematical reasoning. As a first step, we define a formal representation of a tree. The AVL method will be described in terms of a Haskell data type called `SearchTree`.

To avoid unnecessary details, the definition of the `SearchTree` represents folder information generically. The folder contains any type of data, and there is a different type of search tree for each possible type of folder. The identifying number for a folder is an integer. Each node in the tree is either a leaf node (constructed by `Nub`) or an interior node (constructed by `Cel`) containing an identifying number, a folder, and two subtrees.

```
data SearchTree d = Nub  |
                    Cel Integer d (SearchTree d) (SearchTree d)
```

Figure 12.1 shows a formula for a search tree in which the folders are strings. The figure also displays a conventional diagram of the tree the formula represents. The formula is the formal representation of the tree, and the diagram is an informal presentation. This chapter will rely extensively on diagrams to illustrate ideas expressed formally in terms of formulas. To understand the chapter, you will need to develop an ability to convert between diagrams and formulas.

So far, most of the terminology has been cast in terms of the original motivating example of identifying numbers and folders of information. In the usual search tree terminology, the identifying number on which the search is based

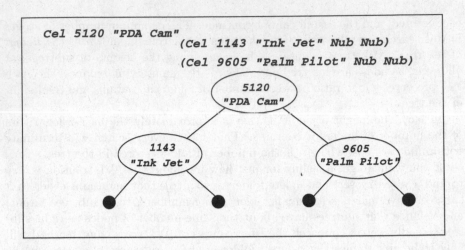

Figure 12.1: Search Tree and Corresponding Diagram

is called the key, and the information associated with the key is called, simply, the data. Most of the following discussion will describe trees in these more commonly used terms.

Formal descriptions of properties of search trees and operations on them depend on subtrees, proper subtrees, concepts of equality between trees, and the occurrence of keys in trees. The following equations provide formal definitions of these predicates. Because the concepts are similar to the ideas of subset and being an element of a set, the usual symbols for those concepts are re-used here in this new context.

The definition of search-tree equality may seem strange because it ignores the data stored in the node. This is because the key is presumed to uniquely identify the data, so if two keys are the same, the data associated with them must be the same. There is no need to compare the data. This provides a subtle advantage: the data may be of a kind for which the equality operator is not defined. For example, it may be an aggregate including some functions, for which the equality operator cannot be defined.

tree equality

$(==) :: \text{SearchTree d} \rightarrow \text{SearchTree d} \rightarrow \text{Bool}$

Nub == Nub = True	$\{N == N\}$
(Cel k d lf rt) == Nub = False	$\{C == N\}$
Nub == (Cel k d lf rt) = False	$\{N == C\}$
(Cel x a xl xr) == (Cel y b yl yr)	
= (x == y) $\wedge$ (xl == yl) $\wedge$ (xr == yr)	$\{C == C\}$

subtree

$$(\subseteq) :: \text{SearchTree d} \rightarrow \text{SearchTree d} \rightarrow \text{Bool}$$

Nub $\subseteq$ s = True $\{N \subseteq\}$

(Cel k d lf rt) $\subseteq$ Nub = False $\{C \subseteq N\}$

(Cel x a xl xr) $\subseteq$ (Cel y b yl yr)

   = ((Cel x a xl xr) == (Cel y b yl yr)) $\vee$ $\{C \subseteq C\}$

      ((Cel x a xl xr) $\subseteq$ yl) $\vee$ ((Cel x a xl xr) $\subseteq$ yr)

proper subtree

$$(\subset) :: \text{SearchTree d} \rightarrow \text{SearchTree d} \rightarrow \text{Bool}$$

s $\subset$ Nub = False $\{\subset N\}$

s $\subset$ (Cel k d lf rt) = (s $\subseteq$ lf) $\vee$ (s $\subseteq$ rt) $\{\subset C\}$

key occurs in tree

$$(\in) :: \text{Integer} \rightarrow \text{SearchTree d} \rightarrow \text{Bool}$$

k $\in$ Nub = False $\{\in N\}$

k $\in$ (Cel x a xl xr) = (k == x) $\vee$ (k $\in$ xl) $\vee$ (k $\in$ xr) $\{\in C\}$

12.5.1 Ordered Search Trees and Tree Induction

A search tree is **ordered** if the key in each non-leaf node is greater than all the keys that occur in the left subtree of the node and is less than all the keys that occur in the right subtree. A leaf is ordered by default. That is, the predicate *ordered*, on the domain of discourse consisting of all search trees, satisfies the following equations.

ordered search tree

ordered (Nub) = True $\{ord\ N\}$

ordered (Cel k d lf rt)

   = $(\forall x \in \text{lf}.\ x < k) \wedge (\forall y \in \text{rt}.\ y > k) \wedge$ $\{ord\ C\}$

      ordered (lf) $\wedge$ *ordered* (rt)

From the definition of the predicate ordered, it's not a big step to guess that an ordered search tree cannot contain duplicate keys. However, saying exactly what that means turns out to be tricky. One approach is to define a function that extracts from a search tree a sequence containing all the data elements in the tree that are associated with a given key, and then to prove that the sequence contains exactly one element if the key occurs in an ordered tree. If the key doesn't occur in the tree, the sequence is empty.

sequence of matching keys

```
dataElems :: SearchTree d -> Integer -> [d]
dataElems Nub x = []                               {dataElems N}
dataElems (Cel k d lf rt) x                        {dataElems C}
  = if k == x
      then (dataElems lf x) ++ [d] ++ (dataElems rt x)
      else (dataElems lf x) ++ (dataElems rt x)
```

Theorem 81. (unique keys, part 1).
$\forall s.(ordered\ (s)\ \wedge\ k \in s)\ \rightarrow$ (length (dataElems s k) == 1)

Theorem 82. (unique keys, part 2).
$\forall s.(ordered\ (s)\ \wedge\ k \notin s)\ \rightarrow$ ((dataElems s k) == [])

We will prove Theorem 82 first, and we will use a new form of induction that we will call **tree induction**.

Principle of Tree Induction.

$$(\forall\ t.\ (\forall\ s \subset t.\ P(s))\ \rightarrow\ P(t))\ \rightarrow\ (\forall t.\ P(t))$$

Note: The domain of discourse of the for-alls is the set of all search trees.

The principle of induction derives from basic elements of set theory, and all forms of inductive proof are equivalent when taken back to this basic framework. In practice, the form of induction used in a particular proof depends on the domain of discourse. Verifying the equivalence of various inductive forms to ordinary, mathematical induction on the natural numbers would require a major digression into details of set theory.

Using the principle of tree induction, we can prove that a predicate is true for every search tree if we can prove a certain implication. The implication is this: if the predicate is true for every proper subtree of a particular, chosen tree, then it must also be true for the chosen tree. The implication must be proved for an arbitrarily chosen tree, but once this implication is proved, the principle of tree induction delivers the conclusion that the predicate is true for every search tree.

The statement of the principle of tree induction is identical to the statement of strong induction for the domain of natural numbers, except that where the relation "less than" appears in the principle of strong induction, the relation "proper subset" appears in the principle of tree induction. The two forms of induction share the implicit requirement that the predicate must be proved directly for the simplest element.

In the case of strong induction, the simplest element is zero. There are no natural numbers smaller than zero. Therefore, when the chosen element is zero, the universe of discourse for the for-all in the hypothesis of the implication $(\forall\ s \lessdot 0.\ P(s))$ is empty. A for-all over an empty universe of discourse is true by default, so for the case when the chosen element is zero, the implication to be proved is (True $\rightarrow$ $P(0)$). The hypothesis in this implication can be of no help in proving its conclusion.

The same is true for tree induction. When the chosen element is a Nub, the universe of discourse for the hypothesis of the implication to be proved is empty. That is, we must prove $((\forall\ s \subset \text{Nub}.\ P(s)) \rightarrow P(\text{Nub}))$, which is the same as $(\text{True} \rightarrow P(\text{Nub}))$. The hypothesis of this implication (namely, "true") cannot help in arriving at its conclusion (namely, $P(\text{Nub})$). We will use tree induction to prove many properties of software that operates on search trees, and a few properties of the search trees themselves. For starters, we use tree induction to prove Theorem 82.

Proof. of Theorem 82
(unique keys, part 2):
$\forall s.(ordered\ (s)\ \wedge\ \text{k} \notin s) \rightarrow ((\text{dataElems s k}) == [])$

Proof: by tree induction.

Base Case.

$(ordered\ (\text{Nub})\ \wedge\ \text{k} \notin \text{Nub}) \rightarrow ((\text{dataElems Nub k}) == [])$
$= \{ord\ N, \in N\}$
$(\text{False} \wedge \text{False}) \rightarrow ((\text{dataElems Nub k}) == [])$
$= \{\wedge\ idem\}$
$\text{False} \rightarrow ((\text{dataElems Nub k}) == [])$
$= \{\text{False} \rightarrow any = \text{True}\}$
True

Inductive Case.

First, we work with just the hypothesis of the implication we're trying to prove.

$(ordered\ (\text{Cel x a lf rt})\ \wedge\ \text{k} \notin (\text{Cel x a lf rt}))$
$= \{\in\ C\}$
$(ordered\ (\text{Cel x a lf rt})\ \wedge\ \neg(\text{x} = \text{k} \vee \text{k} \in \text{lf} \vee \text{k} \in \text{rt}))$
$= \{DM\}$
$(ordered\ (\text{Cel x a lf rt})\ \wedge\ (\text{x} \neq \text{k})\ \wedge\ (\text{k} \notin \text{lf}) \vee (\text{k} \notin \text{rt}))$

We are trying to prove that when the above formula is true, the formula in the conclusion of the theorem is also true. That is, we want to prove that the dataElems function delivers an empty sequence in this case.

$\text{dataElems (Cel x a lf rt) k}$
$= \{\text{dataElems}\ C, \text{x} \neq \text{k}\}$
$(\text{dataElems lf k}) +\!\!+ (\text{dataElems rt k})$
$= \{ord\ C, \text{k} \notin \text{lf}, induction\ hypothesis,\ applied\ twice\}$
$[] +\!\!+ []$
$= \{+\!\!+ []\}$
$[]$

The induction step in the proof occurred when we observed that with respect to the formula (dataElems lf k), the hypotheses of the theorem are true. That is, the tree lf is ordered (by the definition of ordered, since lf is a subtree of an ordered tree) and the key k does not occur in that tree. As lf is a proper subtree of the tree we started with, the principle of induction allows us to assume that the theorem is true for the tree lf. (Remember, induction doesn't require a direct proof in the inductive case. It only requires that you prove an implication whose hypothesis is that the theorem is true for every proper subtree of the one you started with.) In this case, we apply the induction hypothesis twice: once for the left subtree and again for the right subtree. □

Now, what about Theorem 81? Induction also provides the mechanism for its proof.

Proof. of Theorem 81 (unique keys, part 1), by tree induction.

Base Case.

$(ordered$ (Nub) $\land$ k $\in$ Nub) $\to$ (length (dataElems Nub k) $==$ 1)
    $= \{\in N\}$
$(ordered$ (Nub) $\land$ False) $\to$ (length (dataElems Nub k) $==$ 1)
    $= \{\land\ null\}$
False $\to$ (length (dataElems Nub k) $==$ 1)
    $= \{$False $\to\ any\ =\$ True$\}$
True

Inductive Case.

First, we work with just the hypothesis of the implication we're trying to prove.

$(ordered$ (Cel x a lf rt) $\land$ k $\in$ (Cel x a lf rt))
    $= \{\in C\}$
$(ordered$ (Cel x a lf rt) $\land$ (x $=$ k $\lor$ k $\in$ lf $\lor$ k $\in$ rt))
    $= \{\land\ over\ \lor\}$
$(ordered$ (Cel x a lf rt) $\land$ (x $=$ k)) $\lor$
$(ordered$ (Cel x a lf rt) $\land$ k $\in$ lf) $\lor$
$(ordered$ (Cel x a lf rt) $\land$ k $\in$ rt)

We are trying to prove that when the above formula is true, the formula in the conclusion of the theorem is also true. That is, we want to prove that the dataElems function delivers a sequence with exactly one element in this case. The implication we are trying to verify has the following form: $(a\ \lor\ b\ \lor\ c)\ \to\ d$, where d is the conclusion of the theorem (that is, $d\ =\$ (length (dataElems s k) $==$ 1)), and a, b, and c are the terms in the above formula. For example, $a = (ordered\ (\texttt{Cel x a lf rt})\ \wedge\ (\texttt{x = k}))$. Using the Boolean algebra of propositions, one can verify that

$$((a \vee b \vee c) \rightarrow d) = ((a \rightarrow d) \wedge (b \rightarrow d) \wedge (c \rightarrow d))$$

That is, the formula $((a \vee b \vee c) \rightarrow d)$ can be verified by proving that each of the terms, $(a \rightarrow d)$, $(b \rightarrow d)$, and $(c \rightarrow d)$, is true.

Proof of $(a \rightarrow d)$:

$(ordered\ (\texttt{Cel x a lf rt})\ \wedge\ (\texttt{x = k})) \rightarrow$

    $(\texttt{length (dataElems (Cel x a lf rt) k) == 1})$

Again, we work with the hypothesis of the implication first. Since the tree (`Cel x a lf rt`) is ordered, $\texttt{x} \notin \texttt{lf}$ and $\texttt{x} \notin \texttt{rt}$. (All the keys in the left subtree must be smaller than $\texttt{x}$, and all those in the right subtree must be larger than $\texttt{x}$. Because $\texttt{x = k}$, we conclude that $\texttt{k} \notin \texttt{lf}$ and $\texttt{k} \notin \texttt{rt}$.) These observations take us to the conclusion of the theorem by the following logic.

```
length (dataElems (Cel x a lf rt) k)
   = {dataElems C, x = k}
length ((dataElems lf k) ++ [a] ++ (dataElems rt k))
   = {Thm 82, k ∉ lf}
length ([] ++[a] ++ (dataElems rt k))
   = {Thm 82, k ∉ rt}
length ([] ++ [a] ++ [])
   = {++.1, ++[]}
length ([a])
   = {Thm len}
1
```

It turns out that the induction hypothesis was not needed for the proof of $(a \rightarrow d)$. It will be needed for the other two proofs, however.

Proof of $(b \rightarrow d)$:

$(ordered\ (\texttt{Cel x a lf rt})\ \wedge\ \texttt{k} \in \texttt{lf}) \rightarrow$
    $(\texttt{length (dataElems (Cel x a lf rt) k) == 1})$

Again, we work with the hypothesis of the implication first.

$$(ordered \text{ (Cel x a lf rt)} \land \text{k} \in \text{lf})$$
$$\rightarrow \{definition\ of\ ordered\}$$
$$(ordered \text{ (Cel x a lf rt)} \land \text{k} \in \text{lf} \land \text{k} < \text{x})$$
$$\rightarrow \{def\ ord, since\ \text{k} < \text{x} \rightarrow \text{k} \notin \text{rt}\}$$
$$(ordered \text{ (Cel x a lf rt)} \land \text{k} \in \text{lf} \land \text{k} < \text{x} \land \text{k} \notin \text{rt})$$
$$\rightarrow \{\text{dataElems } C, \text{k} < \text{x}\}$$
$$((\text{dataElems (Cel x a lf rt) k)} =$$
$$\quad (\text{dataElems lf k)} +\!\!+ (\text{dataElems rt k}))$$
$$\land \ (ordered \text{ (Cel x a lf rt)} \land \text{k} \in \text{lf} \land \text{k} \notin \text{rt})$$
$$\rightarrow \{Thm\ 82\}$$
$$((\text{dataElems (Cel x a lf rt) k)} = (\text{dataElems lf k)} +\!\!+ [])$$
$$\land \ (ordered \text{ (Cel x a lf rt)} \land \text{k} \in \text{lf} \land \text{k} \notin \text{rt})$$
$$\rightarrow \{+\!\!+ []\}$$
$$((\text{dataElems (Cel x a lf rt) k)} = (\text{dataElems lf k)})$$
$$\land \ (ordered \text{ (Cel x a lf rt)} \land \text{k} \in \text{lf} \land \text{k} \notin \text{rt})$$

Now, because lf is a subtree of (Cel x a lf rt), k ∈ lf, and lf is ordered, the induction hypothesis leads to the desired conclusion.

```
length (dataElems (Cel x a lf rt) k)
   = length (dataElems lf k)
       = 1  {induction hypothesis}
```

The proof of $(c \rightarrow d)$ is similar to the proof of $(b \rightarrow d)$, except that the induction goes down the right side of the tree instead of the left.   □

12.5.2   Retrieving Data from a Search Tree

According to Theorem 81, a key that occurs in an ordered search tree occurs exactly once. It occurs as one of the parameters of a `Cel` constructor, and that constructor will also have a data item as a parameter. Retrieving data from an ordered search tree amounts to finding the `Cel` constructor where a specified key occurs, then delivering the data item from that same `Cel` constructor.

The retrieval operation needs a way to signal whether or not the specified key is present in the tree. In our implementation, this will be done using the `Maybe` data type, which has two constructors: `Just` and `Nothing`. The `Just` constructor will be used to deliver the data item associated with the given key in the tree. For example, (`Just d`) delivers the data item `d`.

If the key is not present, the `Nothing` constructor will be used to signal that it is missing. So, if the retrieval function delivers `Nothing`, the specified key is not present in the tree.

```
getItem :: SearchTree d -> Integer -> Maybe d
getItem (Cel k d lf rt) x  =
  if x < k
     then (getItem lf x)
     else if x > k then (getItem rt x)
```

```
          else (Just d)
getItem Nub x = Nothing
```

We can conclude that getItem works properly if we can prove that whenever the key specified in its second argument is present in the tree given in its first argument, then getItem delivers the data item associated with that key and that if the key is not present, getItem delivers Nothing. We will use tree induction to prove both of these facts.

Theorem 83. $\forall s.(ordered\ (s)\ \wedge\ k \in s)\ \rightarrow\ ((\text{getItem}\ s\ k)\ =\ (\text{Just d}))$, where d is the data item parameter of the Cel constructor in s for which k is the key parameter.

Proof.

$$(k \in s)\ \rightarrow\ (s\ =\ (\text{Cel x a lf rt}))\wedge((k\ =\ x)\vee k \in \text{lf}\vee k \in \text{rt}) \qquad \{\in C\}$$

Now we are in the same situation as in the proof of Theorem 81 (not surprising, as the hypotheses of Theorem 83 are the same as those of Theorem 81). We want to prove an implication whose hypothesis is a three-way disjunction. We follow the same strategy: separate the implication into a conjunction of three implications, and prove each of them separately.

1. Proof of

   $(ordered\ (\text{Cel x a lf rt})\ \wedge\ k \in (\text{Cel x a lf rt})\ \wedge\ (k\ =\ x))\ \rightarrow$

   $((\text{getItem}\ (\text{Cel x a lf rt})\ k)\ =\ (\text{Just d}))$

   where d is the data item parameter of the Cel constructor in the tree (Cel x a lf rt) for which k is the key parameter.

   ```
   (getItem (Cel x a lf rt) k
     = {k = x}
   (getItem (Cel k a lf rt) k
     = {getItem C}
   (Just a)
   ```

   Since a is the data item parameter of the Cel constructor in the tree (Cel x a lf rt) for which k = x is the key parameter, the desired conclusion has been reached.

2. Proof of

   $(ordered\ (\text{Cel x a lf rt})\ \wedge\ k \in (\text{Cel x a lf rt})\ \wedge\ (k \in \text{lf}))\ \rightarrow$

   $((\text{getItem}\ (\text{Cel x a lf rt})\ k)\ =\ (\text{Just d}))$

   where d is the data item parameter of the Cel constructor in the tree (Cel x a lf rt) for which k is the key parameter.

($ordered$ (Cel x a lf rt) $\land$ k $\in$ (Cel x a lf rt) $\land$ (k $\in$ lf))
   = {$def\ ordered$}
($ordered$ (Cel x a lf rt) $\land$ k $\in$ (Cel x a lf rt) $\land$ (k $<$ x))
   $\rightarrow$ {getItem C}
((getItem (Cel x a lf rt) k) = (getItem lf k))
   = {$induction\ hypothesis$}
((getItem (Cel x a lf rt) k) = (Just d))

where d is the data item parameter of the Cel constructor in the tree
(Cel x a lf rt) for which k is the key parameter.

3. The proof of the third implication required by the three-part disjunction
is like this last proof, except that the induction goes down the right-hand
side of the tree instead of the left.

□

12.5.3   Search Time in the Equational Model

We can use the equations for getItem as a prescription for computation. To
compute the value represented by a formula involving getItem, we simply
scan the formula for subformulas that match the left-hand side of one of the
equations. When we find a match, we replace the subformula by the right-hand
side of the equation (with appropriate substitutions for the parameters), and
we continue this procedure until no subformulas match the left-hand side of
either equation. This procedure of repeated substitution for subformulas is the
equational model of computation.

If we want to figure out how many computational steps the equational model
of computation requires to deliver the value of a formula, we count the process
of matching a formula with an equation as a single computational step. We
also count any use of an intrinsic operator, such as logical *and* ($\land$), if-then-else
selection, the Just and Nothing constructors, and the like, as computational
steps.

Consider the computation of the formula (getItem s k). At each step, the
equations either deliver the value directly, requiring only the computational
step of matching the formula with the second equation (this occurs if s is Nub),
or the formula is replaced by an if-then-else selection. In the latter case, the
process of matching the formula with the first equation counts as one computational step, and the if-then-else selection counts as another computational
step.

In the case in which the first equation is the one that matches, the tree s
must have the form (Cel x a lf rt), and the matching subformula is replaced
by the if-then-else selection on the right-hand side of the equation. Where the
computation goes after this substitution depends on the result of the test in the
if-then-else selection. If k, the key specified in (getItem s k), is less than x (the
key in (Cel x a lf rt), which constructs the tree s), the formula (getItem s k)
is replaced by (getItem lf k), and the computation proceeds from there by

again figuring out whether that formula matches the first equation or the second equation. So, the number of computational steps will be one for matching the formula (getItem s k) to the first equation, plus one step for the if-then-else selection, plus the number of steps it takes to compute (getItem lf k).

On the other hand, if the tree s has the form (Cel x a lf rt) and k is greater than x, then the computation requires one step to match the first equation, plus one step for the if-then-else selection that finds out that (k < x) is false, plus one step for the if-then else selection that finds out that (k > x) is true, plus the number of steps required to compute (getItem rt k).

Finally, if k is the same as x, then the computation requires one step to match the first equation, plus one for the if-then-else selection that finds out that (k < x) is false, plus one for the if-then else selection that finds out that (k > x) is false, plus one to construct the result (Just a).

If we let the symbol $gSteps$ (s, k) stand for the number of computational steps required to compute the value represented by the formula (getItem s k), the above reasoning leads us to the following numerical relationships:

$$gSteps\ (\text{Nub}, k)\ =\ 1 \qquad\qquad \{gSteps\ N\}$$
$$gSteps\ (\text{Cel x a lf rt}, k)\ \leq \qquad\qquad \{gSteps\ C\}$$
$$max\ (gSteps\ (\text{lf}, k)\ +\ 2,\ gSteps\ (\text{rt}, k)\ +\ 3, 4)$$

The important point to notice is either the computation is complete in one step (if the tree is Nub), or in four steps (if k == x), or in no more than three more than the number of steps required to search for k in a proper subtree of s. Since the proper subtree is "shorter" than the tree the search started with, the number of steps remaining in the computation is reduced at each stage. Therefore, the total number of steps will be a sum with one term for each stage in the computation, and the number of stages in the computation will not exceed the "height" of the tree.

To turn this informal idea into a sound mathematical analysis, we need to say what we mean by the height of a tree. The idea is that the height of a tree is the longest route from the root through the subtrees to a Nub. That means that the height of a tree of the form (Cel k d lf rt) is one more than the height of the taller of the two subtrees, lf and rt. Here are some equations the height measurement satisfies:

```
height :: SearchTree d -> Integer
height Nub = 0                              {height N}
height(Cel kd lf rt) =
    1 + (max (height lf) (height rt))       {height C}
```

From these equations, it follows that the height of every tree is non-negative. This fact can be guessed from the observation that height is computed by adding non-negative numbers to zero and can be proved by tree induction. In the same vein, it follows that $gSteps$ (s) is non-negative. We are going to imagine that we have carried out a proof of Theorem 84.

Theorem 84. $\forall s.$ height $(s) \geq 0$

We reasoned informally that the number of computation steps required to retrieve a data item from a tree is a sum whose number of terms is the height of the tree. It turns out that each of the terms in that sum is about the same size. Therefore, the number of steps in retrieval is proportional to the height of the tree. Theorem 85 makes this notion precise. We will use tree induction to prove it.

Theorem 85. $\forall s.\ \forall k.$ gSteps $(s, k) \leq 4 *$ (height $(s) + 1$)

Proof. By tree induction.
     Base Case.

> gSteps (Nub, k)
>   = {gSteps N}
> 1
> $\leq$ {arithmetic}
> 4 * (0 + 1)
>   = {height N}
> 4 * (height (Nub) + 1)

Inductive Case.

> gSteps (Cel x a lf rt, k)
> $\leq$ {gSteps C}
> max (2 + gSteps (lf, k), 3 + gSteps (rt, k), 4)
> $\leq$ {induction hypothesis, applied twice}
> max (4 * (height (lf) + 1) + 2, 4 * (height (rt) + 1) + 3, 4)
> $\leq$ {arithmetic}
> max (4 * (height (lf) + 1) + 4, 4 * (height (rt) + 1) + 4, 4)
> $\leq$ {Thm 84, arithmetic}
> max (4 * (height (lf) + 1) + 4, 4 * (height (rt) + 1) + 4)
>   = {arithmetic}
> 4 * max ((height (lf) + 1) + 1, (height (rt) + 1) + 1)
>   = {arithmetic}
> 4 * ((max (height (lf), height (rt)) + 1) + 1)
>   = {arithmetic}
> 4 * ((1 + max (height (lf), height (rt))) + 1)
>   = {height C}
> 4 * (height (Cel x a lf rt) + 1)

$\square$

Theorem 85 guarantees that retrieval time will be short, even for trees that contain a very large number of data items, as long as the tree has a compact shape (not too tall, compared to the number of data items in the tree). In the most compact trees, the number of data items doubles every time the height increases by one. For example, a tree of height one must have the form (Cel k d Nub Nub). This tree contains exactly one data item. We can add data items to this tree by replacing its two subtrees by trees that contain a data item. The new tree would have height two and contain three data items. It would look like this:

(Cel k d (Cel x a Nub Nub) (Cel y b Nub Nub)).

Now the two subtrees, (Cel x a Nub Nub) and (Cel y b Nub Nub) can be modified in the same way to build a tree of height three with seven data items. It is possible to prove, by ordinary mathematical induction, that a tree of height n constructed by this procedure contains $2^n - 1$ data items.

Trees containing a number of data items falling between the gaps in the sequence $1, 3, 7, \ldots 2^n - 1, \ldots$ can be constructed by leaving out some of the data items on the most interior level.

Following this pattern, it is possible to put n data items in a tree of height $log_2(n)$. For trees constructed in a shape like this, the retrieval time required by the getItem function becomes relatively shorter (relative, that is, to the number of data items in the tree) as the number of data items increases. It is this fact that makes it possible for retrieval of data in trees to be way faster than retrieval of unordered data from array-like structures, such as rows of filing cabinets. For a million items in a compact tree, retrieval, according to Theorem 85, takes no more than $4 * (log_2 1,000,000 + 1)$, which is about 80 steps. Compare this to about 500,000 steps, on the average, to retrieve an item from an unordered array.

On the other hand, retrieval from data arranged in order in an array can be as fast as retrieval from a compact tree. However, the number of steps required to insert or delete an item from an ordered array is, on the average, proportional to the number of items stored in the array. If there are a million items in the array, arranged in increasing order, inserting a new one in the middle will require, on the average, moving the half-a-million-or-so items beyond the insertion point. There is room for lots of improvement in the average speed of insertion. It turns out that insertion and deletion in a search tree can be accomplished in an amount of time that increases at the same rate as the height of the tree.

In other words, insertion and deletion in search trees can be done as efficiently as retrieval. It's not easy to figure out how to do this. The rest of this chapter is devoted to explaining the AVL method of efficient insertion and deletion in search trees.

12.6 Balanced Trees

Retrieval, insertion, and deletion in a search tree can be done in an amount of time proportional to the height of the tree. Therefore, it is important to keep a lid on height as items are added or deleted from the tree. The height of a

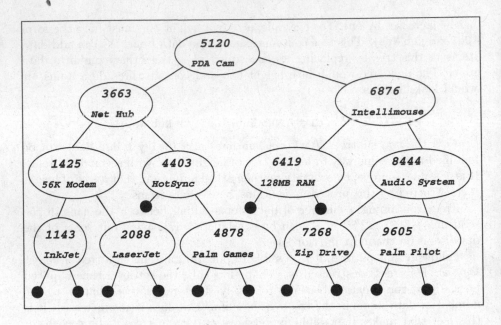

Figure 12.2: Diagram of Height Balanced Search Tree

search tree must grow at least as fast as the logarithm of the number of data items in the tree (another fact that can be proved by induction). Height will grow faster than the logarithm of the number of data items unless the two sides of the tree, left and right, remain balanced as items are inserted.

When we discussed constructing a tree whose height was proportional to the logarithm of the number of data items it contained, we were careful, at each stage, to put the same number of items in the left-hand side of the tree as in the right-hand side. This is known as node balancing. It is the most straightforward way to maintain enough balance to keep a tight lid on height as nodes are inserted, but it turns out to be expensive to maintain node-balancing in trees as items are inserted.

Another form of balancing, known as height balancing, is cheaper to accomplish. Fortunately, height-balanced trees also have the property that height grows at the same rate as the logarithm of the number of data items. This makes it possible to carry out retrieval, insertion, and deletion efficiently. A tree is height balanced if, at all levels, the height of the left subtree is either the same as the height of the right subtree, or taller by one, or shorter by one. This notion is made precise by the following definition.

$$balanced\ (s)\ = \hspace{5cm} \{bal\}$$
$$\forall\ (\texttt{Cel k d lf rt}) \subseteq\ s.\ |\ \texttt{height (lf)}\ -\ \texttt{height (rt)}\ |\ \leq\ 1$$

Figure 12.2 displays a height balanced tree, and Figure 12.3 displays a tree

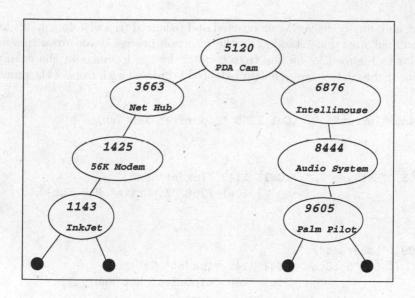

Figure 12.3: Diagram of Unbalanced Search Tree

that is not height balanced. Neither tree is node balanced.

Our goal is to define an insertion operator that preserves order, balance, and the items in the tree. That is, we want the insertion operator, for which we use the symbol (^:), to satisfy the following properties.

$$\forall s.\ \forall k.\ \forall d.\ ordered\ (s)\ \rightarrow\ ordered\ ((k,\, d)\ \hat{}:\ s) \qquad \{\hat{}:\ ord\}$$

$$\forall s.\ \forall k.\ \forall d.\ balanced\ (s)\ \rightarrow\ balanced\ ((k,\, d)\ \hat{}:\ s) \qquad \{\hat{}:\ bal\}$$

$$\forall s.\ \forall k.\ \forall d.\ \forall x.\ (x \in s)\ \rightarrow\ (x \in ((k,d)\hat{}:\ s)) \qquad \{\hat{}:\ old \in\}$$

$$\forall s.\ \forall k.\ \forall d.\ k \in ((k,d)\hat{}:\ s) \qquad \{\hat{}:\ new \in\}$$

$$\forall s.\ \forall k.\ \forall d.\ \forall x.\ ((x \notin s)\ \wedge\ (x\ \neq\ k))\rightarrow (x \notin ((k,d)\hat{}:\ s)) \qquad \{\hat{}:\notin\}$$

Furthermore, the number of steps required to carry out an insertion must grow in the same proportion as the height of the tree.

It is not difficult to find ways to insert new items into small search trees, while maintaining order and balance. To preserve order, simply find the place at the most interior part of the tree where the new key goes, moving left or right down the tree according to whether the new key is smaller or larger than the key being considered. When you arrive at a Nub, replace it with a Cel- constructed tree containing the new key, its associated data, and two Nub subtrees. This automatically preserves order. If you're lucky, it may also preserve balance. However, the replacement subtree has height 1, while the subtree it replaced (a Nub tree) had height zero. If this happens to occur at a point where the tree was already tall, this can make it too tall at that point. If this happens, you need to figure out a way to rearrange the tree to get it back into balance.

The following example starts with a tree containing one item, then inserts three new items, one at a time, producing a sequence of ordered and balanced trees, and, finally delivering an ordered and balanced tree with four items in it. (It will aid your understanding of the insertion process if you draw diagrams similar to Figure 12.2 for the trees denoted by the formulas in the example. Notice that each tree is also height balanced is ordered with respect the numeric key.)

```
(1143, "InkJet")  : (Cel 7268 "ZipDrive" Nub Nub)

    ⇒

(4403, "HotSync") ^: (Cell 1143 "InkJet" Nub
                           (Cel 7268 "ZipDrive" Nub Nub))

    ⇒

(2088, "LaserJet") ^:
    (4403 "HotSync" (Cel 1143 "InkJet" Nub Nub)
                      (Cel 7268 "ZipDrive" Nub Nub))

    ⇒

(Cel 2088 "LaserJet"
        (Cel 1143 "InkJet" Nub Nub)
        (Cel 4403 "HotSync" Nub (Cel 7268 "ZipDrive" Nub Nub)))
```

Although it is not difficult to find a way to rearrange a small tree to get it back into balance when the insertion causes it to go out of balance, finding a way to rebalance a large tree can be tricky. We are going to consider four special cases of the general problem. Then, we are going to build a solution to the full problem based on these four special cases. Two of the special cases admit a straightforward solution to the rebalancing problem. The other two require an ingenious insight. We'll start with the easy cases.

12.6.1 Rebalancing in the Easy Cases

The first point to notice is that if we start with a tree that is ordered and balanced, then insert a new key in such a way that order is preserved, the worst that can happen with respect to balance is that the height of the subtree where the insertion occurred is two greater than the height of the other subtree.

Let's first consider the case where the left subtree is too tall. Since the height of the left subtree is two more than the height of the right subtree, the height of the left subtree must be at least two. Therefore, it cannot be the Nub tree. It must be Cel-constructed.

That is, in this special case, the out-of-balance tree is Cel-constructed, its right subtree is an ordered and balanced tree of height n, and its left subtree is a Cel-constructed tree of height $n + 2$. Additionally, we assume that this left subtree is ordered, that its left subtree is an ordered, balanced tree of height

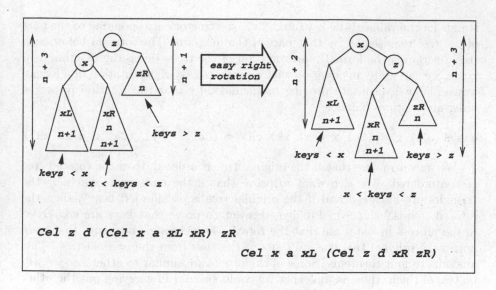

Figure 12.4: Easy-Right Rotation of Outside Left-Heavy Tree

$n + 2$, and that its right subtree is an ordered, balanced tree whose height is either n or $n + 1$. We call a tree configured in this way "outside left-heavy" because its tallest part is on the left side of the left subtree.

It turns out that a tree with these properties can be transformed into an ordered, balanced tree by performing what we call an easy-right rotation. This is accomplished by constructing a new tree whose root contains the key/data pair from the left subtree of the tree we started with and whose right subtree contains the key/data pair that was at the root.

Of course, we have to figure out what to do with the right subtree of the original tree and also what to do with the left and right subtrees of the left subtree of the original tree. These we place in the only slots that will preserve order, and it just happens that this rearrangement restores balance. In other words, we get lucky.

Figure 12.4 diagrams the easy-right rotation. In the diagram, triangles represent ordered, balanced trees that remain unchanged in the rotated tree. A label placed near the top of the triangle is simply used to identify the subtree. Numeric formulas written near the bottom of the triangle are potential heights of the subtree. For example, if the numeric formulas n and $n + 1$ appear near the bottom of the triangle, the subtree may have height n, or it may have height $n + 1$, but it cannot have any other height. Heights for portions of the tree are also sketched in near the sides of the tree diagrams.

All keys in the tree represented by a triangle will be in the range indicated on the diagram. The keys associated with particular Cel constructors are named in circles. For example, if the name x appears in a circle in the diagram, it stands for the value of the key in the `Cel` constructor corresponding to the part of the tree represented by that part of the diagram. The formula below each tree diagram is the formal representation of the tree. Using the diagram as a guide, we can write an equation that represents easy-right rotation in a formal way. This will facilitate proving mathematically that the rotation preserves order and restores balance.

`easyR (Cel z d (Cel x a xL xR) zR) = (Cel x a xL (Cel z d xR zR))`

We want to prove that if the original tree is ordered, then the rotated tree is also ordered. We also want to prove that if the trees represented by the triangles are balanced and if the original tree is outside left-heavy, then the rotated tree is balanced. Finally, we want to prove that keys are conserved in the process in the sense that the rotation neither adds new keys that were not in the original tree nor loses any of the keys from the original tree. This amounts to four theorems. Some of the proofs are similar to others, so we will do two of them, then assume that we could succeed in carrying out the other two proofs if we made the attempt.

Here is a formal definition of what it means to be outside left-heavy:

$outLeft$ `(Cel z d (Cel x a xL xR) zR)` =
   `(height (xL)` $\geq$ `height (xR))` $\wedge$ `(height (xL)` $\leq$ `height (xR)` $+$ `1)` $\wedge$
   `(height (xR)` $\geq$ `height (zR))` $\wedge$ `(height (xL)` $=$ `height (zR)` $+$ `1)`

Theorem 86. (`easyR` preserves order)
   $\forall z. \ \forall d. \ \forall x. \ \forall a. \ \forall xL. \ \forall xR. \ \forall zR.$
      $ordered$ `(Cel z d (Cel x a xL xR) zR)` $\rightarrow$
         $ordered$ `(easyR (Cel z d (Cel x a xL xR) zR))`

Theorem 87. (`easyR` restores balance to outside left-heavy trees)
   $\forall z. \ \forall d. \ \forall x. \ \forall a. \ \forall xL. \ \forall xR. \ \forall zR.$
   $(balanced$ `(xL)` $\wedge \ balanced$ `(xR)` $\wedge \ balanced$ `(zR)`
      $\wedge \ outLeft$ `(Cel z d (Cel x a xL xR) zR))`
   $\rightarrow \ balanced$ `(easyR (Cel z d (Cel x a xL xR) zR))`

Proof. One way to prove that an implication formula is true is to prove that its conclusion is true whenever its hypothesis is. Therefore, it will be sufficient to prove that the formula

$$balanced \ \texttt{(easyR (Cel z d (Cel x a xL xR) zR))}$$

is true whenever the formulas

$$(balanced \ \texttt{(xL)} \ \wedge \ balanced \ \texttt{(xR)} \ \wedge \ balanced \ \texttt{(zR)})$$

and

$$outLeft \ \texttt{(Cel z d (Cel x a xL xR) zR)}$$

are both true.

balanced (easyR (Cel z d (Cel x a xL xR) zR))
  = {easyR}
balanced (Cel x a xL (Cel z d xR zR))
  = {*bal*}
| height (xL) − height (Cel z d xR zR) | ≤ 1 ∧
    balanced (Cel z d xR zR)
    ∧ *balanced* (xL) ∧ *balanced* (xR) ∧ *balanced* (zR)

The last three terms in this formula are part of the hypothesis. So, that part is true. We now only need to verify that the other terms are also true. Since we know that xR and zR are balanced, the term *balanced* (Cel z d xR zR) is true if xR and zR differ in height by one or less.

This fact can be deduced from the *outLeft* (Cel z d (Cel x a xL xR) zR) hypothesis because of the constraints it puts on the heights of the various subtrees. In addition this same *outLeft* hypothesis implies that the following equation is true:

$$max \text{ (height (xR), height (zR))} = \text{height (xR)} \quad \{*\}$$

Using this equation, we can derive another one as follows.

height (Cel z d xR zR)
  = {height C}
1 + *max* (height (xR), height (zR))
  = {*}
1 + height (xR)

That is, the following equation is true.

$$\text{height (Cel z d xR zR)} = \text{height (xR)} + 1 \quad \{**\}$$

This equation allows us to complete the proof of Theorem 87 by verifying that the first term in the formula is true.

| height (xL) − height (Cel z d xR zR) |
  = {**}
| height (xL) − (height (xR) + 1) |
  ≤ {*outLeft* (Cel z d (Cel x a xL xR) zR)}
1

□

Theorem 88. (easyR loses no keys)
∀k. ∀z. ∀d. ∀x. ∀a. ∀xL. ∀xR. ∀zR.
  k ∈ (Cel z d (Cel x a xL xR) zR) → k ∈ easyR (Cel z d (Cel x a xL xR) zR)

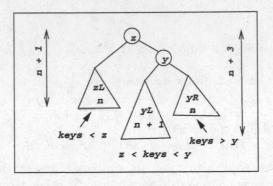

Figure 12.5: Inside Right-Heavy Tree

Proof.

$k \in$ (Cel z d (Cel x a xL xR) zR)
$\quad = \{\in C\}$
$(k = z) \lor (k = x) \lor (k \in xL) \lor (k \in xR) \lor (k \in zR)$
$\quad = \{\in C\}$
$k \in$ (Cel x a xL (Cel z d xR zR))
$\quad = \{easyR\}$
$k \in$ easyR (Cel z d (Cel x a xL xR) zR)

$\square$

Theorem 89. (easyR adds no keys)
$\quad \forall k.\ \forall z.\ \forall d.\ \forall x.\ \forall a.\ \forall xL.\ \forall xR.\ \forall zR.$
$\quad k \notin$ (Cel z d (Cel x a xL xR) zR) $\rightarrow$ k $\notin$ easyR (Cel z d (Cel x a xL xR) zR)

Theorems 86 through 89 confirm that the easy right rotation does the right thing to outside left-heavy trees. The outside right-heavy case is so similar that we are going to assume we could state and prove the corresponding theorems for that case, and will use them as if we had done so. For the record, here is the rotation function for outside right-heavy trees:

easyL (Cel z d zL (Cel y b yL yR)) = (Cel y b (Cel z d zL yL) yR)

12.6.2   Rebalancing in the Hard Cases

Since the easy cases occur when the tallest part of the tree is on the outside, you can guess that the hard cases occur when the tallest part is on the inside. The inside right-heavy case occurs when the height of the right subtree is two more than the height of the left subtree, and it is the left subtree of that right subtree where the tallest part of the tree resides.

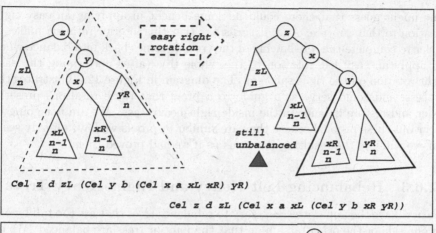

Cel z d zL (Cel y b (Cel x a xL xR) yR)

Cel z d zL (Cel x a xL (Cel y b xR yR))

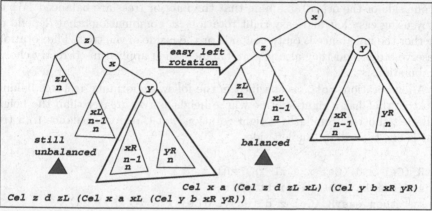

Cel x a (Cel z d zL xL) (Cel y b xR yR)

Cel z d zL (Cel x a xL (Cel y b xR yR))

Figure 12.6: Rebalancing an Inside Right-Heavy Tree

Because the easy cases (outside heavy trees) include the possibility that the left and right subtrees of the tallest part have the same height, we can be more explicit about subtree heights in the inside heavy cases. Figure 12.5 uses the same conventions as our early tree diagrams to display an inside right-heavy tree.

The tallest part of the tree in Figure 12.5 occurs at the yL subtree. It has height $n + 1$, which is at least one. Therefore, it is not a Nub tree. Its Cel-constructed structure can be further exposed as a key, x, with two subtrees, xL and xR, as shown in Figure 12.6. At least one of the subtrees of yL must have height n. Otherwise, the height of yL could not be $n + 1$. The other subtree of yL could have either height n or height $n - 1$. The diagram indicates that these subtrees can have either height.

The right subtree of an inside right-heavy is not out of balance, but the height of its left subtree is greater (by one) than that of its right subtree.

One might guess that there could be some benefit in applying an easy right
rotation to that subtree. And, it turns out that this interior rotation makes it
easier to complete the rebalancing of the tree. It comes back into balance when
we apply an easy left rotation to the whole tree (after performing the easy
right rotation on the right subtree). The diagram in Figure 12.6 illustrates the
process, and could serve as a guide to a proof that these rotations preserve
order and restore balance in the inside right-heavy case. We are not going to
carry out these proofs because they are similar to proofs we have already seen,
but we will cite the resulting theorems as if we had proved them.

12.6.3 Rebalancing Left-Heavy and Right-Heavy Trees

We now have the apparatus necessary to rebalance trees that are too tall by one
on one side or the other, assuming that the interior trees are balanced. We do
so by using easy left and easy right rotations in combinations that depend on
whether the imbalance is on the outside or the inside of the tree. The rotations
preserve order, and they also restore balance when applied in correctly chosen
combinations.

A left rotation, `rotL`, as specified in the following formula, restores balance
to a tree that has a right subtree whose height is two greater than the height
of its left subtree. Similarly, a right rotation, `rotR`, restores balance to a tree
that is one too tall on the left side.

```
rotR (Cel z d (Cel x a xL xR) zR) =
   if (height xL) < (height xR)
      then easyR (Cel z d (easyL (Cel x a xL xR)) zR)
      else easyR (Cel z d (Cel x a xL xR) zR)

rotL (Cel z d zL (Cel y b yL yR)) =
   if (height yR) < (height yL)
      then easyL (Cel z d zL (easyR (Cel y b yL yR)))
      else easyL (Cel z d zL (Cel y b yL yR))
```

The following theorems are straightforward consequences of theorems we
already know about the easy rotations.

Theorem 90. (rotations preserve order)
$\forall$z. $\forall$d. $\forall$x. $\forall$a. $\forall$xL. $\forall$xR. $\forall$zR.
ordered (Cel z d (Cel x a xL xR) zR) $\rightarrow$

> *ordered* (rotR (Cel z d (Cel x a xL xR) zR))

$\forall$z. $\forall$d. $\forall$x. $\forall$a. $\forall$xL. $\forall$xR. $\forall$zR.
ordered (Cel z d zL (Cel y b yL yR))
> $\rightarrow$ *ordered* (rotL (Cel z d zL (Cel y b yL yR)))

Theorem 91. (rotations restore balance)

∀z. ∀d. ∀x. ∀a. ∀xL. ∀xR. ∀zR.

balanced (xL) ∧ *balanced* (xR) ∧ *balanced* (zR) ∧ height (Cel x a xL xR) >
height (zR) + 1 → *balanced* (rotR (Cel z d (Cel x a xL xR) zR))

∀z. ∀d. ∀y. ∀b. ∀zL. ∀yL. ∀yR.

balanced (zL) ∧ *balanced* (yL) ∧ *balanced*(yR) ∧ height (Cel y b yL yR) >
height (zL) + 1 → *balanced* (rotL (Cel z d zL (Cel y b yL yR)))

Theorem 92. (rotations conserve keys)

∀z. ∀d. ∀x. ∀a. ∀xL. ∀xR. ∀zR.

k ∈ (Cel z d (Cel x a xL xR) zR))

→ k ∈ rotR (Cel z d (Cel x a xL xR) zR)

and

∀z. ∀d. ∀x. ∀a. ∀xL. ∀xR. ∀zR.

k ∉ (Cel z d (Cel x a xL xR) zR)))

→ k ∉ rotR (Cel z d (Cel x a xL xR) zR))

We are going to assume that we could prove these theorems if someone asked us to, and we will cite them as if we had constructed their proofs.

12.6.4 Inductive Equations for Insertion

Insertion of a new element in a tree is a matter of deciding which side of the tree it goes in and inserting it on that side. If there is no room at the top level, simply apply the same idea, inductively, to the subtree where the insertion is to take place.

After the insertion, it may happen that the tree is out of balance. If so, the height of the side where the insertion occurred will be exactly two more than the height of the other side, and a rotation will restore balance.

In formal terms, the operation (ˆ:) satisfies the following equations.

```
(k, d) ^: Nub = (Cel k d Nub Nub)                    {^: N}
(k, d) ^: (Cel x a lf rt) =                          {^: C}
if z < x then
        if (height newL) > (height rt) + 1
          then rotR (Cel k d newL rt)
          else (Cel x a newL rt)
    else if z > x then
        if (height newR) > (height lf) + 1
          then rotL (Cel x a lf newR)
          else (Cel x a lf newR)
    else (Cel x d lf rt)
    where
    newL = (k, d) ^: lf
    newR = (k, d) ^: rt
```

The "where" clause in the second equation names the tree delivered by insertion into the subtree that preserves order. Only one of these insertions will take place, of course. The new (key, data) pair goes into the left subtree if the key is smaller than the key at the root. If it is greater, it goes into the right subtree.

Otherwise, the key to be inserted must be the same as the key at the root. There are many choices for things to do at this point. It might be considered an error to attempt to insert a duplicate key into a tree, in which case it would be best to deliver a value that could signal the problem. One choice of such a value would be the Nub tree, since it could not be the result of inserting a new key into a tree. On the other hand, it might be appropriate to simply associate the new data with the key that matches the key that was to have been inserted. We made the latter choice because it makes our theorems about conservation of keys easier to state.

The insertion equations imply three important properties of the operation: insertion preserves order, preserves balance, and conserves keys. That is, if the right-hand operand of an insertion is an ordered and balanced tree, then insertion will deliver a tree that is ordered and balanced. Furthermore, the tree delivered by insertion will contain all the keys that occur in the right-hand operand, plus the key in the left-hand operand, and it will contain only those keys and no others.

Theorems 93, 94, and 95 state these properties. Tree induction can be used to prove all of these theorems. We are going to carry out only one of them. The pattern of the others is similar, and we will assume we could construct those proofs if someone asked.

Theorem 93. (insertion preserves order).

$$\forall s.\ \forall k.\ \forall d.\ ordered\ (s)\ \rightarrow\ ordered\ ((k,d)\ \hat{}:\ s)$$

Theorem 94. (insertion preserves balance).

$$\forall s.\ \forall k.\ \forall d.\ balanced\ (s)\ \rightarrow\ balanced\ ((k,d)\ \hat{}:\ s)$$

Theorem 95. (insertion conserves keys)

$$\forall s.\ \forall k.\ \forall d.\ k \in ((k,\ d)\ \hat{}:\ s)$$
$$\forall s.\ \forall k.\ \forall d.\ \forall x.\ x \in s\ \rightarrow\ x \in ((k,\ d)\ \hat{}:\ s)$$
$$\forall s.\ \forall k.\ \forall d.\ \forall x.\ (x \notin s\ \land\ x \neq k)\ \rightarrow\ x \notin ((k,\ d)\ \hat{}:\ s)$$

Proof. of Theorem 94 (by tree induction)

Base Case.

∀ k. ∀d. *balanced* (Nub) → *balanced* ((k, d) ˆ: Nub)
  = {ˆ: *N*}
∀k. ∀d. *balanced* (Nub) → *balanced* (Cel k d Nub Nub)
  = {*bal N*}
∀k. ∀d. True → *balanced* (Cel k d Nub Nub)
  = {*bal C*}
∀k. ∀d. True → True
  = {→}
True

Inductive Case.

In this part of the proof, we will apply the {ˆ: *C*} equation, which takes three forms, depending on the relationship between the key being inserted and the key associated with the tree's cell constructor (k < x) ∨ (k > x) ∨ (k = x). To prove an implication with a disjunction in the hypothesis, it is sufficient to prove the conclusion separately for each of the terms in the disjunction. (We know this from the following equation of propositional logic: $((a ∨ b ∨ c) → d) = ((a → d) ∧ (b → d) ∧ (c → d))$

The proof takes different forms for each of these separate parts, but the reasoning is similar enough that we are going to carry out just one of the parts, and leave the others to the imagination.

The part we are going to prove is when the key to be inserted is smaller than the key associated with the Cel constructor for the tree. A proof strategy for an implication is to start from the hypothesis and derive the conclusion. The hypothesis of the implication, in the inductive case, is the formula *balanced* (Cel x a lf rt). Since lf is a proper subtree of (Cel x a lf rt), the induction hypothesis leads to the conclusion that the formula *balanced* ((k, d)ˆ: lf) is true.

We are proving three implications, and in the first of these we assume that the key, k, to be inserted is less than the key, x, associated with the Cel constructor for the tree. To complete this part of the proof, we must verify that the conclusion of the implication is true when the hypothesis is true and k < x.

Here, there are two cases because the equation {ˆ: *C*} specifies that (Cel x a ((k,d) ˆ: lf) rt) is the value of ((k, d) ˆ: (Cel x a lf rt)) when | height ((k,d) ˆ: lf) − height (rt) | ≤ 1 and specifies rotR ((k,d) ˆ: (Cel x a lf rt)) as the value when height ((k,d) ˆ: lf) > height (rt) + 1.

We reason as follows:

balanced ((k, d) ˆ: (Cel x a lf rt))
    = {ˆ: C, k < x}
balanced (Cel x a ((k, d) ˆ: lf) rt)
    = {*bal*}
∀(Cel y a yL yR) ⊆ (Cel x a ((k, d) ˆ: lf) rt).
       | height (yL) − height (yR) | ≤ 1

This "forall" predicate can be partitioned into three logic formulas, all of which must be true:

| height ((k, d) ˆ: lf) − height (rt) | ≤ 1   {*formula 1*}
∀(Cel y a yL yR) ⊆ ((k, d) ˆ: lf).
       | height (yL) − height (yR) | ≤ 1   {*formula 2*}
∀(Cel y a yL yR) ⊆ rt.
       | height (yL) − height(yR) | ≤ 1   {*formula 3*}

We have already verified, using the induction hypothesis, that ((k, d) ˆ: lf) is balanced. Therefore, formula 2 must be true. With regard to formula 3, the hypothesis of the implication we are trying to prove is that (Cel x a lf rt) is balanced. Therefore, all its subtrees are balanced, and since rt is one of those, all of the non-empty subtrees of rt must have left and right subtrees with heights that differ by one or less.

With regard to formula 1, we must consider two situations because the equation {ˆ: C} specifies (Cel x a ((k, d) ˆ: lf) rt) as the value of ((k, d) ˆ: (Cel x a lf rt)) when | height ((k, d) ˆ: lf) − height (rt) | ≤ 1 and specifies rotR ((k, d) ˆ: (Cel x a lf rt)) as the value when height ((k, d) ˆ: lf) > height (rt) + 1. The first situation verifies formula 1 directly. In the second situation, Theorem 91 (rotations restore balance) implies that formula 1 is true.

This completes the proof of the inductive case of Theorem 94 when k < x. When k > x, the reasoning goes the same way, but down the right-hand side of the tree instead of the left. When k = x, the insertion operation delivers a tree identical to the input, except for the data item in the Cel constructor for the tree. The input tree is balanced, and swapping the data item in its Cel constructor for a different one cannot affect the balance of the tree. By this route, the proof of the inductive case can be completed.   □

12.6.5   Insertion in Logarithmic Time

In the same way that we reasoned, by induction, that the number of computational steps required for retrieval of a data item from a search tree is proportional to the height, we can prove that the number of steps required for insertion is also proportional to the height. The induction for insertion proceeds in the same way as the induction for retrieval, except that some additional basic operations are encountered, such as the Cel and Nub constructors, the rotation operators, and the height computation. If we can convince ourselves that each of these operations takes a fixed amount of time, regardless of what their operands are, the outcome of the induction will be that the number of steps required for insertion is proportional to tree height.

Nobody will have difficulty believing that the Nub constructor takes a fixed amount of time, as it has no operands. The Cel constructor calls for more careful reasoning because it involves building a tree from components, and we want to argue that building the tree does not depend on the size of the components. We can assume that the components have already been constructed, so we just need to verify that if the computer is given a key k, data item d, and two pre-constructed subtrees lf and rt, the number of computational steps required for the computer to deliver the tree (Cel k d lf rt) does not depend on the size of the key, data item, or subtrees.

Compilers for languages like Haskell refer to constructed entities by address. For example, if the subtree lf, involved in the construction of (Cel k d lf rt), has been previously constructed, then it has been recorded somewhere in the computer's memory. To construct the tree (Cel k d lf rt), the computer simply places, into the structure delivered by the Cel constructor, the address where lf has been recorded. Similarly, it puts the addresses for k, d, and rt in the structure. Because addresses have a fixed size, all of this construction is completed in a fixed amount of time, regardless of the values represented by k, d, lf, and rt. As for the rotation operators, they simply combine a fixed number of Cel and Nub constructions, so they also can be completed in a fixed number of steps.

The height operator is another matter. The equations we have given for it, when used according to the procedures of the equational model of computation, produce a computation in which the number of steps is proportional to the height of the tree. If we were to use a version of the height operator satisfying equations {height N} and {height C}, we would fail to accomplish our goal of producing an insertion operator that can be carried out in a number of steps proportional to the height. Therefore, we need a different height operator.

The trick is to record the height directly in the tree, and insert in incrementally, as the tree is being constructed. This requires a new definition for search trees:

```
data SearchTree d = Nub    |
                    Cel Integer Integer d
                        (SearchTree d) (SearchTree d)
```

This new definition has a extra Integer parameter for the Cel constructor, and this parameter is used to record the height of the tree. To construct a tree, its height must be specified, along with the key, data item, and two subtrees. The construction (Cel h k d lf rt) denotes a tree of height h with the key k, data item d, and left and right subtrees lf and rt. The new height operator could be specified by the following equations.

```
height Nub = 0
```

```
height (Cel h k d lf rt) = h
```

Delivering the height of a tree simply amounts to pulling it out of an existing structure, an operation that requires a fixed amount of effort, regardless of the size of the tree.

Of course, all of the `Cel` constructions that appear in equations must be modified to include the height, but it turns out that this can be done because the height to be recorded by a `Cel` construction can be computed from the heights of the subtrees supplied for the construction. For example, the `easyL` rotation

```
easyL(Cel hz z d  zL (Cel hy y b yL yR)) =
    (Cel hNewy y b (Cel hNewz z d zL yL) yR)
        where
        hNewy = max hNewz (height yR)
        hNewz = max (height zL) (height yL)
```

Similar modifications to the other rotations and to the `Cel` constructions in the insertion operator lead to an insertion operator that can be computed in an amount of time proportional to the height of the tree in which the insertion occurs.

Since the height of a balanced tree is proportional to the logarithm of the number of data items in the tree (together with all subtrees contained in the tree), insertion can be done in an amount of time proportional to the logarithm of the number of data items in the tree. This is the insertion part of the AVL miracle. The deletion part, which we are going to turn to now, is a little more complicated, but not much.

12.6.6   Deletion

The process of deleting a key and its associated data item from a search tree, while maintaining order and balance, depends on where the deleted item occurs in the tree. Wherever it occurs, the `Cel` constructor at that point forms an entire tree (a subtree of the one we started with). If we can find a way to delete the item associated with a particular `Cel` constructor, we can build a new subtree containing all the items of the original subtree except the one associated with its `Cel` constructor. We can then replace the portion of the original tree corresponding to that `Cel` constructor with the new subtree.

If the new subtree occurs as a proper subtree of the tree as a whole, we need to make sure the resulting tree is balanced. (It will automatically be ordered because all of the keys occurring in the subtree occurred in the original subtree, so they must have stood in the proper relationship to other keys in the tree.) We have deleted a node, so it may happen that the new subtree is shorter than the old one, and this could disrupt the balance of the whole tree. Therefore, we compare the height of the new subtree to the height of the other subtree in the `Cel` constructor where the new subtree resides. If the heights differ by

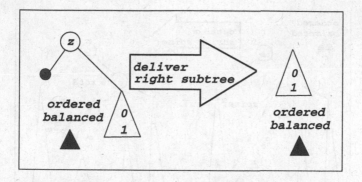

Figure 12.7: Deleting Root When Left Subtree Is Nub

more than one, we rotate right or left as necessary, then carry out a similar check and possible rotation one level up in the tree. We continue this process until we come to the Cel constructor for the tree as a whole.

That is the plan. The first step is to figure out how to delete an item associated with the Cel constructor for a tree. That is, how to delete a key that occurs at the root of a search tree. We break this into two cases. As usual, one of the cases is easy to deal with, and the other one isn't. We will take care of the easy case first.

The easy case is when the left subtree is a Nub. Since the tree is balanced, the height of the right subtree must be either zero or one. There are no items in the left subtree to worry about, so all items (except the one to be deleted) are retained if we deliver the right subtree as the result. Furthermore, the right subtree is ordered and balanced. It has to be because it is a subtree of the original tree, which was ordered and balanced. Figure 12.7 charts the transformation. As in previous diagrams, a triangle represents an ordered and balanced tree, and the numbers near the bottom of the triangle specify the possibilities for the height of the tree represented by the triangle.

When the left subtree is not a Nub tree, the tree may have any height, and there is more to do. Visualizing the tree transformations as diagrammed in Figure 12.8 may make it easier to follow the deletion procedure in this case. The idea is to march down the spine on the right side of the left subtree until we come to a node whose right subtree is Nub. In keeping with the spine metaphor, we call this node the sacrum. (That is the tailbone. You can look it up.) The sacrum is the node with the key x in the leftmost tree shown in Figure 12.8, and the spine consists of the Cel constructor of the left subtree, the Cel constructor of its right subtree, the Cel constructor of the right subtree of that subtree, and so on down to the sacrum.

The left subtree of the sacrum will have height zero or one (otherwise, the tree would have been out of balance). We cache the key and data item of the sacrum, replace it by its left subtree, and perform right rotations as needed for

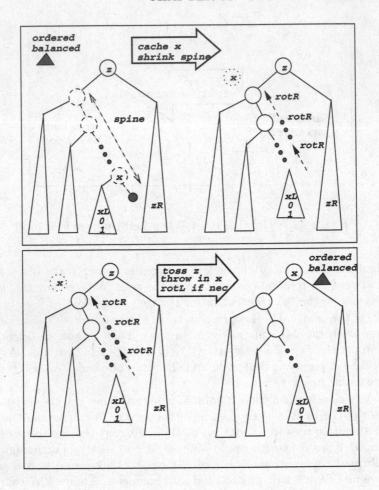

Figure 12.8: Deleting Root When Left Subtree Is `Cell`

balance all the way back up the spine.

Right rotations are needed where the height of the left subtree is two more than that of the right subtree. The left subtrees along the spine will always have heights exceeding that of the corresponding right subtrees by no more than two because the tree was originally balanced, which implies height differences of one or less. When we replace the subtree rooted at the sacrum (a subtree that has a `Nub` tree as its right subtree) by its left subtree, the height of the new tree at the position of the sacrum may be one less than it was before. This new tree is the right subtree of another subtree, so that subtree may be too tall by one on the left. If it is, we know from Theorem 91 that a right rotation will rebalance it. We also know, from Theorem 90, that a right rotation will preserve order. Therefore, at each stage the subtree produced is ordered and balanced.

When we reach the top of the spine, we deliver a new tree with key, data item, left subtree, and right subtree specified as follows: (1) use the key and data item cached key from the sacrum, (2) for the left subtree, use the tree we constructed as we rotated our way back up the spine, and (3) for the right subtree, use the right subtree from the original tree.

Finally, we rotate the tree left, if necessary to get it back in balance. This may be necessary because the height of the left subtree may now be one less than it was originally, and this may have caused the tree to get out of balance by one. Placing the cached key (the one that was in the sacrum) at the root preserves order because it was the largest key in the left subtree, so all keys in the left subtree will be smaller than it is, as they should be in an ordered search tree. Also, it is smaller than any key in the original right subtree because it was smaller than the key at the root, and the key at the root was larger than any key in the original left subtree.

To make these informal ideas amenable to mathematical reasoning, we need formal equations to specify the relationships among the trees and subtrees involved. It will help to consider parts of the transformation separately.

12.6.7 Shrinking the Spine

One major component in the transformation is the shrinking of the spine when the left subtree is not a Nub. In Figure 12.8, this is the part of the transformation diagrammed in the top box of the figure. At the same time, we cache the key and data item at the sacrum. So, this part of the transformation needs to deliver a new left subtree, with a shrunken spine, together with the key and data item from the sacrum. We will implement this delivery with a three-tuple in which the first component is the key from the sacrum, the second component is the data item from the sacrum, and the third component is the new, shrunken, left subtree. With these conventions, the following equations specify the relationships among the components that the diagram in Figure 12.8 displays.

```
shrink(Cel y b lf rt) =                                      {shrink}
   if rt == Nub then (y, b, lf)
   else if (height lf) > (height shrunken) + 1
           then (x, a, rotR (Cel y b lf shrunken))
   else (x, a, Cel y b lf shrunken)
   where
   (x, a, shrunken) = shrink rt
```

The shrink operation preserves order and balance, and it delivers a key that exceeds all the keys remaining in the shrunken tree. The following theorems, all of which can be proved using tree induction, make these claims precise. We will prove only one of them and, as usual, assume it is possible to construct the other proofs.

Theorem 96. (shrink preserves balance)

$\forall$y. $\forall$b. $\forall$lf. $\forall$rt.

   (*balanced* (Cel y b lf rt) $\wedge$ ((x, a, sh) = shrink (Cel y b lf rt))) $\rightarrow$
balanced (sh)

Theorem 97. (shrink loses no keys)

$\forall$y. $\forall$b. $\forall$lf. $\forall$rt.

   (*ordered* (Cel y b lf rt) $\wedge$ ((x, a, sh) = shrink (Cel y b lf rt)) $\wedge$ (k $\in$
(Cel y b lf rt)) $\wedge$ (k $\neq$ x)) $\rightarrow$ (k $\in$ sh)

Theorem 98. (shrink adds no keys)

$\forall$y. $\forall$b. $\forall$lf. $\forall$rt.

   ordered (Cel y b lf rt) $\wedge$ ((x, a, sh) = shrink (Cel y b lf rt)) $\wedge$ (k $\notin$
(Cel y b lf rt)) $\rightarrow$ (k $\notin$ sh)

Theorem 99. (shrink preserves order)

$\forall$y. $\forall$b. $\forall$lf. $\forall$rt.

   ordered (Cel y b lf rt) $\wedge$ ((x, a, sh) = shrink (Cel y b lf rt)) $\rightarrow$
(*ordered* (sh) $\wedge$ ($\forall$k $\in$ sh. x $>$ k))

Proof. of Theorem 99 (by tree induction).

   Base Case.

   ordered (Cel y b lf Nub) $\rightarrow$ (*ordered* (lf) $\wedge$ ($\forall$k $\in$ lf.y $>$ k))

   by the definition of *ordered*, and

   ((y, b, lf) = shrink (Cel y b lf Nub))

   by the definition of shrink.

.   Therefore, as sh = lf and x = y,

   (*ordered* (Cel y b lf Nub) $\wedge$ ((x, a, sh) = shrink (Cel y b lf Nub)))
      $\rightarrow$ (*ordered* (sh) $\wedge$ ($\forall$k $\in$ sh. x $>$ k))

   Inductive Case.

   We are assuming, in the inductive case that (Cel y b lf (Cel k d kL kR)) is
ordered and that (x, a, sh) = shrink (Cel y b lf (Cel k d kL kR)). In this case
the else-branch will be the applicable formula in the {shrink} equation. Therefore, sh will have one of the following two values: (Cel y b lf shrunken) or
rotR (Cel y b lf shrunken), where (x, a, shrunken) = shrink (Cel k d kL kR).
Also, we know from the conservation-of-keys theorems for rotR (Theorem 92)
and for shrink (Theorems 97 and 98) that in either case sh contains all the
keys of lf plus all the keys of (Cel k d kL kR) except x.

From the induction hypothesis, we know that x exceeds all the keys of shrunken. This means that x exceeds all the keys of (Cel k d kL kR) except x, itself, and that x ∈ (Cel k d kL kR). Therefore, since (Cel y b lf (Cel k d kL kR)) is ordered, it must be true that x > y and that x exceeds all the keys in lf.

This accounts for all the keys in sh, so it must also be true that x exceeds all the keys in sh. Furthermore, if sh = (Cel y b lf shrunken), then it must be ordered because lf is ordered (a consequence of the fact that (Cel y b lf (Cel k d kL kR)) is ordered), shrunken is ordered (a consequence of the induction hypothesis), and y exceeds all the elements in lf and precedes all the elements in (Cel k d kL kR) (a consequence of the fact that (Cel y b lf (Cel k d kL kR)) is ordered). Finally, if it turns out that sh = rotR (Cel y b lf shrunken), then sh must be ordered because rotations preserve order (Theorem 90). □

12.6.8 Equations for Deleting Root

Figure 12.8 describes the deletion process, for the case when the left subtree is not Nub, in two steps: (1) cache the sacrum and shrink the spine in the left subtree and (2) form a new tree (with the cached key/data at the root, the shrunken left subtree, and the right subtree that came with the original). Finally, rotate left if the constructed tree is out of balance. The following equation expresses these relationships in a formal way.

```
delRoot (Cel z d Nub zR) = zR                    {delRoot N}
delRoot (Cel z d (Cel y b lf rt) zR) =           {delRoot C}
  if (height zR) > (height shrunken) + 1
    then rotL (Cel x a shrunken rt)
    else (Cel x a shrunken rt)
    where
    (x, a, shrunken) = shrink (Cel y b lf rt)
```

These equations imply that delRoot preserves order and balance and conserves keys (except for the one at the root). The following theorems, which can be proved using tree induction, state these facts.

Theorem 100. (delRoot preserves order)
∀y. ∀b. ∀lf. ∀rt.
order (Cel y b lf rt) → *order* (delRoot (Cel y b lf rt))

Theorem 101. (delRoot preserves balance)
∀y. ∀b. ∀lf. ∀rt.
balanced (Cel y b lf rt) → *balanced* (delRoot (Cel y b lf rt))

Theorem 102. (delRoot conserves keys)
∀y. ∀b. ∀lf. ∀rt.
(*ordered* (Cel y b lf rt) ∧ (k ∈ (Cel y b lf rt)) ∧ (k ≠ y)) →

$$(k \in \text{delRoot (Cel y b lf rt)})$$
and
$$(\textit{ordered} \text{ (Cel y b lf rt)} \land (k \in \text{delRoot (Cel y b lf rt)})) \rightarrow$$
$$(k \in \text{(Cel y b lf rt)})$$

12.6.9 Equations for Deletion

With equations for shrinking the spine and deleting the root in place, we are now in a position to discuss the equations for deleting a key from a search tree. The key to be deleted will be located in the root of a subtree of the tree from which the deletion is to occur. Part of the process is to locate that subtree. Since we are assuming the tree is ordered, these parts of the equations will be similar to the equations for retrieval from search trees. Once we have located the key at the root of a subtree, we apply delRoot to delete it and use the tree it delivers in place of the subtree where the key was located. Of course, the height of this tree may now be one less than it was before, and that might unbalance the tree. If so, we apply a rotation to rebalance it.

We have been assuming that the key to be deleted occurs somewhere in the tree. If it doesn't, of course it can't be deleted. In that case, we simply deliver the tree we started with. Other alternatives for this case might be appropriate for certain applications, but we have chosen this one because it makes the theorems straightforward.

The following equations specify the relationships among search trees, keys, and the results of deleting a key from a tree, following the outline we have been discussing informally.

```
Nub ^- k = Nub                                    {^- N}

(Cel z d lf rt)   ^- 'k =                         {^- C}
    if k = z then delRoot (Cel x a lf rt)
    else if k < z && (height rt) > (height newL) + 1
            then rotL (Cel z d newL rt)
            else (Cel z d newL rt)
    else if (height lf) > (height newR) + 1
            then rotR (Cel z d lf newR)
            else (Cel z d lf newR)
    where
    newL = lf ^- k
    newR =rt ^- k
```

These equations imply that deletion satisfies order and balance preservation laws and key conservation laws similar to those of insertion. The following theorems, which are provable using tree induction, express these laws.

Theorem 103. (deletion preserves order).
$$\forall s. \forall k. \textit{ordered } (s) \rightarrow \textit{ordered } (s \text{ } \hat{} \text{- } k)$$

Theorem 104. (deletion preserves balance).

$\forall s. \ \forall \mathrm{k}. \ balanced \ (s) \ \rightarrow \ balanced \ (s \ \hat{} \text{-} \ \mathrm{k})$

Theorem 105. (insertion conserves keys).

$\forall s. \ \forall \mathrm{k}. \ ordered(s) \ \rightarrow \ (\mathrm{k} \notin (s \ \hat{} \text{-} \ \mathrm{k}))$

$\forall s. \ \forall \mathrm{k}. \ \forall \mathrm{x}. \ (ordered(s) \ \wedge \ (\mathrm{x} \neq \mathrm{k}) \ \wedge \ (\mathrm{x} \in s)) \ \rightarrow \ (\mathrm{x} \in (s \ \hat{} \text{-} \ \mathrm{k}))$

$\forall s. \ \forall \mathrm{k}. \ \forall \mathrm{x}. \ (ordered(s) \ \wedge \ (\mathrm{x} \notin s)) \ \rightarrow \ (\mathrm{x} \notin (s \ \hat{} \text{-} \ \mathrm{k}))$

12.6.10 Deletion in Logarithmic Time

The theorems about retrieval (Theorem 83), insertion (Theorems 93-95), and deletion (Theorems 103-105) confirm the properties that are important for confirming the correctness of retrieval, insertion, and deletion operations. The theorems are consequences of the assumption that these operations satisfy the equations labeled {getItem N}, {getItem C}, {$\hat{}$: N}, {$\hat{}$: C}, {$\hat{}$- N}, and {$\hat{}$- C}. No other assumptions are needed.

The number of computational steps required to delete a key and its associated data item from a search tree depends on how the equations are interpreted. We assume that the interpretation follows the equational model of computation.

In this case, at each level in a tree from which a key is being deleted, from the root to the point where the deletion occurs, the computation will require at least the construction of a tree by a Cel constructor, and possibly also a rotation. Each of these operations requires a fixed number of steps, independent of the size of the tree. Because the height is the maximum number of levels in the tree, the total number of steps required for this part of the deletion operation is proportional, at worst, to the height of the tree.

In the subtree where the deletion actually takes place, the key to be deleted occurs in the root, and the delRoot operation is applied. How many computational steps will this require? It will require a Cel construction at the root level, and possibly a rotation. It also calls for shrinking the spine.

The rotation and Cel construction, together, are a fixed amount of computation, regardless of the size of the tree. The shrinking operation, on the other hand, requires examining each level of the tree down the spine and performing a Cel construction and possibly a rotation at each of those levels. Nevertheless, the amount of computation at each level is independent of the size of the tree, so the total number of steps will be proportional to the height of the tree to which the delRoot operation is being applied. Since this height is no more than the height of the tree, the total number of computational steps in a deletion operation is proportional to the height of the tree.

This argument about the number of computational steps in deletion has been an informal one, but it can be made precise through tree induction, along the same lines as our proof that the number of computational steps in retrieval and in insertion is proportional to tree height.

As with insertion, the height operation will slow down the deletion process significantly unless the height is imbedded in the tree structure and updated incrementally. This means that everywhere in the equations for deletion that a Cel construction occurs, the value of the height must be recorded. Rewriting the equations to take care of this detail is straightforward. The following version of the equation {^- C} show how it can be done in one case. To complete the process, the equations for deletion at the root and for shrinking the spine would have to be similarly rewritten.

```
(Cel z d h lf rt) ^- k =
    if k = z then delRoot (Cel h x a lf rt)
    else if k < z && (height rt) > (height newL) + 1
         then rotL (Cel ((max (height newL) (height rt)) + 1)
                        z d newL rt)
         else (Cel ((max (height newL) (height rt)) + 1)
                    z d newL rt)
    else if (height lf) > (height newR) + 1
         then rotR (Cel ((max (height lf) (height newR)) + 1)
                        z d lf newR)
         else (Cel z ((max (height lf) (height newR)) + 1)
                    d lf newR)
    where
    newL = lf ^- k
    newR = rt ^- k
```

12.7   Things We Didn't Tell You

It is not difficult to prove, using tree induction, that the height of a search tree s is the next integer larger than the base-2 logarithm of the number of keys that occur in the search tree if you assume that all of the subtrees of s have about the same number of nodes in their left subtree as in their right. This is a strict kind of balancing, but it is not the kind of balancing we have used in writing equations for insertion and deletion.

The kind of balance we have used has to do with the heights of subtrees, not the number of keys they contain. It is not so easy to verify that the height of a search tree that is balanced in the sense of the equations {*bal N*} and {*bal C*} is proportional to the logarithm of the number of keys in the tree. It turns out that it is (in fact, the height of such a tree is at most 44% more than the base-2 logarithm of the number of keys that occur in the search tree), but we haven't proved it. It wouldn't be hard for you to track down a proof of this fact if you wanted to, but this is one of the details we are going to skip in our discussion.

A more troubling point is that we have left out the proofs of many theorems, and we haven't been entirely formal in most of the proofs we have taken the trouble to write down. Our proofs are probably right in broad outline, but it is certain that if we tried to carry them down to the last detail, we would make mistakes. People simply are not attentive enough and patient enough to work out, correctly, all the details in the thousands of logic formulas that would be required to complete these proofs in an entirely formal manner.

For such an undertaking, a mechanized logic is required. That is, running a computer program that verifies that our proofs are correct in every detail is the only practical way to be sure. We don't know of such a program for Haskell programs, but practical tools of this kind do exist for other notations based on the same computation model. For example, ACL2 [1] is a mechanized logic based in a programming language called Common Lisp.

ACL2 is able not only to verify that a proof is correct, but is actually able to construct many inductive proofs on its own. The strategy one would use, if the goal were to make sure a set equations implies a certain property of the operators involved, would be to use a tool like ACL2 to derive a proof of the property by stating a sequence of simple theorems that lead to the proof. The proof of each of the simple theorems would cite, somewhere along the line, one or more of the previous theorems in the sequence.

The tool would prove the simple theorems automatically, and the result would be a fully verified proof. The hard part is stating just the right sequence of theorems simple enough for the tool to prove automatically. This requires the same kinds of skills as constructing proofs at the level of detail we have been using in our discussion of AVL trees.

Proof assistants such as ACL2 have become, over the past few years, after decades of effort by their developers, practical to use as a regular part of the software development process. By attempting to do proofs of properties of software without the use of such tools, students learn how to state relevant theorems and how to build a proof of a theorem through a sequence of simpler theorems. In other words, students learn what they need to know to make use of proof assistants in the practice of software development.

It is our hope that by now you have gained enough experience with this sort of proof to enable you to make use of well-established proof assistants. We also hope that you will be motivated to do so by a certain level of confidence, born of your experience in working through problems, that the phrase "defect-free software" is not an oxymoron.

Chapter 13

Discrete Mathematics in Circuit Design

The techniques of discrete mathematics that you have been studying in this book are used throughout computer science. We have seen many small examples of the application of mathematics to computing, and we have also used programming to help with the mathematics. The previous chapter has examined a sophisticated application, the design of AVL trees.

This chapter looks at another application of formal reasoning and a mathematical approach to computer science: the use of discrete mathematics to help with the process of designing digital circuits. Hardware design is not the real subject here, so we will explain just enough about hardware to explain the particular circuits involved in the discussion.

In addition to applying discrete mathematics, we will use Haskell to specify and simulate circuits. The combination of discrete mathematics and Haskell makes it possible to carry out several useful tasks: precise specification of circuits, simulation, correctness proofs, and circuit derivations.

Digital circuit design is a vast subject area, and there is not space here to cover all of it. Therefore we will consider only one class of digital circuits (*combinational* circuits, which don't contain flip flops). However, that restriction is made *only* to keep the chapter short; discrete mathematics is used heavily throughout the entire subjects of digital circuit design and computer architecture.

You do not need to have any prior knowledge about hardware in order to read this chapter; everything you need to know is covered here. We will begin by defining the basic hardware components, Boolean logic gates, and then will look at how to specify and simulate simple circuits using Haskell. Then we apply the methods of Propositional Logic to circuit design, including reasoning with truth tables and algebraic reasoning about circuits.

One of the principal problems in designing circuits is ensuring that they

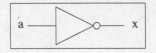

Figure 13.1: Symbol for the Inverter

are correct. It is extremely expensive to debug circuit designs by building and testing them, so getting the design right in the first place is crucial. The last sections in this chapter address this issue: we use mathematics to state precisely what it means for an addition circuit to be correct. We use recursion and higher order functions to design an adder, and we use induction to prove that it works correctly.

13.1   Boolean Logic Gates

Digital circuits are constructed with primitive circuits called *logic gates*. There are logic gates that implement the basic operations of propositional logic, $\wedge$, $\vee$, $\neg$, as well as a few other similar operations.

The simplest logic gate is the *inverter*, which implements the logical not ($\neg$) operation. It takes one input and produces one output; Figure 13.1 shows the standard symbol for the inverter with input a and output x. Instead of using the $\neg$ symbol to specify an inverter, we will use the name *inv*; thus *inv a* means the output of an inverter whose input is a. The inverter's truth table is identical to the truth table for the logical not ($\neg$) operator. Throughout this chapter, we will frequently use 0 and 1 as shorthand notations for False and True. This convention is frequently used in hardware design.

a	$inv\ a$
0	1
1	0

Some of the most commonly used logic gates take two inputs. The logical $\wedge$ operation is performed by the *and2* gate, whose symbol is shown in Figure 13.2. This gate is named *and2* because it takes two inputs; there are similar gates *and3*, *and4* that take 3 and 4 inputs, respectively. The inclusive logical or operation $\vee$ is produced by the *or2* gate (Figure 13.3), and the exclusive or operation, which produces 1 if either argument is 1 *but not both*, is provided by the *xor* gate (Figure 13.4). The following table defines the outputs of these logic gates.

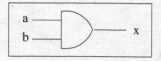

Figure 13.2: And Gate

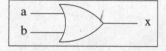

Figure 13.3: Or Gate

a	b	and2 a b	or2 a b	xor a b
0	0	0	0	0
0	1	0	1	1
1	0	0	1	1
1	1	1	1	0

13.2 Functional Circuit Specification

Since a digital circuit produces outputs that depend on its inputs, a circuit can be modelled by a mathematical function. Furthermore, we can implement such functions directly with Haskell functions. This provides several valuable benefits, including error checking, powerful specification techniques, and tools for circuit simulation.

A circuit can be specified in two ways: using a Haskell function definition, or using a schematic diagram. Most of the time we will use both forms, and later in the chapter you will see some of the advantages of each kind of specification. Meanwhile, we will look at how to write functional circuit specifications and what the corresponding schematic diagrams look like.

A circuit's function is applied to its inputs, and the result is the output. For example, to specify that a and b should be connected to the inputs of an

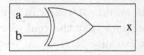

Figure 13.4: Exclusive Or Gate

Figure 13.5: The Circuit y = inv (and2 a b)

and2 gate, and the output should be named *x*, we would write the Haskell specification:

```
x = and2 a b
```

Function applications are also used to make connections between several components. The following specification says that the output of the *and2* gate should be inverted, and the inverted output is named *y*. Figure 13.5 shows the corresponding circuit diagram.

```
y = inv (and2 a b)
```

The type of a circuit indicates what its inputs and outputs are. The value carried by a wire is called a *signal*, and there is a class `Signal` of Haskell types that can be used to represent such a value. Obviously `Bool` is a member of the `Signal` class, since we could use `True` and `False` to represent the values of logic signals. There are other types as well that are useful in various circumstances. The logic gates have the following types:

```
inv :: Signal a => a -> a
and2, or2, xor :: Signal a => a -> a -> a
```

For every type in the `Signal` class, there are constants `zero` and `one` that represent the basic logic values; these correspond to `False` and `True`. In the examples that follow, specialised constants `False` (for logical 0 or False) and `True` (for logical 1 or True) will be used; the advantage of `False` and `True` is that the system knows which signal type to use for them, so you can omit type signatures when evaluating expressions.

13.2.1   Circuit Simulation

A circuit simulator is a computer program that predicts the behaviour of a circuit, without requiring that the circuit be constructed physically. The program behaves just like the circuit would: it reads in a set of inputs to the circuit, and it produces the same outputs that the real circuit would.

Simulation is important because it is much easier, cheaper, and faster to test a design by simulating it with a computer than by constructing it. Just as programs have to be debugged, complex circuit designs also contain errors and must go through an extensive testing and debugging process. With a circuit simulator, it is possible to test the correctness of a design immediately; in contrast, it may take days or weeks to fabricate a physical prototype circuit.

You can simulate any circuit by applying it to suitable signal inputs. For example, we can simulate an `and2` gate by applying `and2` to each of the four possible sets of input values. Compare the results of the following execution with the truth table for `and2`:

```
> and2 False False
False
> and2 False True
False
> and2 True False
False
> and2 True True
True
```

A useful technique is to put a set of test cases in the circuit specification file, right after the circuit itself. This serves as documentation, a set of examples to help a reader to understand what is going on, and it's also useful to check that the circuit is still working when other parts of the system have been modified.

Here is a complete set of test cases for simulating the *and2* gate. It is organised just like a truth table: each line consists of a test for particular data values of the inputs. On each line there is a -- symbol indicating that the rest of the line is a comment, and after the -- we give the expected result.

```
and2 False False    -- False
and2 False True     -- False
and2 True False     -- False
and2 True True      -- True
```

A convenient way to execute the test cases is to put up two windows on the screen: a text editor containing this file, and an interactive session with Haskell. Use the mouse to copy and paste the first line of the test into the Haskell window, and compare the actual result with the expected result. Repeat this for each line in the test suite. The software tools module, `Stdm.hs`, contains more powerful simulation tools; see the accompanying documentation.

13.2.2 Circuit Synthesis from Truth Tables

Hardware design is not a random process (at least, it shouldn't be). There are many systematic techniques for designing robust circuits. For small circuits, the specification is commonly expressed in the form of a truth table, and you need to design a circuit which implements that truth table. This section presents a systematic method for solving this problem. It has two great advantages: the method is *simple*, and it *always works*. Sometimes the method doesn't produce the most efficient solution, but that may not be so important, and if it is, there are also systematic methods for optimising circuits.

Every truth table can be written in a general form, where there is one line for every possible combination of input values, and a variable $p, q, r, s \ldots$ specifies the value of the result. For example, here is the general truth table for a circuit f that takes two inputs:

x	y	$f\,x\,y$
0	0	p
0	1	q
1	0	r
1	1	s

A truth table with k input variables will have 2^k lines. To illustrate how to synthesise a logic function, let's consider the following example with three input variables:

a	b	c	$f\,a\,b\,c$
0	0	0	0
0	0	1	1
0	1	0	0
0	1	1	1
1	0	0	1
1	0	1	0
1	1	0	1
1	1	1	0

The idea is simple: the output $f\,a\,b\,c$ should be 1 whenever the inputs correspond to any of the lines where the rightmost column contains 1. All we need is a logical expression which is True for each such line and False for the others; the value of $f\,a\,b\,c$ is then just the logical *or* of all those expressions.

Because there happen to be four lines with a result of 1, we will need four expressions, one for each of these lines: we don't yet know what these expressions are, so we just call them $expr_1$, $expr_2$, and so on. The required expression has the form

$$f\,a\,b\,c = expr_1 \lor expr_2 \lor expr_3 \lor expr_4.$$

The next step is to figure out what these four expressions are. The first one, $expr_1$, should be 1 if the inputs specify the first line of the table where the output is 1. This happens when $a = 0$, $b = 0$ and $c = 1$; equivalently, it happens when $\neg a$, $\neg b$ and c are all true. Therefore the expression is simply $expr_1 = \neg a \land \neg b \land c$. This expression has the property we were looking for: it is true when the inputs are $a = 0, b = 0, c = 1$, and it is false for all other inputs. The other expressions are worked out the same way:

a	b	c	x	expr
0	0	0	0	
0	0	1	1	$\neg a \wedge \neg b \wedge c$
0	1	0	0	
0	1	1	1	$\neg a \wedge b \wedge c$
1	0	0	1	$a \wedge \neg b \wedge \neg c$
1	0	1	0	
1	1	0	1	$a \wedge b \wedge \neg c$
1	1	1	0	

Now we just plug the expressions into the equation for $f\ a\ b\ c$:

$$f\ a\ b\ c\ =\ (\neg a \wedge \neg b \wedge c) \vee (\neg a \wedge b \wedge c) \vee (a \wedge \neg b \wedge \neg c) \vee (a \wedge b \wedge \neg c)$$

There are several useful refinements of this technique, but those are strictly optional. The important point is that we have a simple method that can be used to synthesise a logical expression—and hence a digital circuit—to implement any truth table.

Often there is a straightforward but inefficient way to design a circuit; an efficient implementation should be used in the final product, but this may be difficult to design. Furthermore, debugging is quicker for easy designs. Because of this, a useful approach is to begin by specifying the simple circuit and then to *transform* it to a more efficient one. The transformation consists of a logical proof that the two circuits have implement exactly the same function. Boolean algebra is a powerful tool for circuit transformation; we can start with a specification expressed as a logical expression and transform it through a sequence of steps until a circuit with satisfactory performance is found.

Exercise 1. Design a circuit that implements the following truth table:

a	b	c	$f\ a\ b\ c$
0	0	0	1
0	0	1	1
0	1	0	0
0	1	1	0
1	0	0	0
1	0	1	1
1	1	0	1
1	1	1	1

Modern digital circuits can be very large and complex; current processor chips contain hundreds of millions of components. Such circuits cannot be designed as one giant diagram, with every component inserted individually. The key to design is *abstraction*. The circuit is organised in a series of levels of abstraction.

At the lowest level are the logic gates and other primitive components. These are used to design the next level up, including circuits like multiplexors,

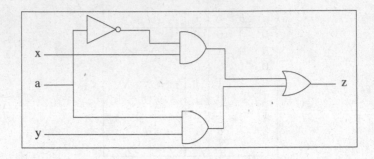

Figure 13.6: Multiplexor

demultiplexors, half and full adders, etc. Such circuits contain around 5 to 10 lower level components, so their specifications are not too complicated. The next level up of the design comprises basic circuits, not primitive logic gates: a design containing 5 or 10 circuits at the level of a multiplexor, for example, would actually contain on the order of 100 logic gates. The same process continues for many levels of abstraction, but at all stages the design is kept reasonably simple through the use of sufficiently high level building blocks.

This section shows just a few simple circuit designs, in order to give some feeling for how abstraction is used. We will be concerned with just three levels: the primitive logic gates; the simplest circuits, including multiplexors and adders for individual bits, and the next level up where binary numbers are added.

13.2.3   Multiplexors

A multiplexor is the hardware equivalent of a conditional (if—then—else) expression. It takes a control input a and two data inputs, x and y. There is one output; if a is 0 then the output is x, but if a is 1 then the output is y. The circuit is implemented using the standard logic gates (see Figure 13.6):

```
mux1 :: Signal a => a -> a -> a -> a
mux1 a x y = or2 (and2 (inv a) x) (and2 a y)
```

A demultiplexor is the opposite of a multiplexor. It has a single data input x and a control input a. The circuit produces two outputs (z_0, z_1). The x input is sent to whichever output is selected by a, and the other output is 0 regardless of the value of x. Figure 13.7 shows the circuit, which is specified as follows:

```
demux1 :: Signal a => a -> a -> (a,a)
demux1 a x = (and2 (inv a) x, and2 a x)
```

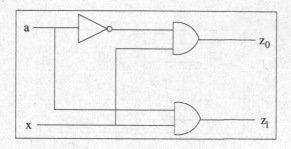

Figure 13.7: Demultiplexor

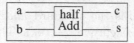

Figure 13.8: Half Adder Black Box

Exercise 2. Recall the informal description of the multiplexor: if a is 0 then the output is x, but if a is 1 then the output is y. Write a truth table that states this formally, and then use the procedure from Section 13.2.2 to design a multiplexor circuit. Compare your solution with the definition of `mux1` given above.

13.2.4 Bit Arithmetic

It's natural to use bits to represent Boolean values and to perform logical calculations with them. An even more common application is to use bits to represent numbers, for example using the binary number system. In fact, the word 'bit' reflects this usage: it originated as an acronym for **B**inary Dig**it**. In this section, we will look at digital circuits for adding individual bits; the following sections extend this to words representing binary numbers.

The most basic addition circuit is the *half adder*, which takes two bits a and b to be added together, and produces a two-bit result (c, s) where c is the carry and s is the sum. Figure 13.8 gives the black box diagram for a half adder, and here is its truth table:

a	b	c	s
0	0	0	0
0	1	0	1
1	0	0	1
1	1	1	0

A circuit implementing the half adder could be synthesised using the method given in Section 13.2.2, but we could also just observe that the carry output

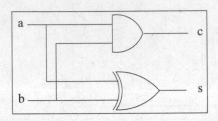

Figure 13.9: Half Adder Circuit

c has the same truth table as the standard *and2* logic gate, while the sum output s is the same as the exclusive or (*xor*). Therefore a half adder can be implemented with just two logic gates (Figure 13.9).

```
halfAdd :: Signal a => a -> a -> (a,a)
halfAdd a b = (and2 a b, xor a b)
```

It is important to be sure that a circuit design is correct before actually fabricating the hardware. One method for improving confidence in correctness is to simulate the circuit. This approach works well for circuits that have a small number of inputs (and no internal state), like `halfAdd`, because it's possible to check every possible combination of input values. Most real-world circuits are too complex for exhaustive testing, and a good approach is to perform some testing and also to carry out a correctness proof. This provides two independent methods for checking the circuit, greatly reducing the likelihood that errors will go unnoticed.

Although `halfAdd` is simple enough to allow complete testing on all possible inputs, we will also consider how to prove its correctness. In order to do this, it's useful to define a *bitValue* function that converts a bit signal into an integer, either 0 or 1. This function requires that the signal arguments be members of the *Static* class, which ensures that they have fixed numeric values. (There are non-static signals used in circuits with flip flops, but those details need not concern us here.)

```
bitValue :: Static a => a -> Int
bitvalue x = if x==zero then 0 else 1
```

The following theorem says that the half adder circuit produces the correct result; that is, if we interpret the output (c, s) as a binary number, then this is actually the sum of the numeric values of the inputs.

Theorem 106. Let $(c, s) = halfAdd\ a\ b$. Then

$$2 \times bitValue\ c + bitValue\ s \ = \ bitValue\ a + bitValue\ b.$$

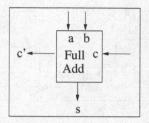

Figure 13.10: Full Adder Black Box

Proof. This theorem is easily proved by checking the equation for each of the four possible combinations of input values. This is essentially the same as using Haskell to simulate the circuit for all the input combinations; the only difference is notational. The details of this proof are left to you to work out. For larger circuits, correctness proofs are not at all like simulators. We will see the difference later in this chapter.   □

In order to add words representing binary numbers, it will be necessary to add three bits: one data bit from each of the words, and a carry input bit. This function is provided by the *full adder* circuit (Figure 13.10); as with the half adder, there is a two-bit result (c', s), where c' is the carry output and s is the sum.

a	b	c	c'	s
0	0	0	0	0
0	0	1	0	1
0	1	0	0	1
0	1	1	1	0
1	0	0	0	1
1	0	1	1	0
1	1	0	1	0
1	1	1	1	1

There are many ways to implement the full adder circuit. The traditional method, given below and in Figure 13.11, uses two half adders. This is an example of the use of abstraction in circuit design: the specification of the full adder is simplified by the use of the `halfAdd` circuit. In general, larger circuits are implemented using a handful of somewhat-smaller circuits, and designers don't implement everything directly using logic gates.

```
fullAdd :: Signal a => (a,a) -> a -> (a,a)
fullAdd (a,b) c = (or2 w y, s)
  where (w,x) = halfAdd a b
        (y,s) = halfAdd x c
```

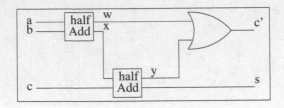

Figure 13.11: Implementation of Full Adder

The following theorem says that the full adder produces the correct result; it will be needed later to prove the correctness of adders for general binary numbers.

Theorem 107 (Correctness of full adder). Let $(c', s) = fullAdd\ (a, b)\ c$, so that c' is the carry output and s is the sum output. Then

$$bitValue\ c' \times 2 + bitValue\ s\ =\ bitValue\ a + bitValue\ b + bitValue\ c.$$

Exercise 3. Use Haskell to test the half adder on the following test cases, and check that it produces the correct results. (You'll need to load the Stdm.hs file.)

```
Test cases for half adder, with predicted results
halfAdd False False   -- 0 0
halfAdd False True    -- 0 1
halfAdd True False    -- 1 0
halfAdd True True     -- 1 1
```

Exercise 4. Prove Theorem 106 using truth tables.

Exercise 5. Prove Theorem 107.

13.2.5   Binary Representation

Binary numbers consist of a sequence of bits called a *word*. This is represented as a list. For example, if you have four individual signals named w, x, y, and z, you can treat them as a word by writing $[w, x, y, z]$, and its type is Static $a \Rightarrow [a]$.

There are, unfortunately, two traditional schemes for numbering the bits in a word: $[x_0, x_1, x_2, x_3]$ and $[x_3, x_2, x_1, x_0]$. We will use the first scheme, where the leftmost bit of a word has index 0 and the rightmost has index $k - 1$, where k is the number of bits in the word. The binary value of the word $[x_0, x_1, x_2, x_3]$ is

$$x_0 \times 2^3 + x_1 \times 2^2 + x_2 \times 2^1 + x_3 \times 2^0.$$

In general, the value of a k-bit word $x = [x_0, \ldots, x_{k-1}]$ is

$$\sum_{i=0}^{k-1} x_i \times 2^{k-(i+1)}.$$

This value is calculated by the *wordValue* function:

```
wordValue :: Static a => [a] -> Integer
wordValue [] = 0
wordValue (x:xs) = 2^k * bitValue x + wordValue xs
    where k = length xs
```

Notice that in the binary number system the smallest value that can be represented in k bits is 0, and the largest value is $2^k - 1$. Negative numbers are not representable at all in the binary system. Most modern computers represent integers using the *two's complement* number system, which allows for negative numbers. One nice property of two's complement is that an ordinary binary addition circuit can be used to perform addition on two's complement numbers. Consequently, we won't worry about negative numbers here, but will proceed to the addition of binary numbers.

Exercise 6. Work out the numeric value of the word $[1, 0, 0, 1, 0]$. Then check your result by using the computer to evaluate:

```
wordValue [True,False,False,True,False]
```

13.3 Ripple Carry Addition

A ripple carry adder (Figure 13.12) is used to calculate the sum of two words. When the word size is four bits, the binary arguments are words containing the bits $[x_0, x_1, x_2, x_3]$ and $[y_0, y_1, y_2, y_3]$. The most significant bits are x_0 and y_0, and they appear on the left of the word; the least significant bits are x_3 and y_3, and they appear at the right of the word. The ripple carry adder also takes a carry input c (this makes it possible to add larger numbers by performing a sequence of additions). The output produced by the circuit is a single carry output bit, and a word of sum bits. We require that the two input words and the sum word all contain the same number of bits.

The following specification (Figure 13.13) uses four full adders to construct a 4-bit ripple carry adder. In bit position i, the data inputs are x_i and y_i, and the carry input is c_{i+1}. The sum produced by position i is s_i, and the carry output c_i will be sent to the bit position to the left (position $i - 1$).

```
add4 :: Signal a => a -> [(a,a)] -> (a,[a])
add4 c [(x0,y0),(x1,y1),(x2,y2),(x3,y3)] =
    (c0, [s0,s1,s2,s3])
```

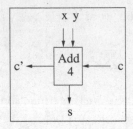

Figure 13.12: 4-Bit Ripple Adder Black Box

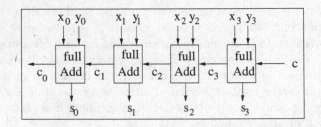

Figure 13.13: 4-Bit Ripple Carry Adder

```
where (c0,s0) = fullAdd (x0,y0) c1
      (c1,s1) = fullAdd (x1,y1) c2
      (c2,s2) = fullAdd (x2,y2) c3
      (c3,s3) = fullAdd (x3,y3) c
```

To use the adder, we must convert the input numbers into 4-bit binary representations. For example, here is the addition of $3 + 8$.

```
Example: addition of 3 + 8
   3 + 8
 =   0011   (  2+1 =  3)
   + 1000   (    8 =  8)
 =   1011   (8+2+1 = 11)
Calculate this by evaluating
   add4 False [(False,True),(False,False),
               (True,False),(True,False)]
The expected result is
   (False, [True,False,True,True])
```

Exercise 7. Use Haskell to evaluate the example above, and check that the result matches the expected result.

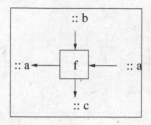

Figure 13.14: Building Block Circuit for mscanr

13.3.1 Circuit Patterns

The `add4` specification in the previous section is not too complicated, but it would be awfully tedious to extend it to handle words containing 32 or 64 bits (which are the sizes most commonly used with current generation processors). The analogous specifications would contain 32 or 64 local equations, and there would be a correspondingly large number of indexed names. Besides the sheer size, such specifications would be highly error-prone. Furthermore, it would be better to define the family of all ripple carry adders, rather than to keep on defining new ones at various different word sizes.

A much better approach is to define the general k-bit ripple carry adder once and for all, so that it works for arbitrary k. To do this, we can't name the individual bits explicitly, like x_0, x_1 and so on. Instead, we need to use a method that works for any word size without referring explicitly to the individual bits. The most intuitive description of the adder would say 'each full adder has its carry input connected to the carry output of its right neighbour,' and this is exactly the idea that needs to be formalised with a function.

A higher order function can be used to express the abstract structure of circuits like `add4`. The idea is to write a function whose argument is a circuit specification; the higher order function connects up as many copies as required of the circuit it is given. Figure 13.14 shows the sort of building block needed for the ripple carry adder; it matches the black box structure of the full adder.

The *mscanr* function takes a building-block circuit with an appropriate type. It creates as many copies of the building block as are required and makes all the internal connections that are needed. Figure 13.15 depicts the structure of the resulting circuit, and Figure 13.16 shows the circuit defined by *mscanr* as a black box.

```
mscanr :: (b->a->(a,c)) -> a -> [b] -> (a,[c])
mscanr f a [] = (a,[])
mscanr f a (x:xs) =
  let (a',ys) = mscanr f a xs
      (a'',y) = f x a'
  in (a'', y:ys)
```

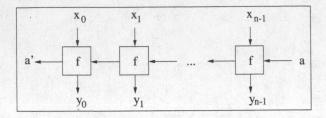

Figure 13.15: Structure of `mscanr` Pattern

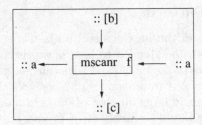

Figure 13.16: Black Box for `mscanr` Pattern

Exercise 8. Let $f :: b \to a \to (a, c)$ be a black box circuit. Draw a diagram showing the structure of the circuit specified by

$$mscanr\ f\ a\ [(x_0, y_0), (x_1, y_1), (x_2, y_2)].$$

13.3.2   The n-Bit Ripple Carry Adder

Now we can use the higher order function *mscanr* to define a general ripple carry adder that works for any word size. The definition is much more intuitive than brute-force definitions like *add4*, once you understand the idea of using higher order functions to express regular circuit patterns.

The *mscanr* function expresses the pattern of the ripple carry adder. It simply says that a ripple carry adder consists of a row of full adders, one for every bit position. The carry input to each full adder is connected to the carry output from the full adder to the right, and the carry input to the rightmost (least significant) bit position is the carry input to the entire addition.

The first argument is a circuit specification with type $b \to a \to (a, c)$. Recall that a full adder has type

$$Signal\ a \Rightarrow (a, a) \to a \to (a, a).$$

This fits the *mscanr* pattern, and a ripple carry adder consists of a row of full adders with the carry input of each connected to the carry output of its right neighbour.

```
rippleAdd :: Signal a => a -> [(a,a)] -> (a, [a])
rippleAdd c zs = mscanr fullAdd c zs
```

This definition works for arbitrary word size. The size of a particular circuit is determined by the size of the input data word. The definition itself doesn't get longer if the words become longer! It's now easy to specify a 6-bit adder, as the following test case demonstrates.

```
Example: addition of 23+11
   23 + 11
 =   010111   (16+4+2+1 = 23)
   + 001011   (   8+2+1 = 11) with carry input = 0
 =   100010   (   32+2 = 34) with carry output = 0
Calculate with the circuit by evaluating
   rippleAdd False [(False,False),(True,False),(False,True),
              (True,False),(True,True),(True,True)]
The expected result is
   (False, [True,False,False,False,True,False])
```

Exercise 9. Work out a test case using the ripple carry adder to calculate 13+41=54, using 6-bit words. Test it using the computer.

13.3.3 Correctness of the Ripple Carry Adder

Theorem 108. Let xs and ys be k-bit words, so $xs, ys :: Signal\, a \Rightarrow [a]$. Define $(c, sum) = rippleAdd\, zero\, (zip\, xs\, ys)$; thus $c :: a$ is the carry output and $ss :: [a]$ is the sum word. Then

$$bitValue\, c \times 2^k + wordValue\, ss\ =\ wordValue\, xs + wordValue\, ys.$$

The left-hand side of the equation is the numeric value of the output of the ripple carry adder circuit, and the right-hand side is the numeric value of its inputs. Thus the equation says that the circuit produces the correct answer.

Proof. Induction over k. For the base case, $k = 0$, and $xs = ys = []$. First we simplify:

$(c, ss)\ =\ rippleAdd\, zero\, []$
$= mscanr\, fullAdd\, zero\, []$
$= (zero, [])$
$c = zero$
$ss = []$
$wordValue\, []\ +\ wordValue\, []$
$\quad = 0 + 0$

$$= 0$$

$$bitValue \; c \times 2^k + wordValue \; ss$$
$$= 0 \times 2^0 + 0$$
$$= 0 \times 2^0 + wordValue \; [\,]$$

For the inductive case, let $k = length \; xs = length \; ys$, and assume

$$bitValue \; c \times 2^k + wordValue \; ss = wordValue \; xs + wordValue \; ys,$$

where $(c, ss) = rippleAdd \; zero \; (zip \; xs \; ys)$. The aim is to prove that

$$bitValue \; c \times 2^{k+1} + wordValue \; ss =$$
$$wordValue \; (x : xs) + wordValue \; (y : ys),$$

where $(c, ss) = rippleAdd \; zero \; (zip \; (x : xs) \; (y : ys))$.
First we simplify:

$$(c, ss) = mscanr \; fullAdd \; zero \; (zip \; (x : xs) \; (y : ys))$$
$$= mscanr \; fullAdd \; zero \; ((x, y) : zip \; xs \; ys)$$
$$= let \; (c', ss) = mscanr \; fullAdd \; zero \; (zip \; xs \; ys)$$
$$= rippleAdd \; zero \; (zip \; xs \; ys)$$
$$(c'', s) = fullAdd \; (x, y) \; c'$$
$$in \; (c'', s : ss)$$

Now the left-hand side of the equation can be transformed into the right-hand side, using equational reasoning:

lhs (numeric value of output from the adder)
$$= bitValue \; c \times 2^{k+1} + wordValue \; ss$$
$$= bitValue \; c'' \times 2^{k+1} + wordValue \; (s : ss)$$
$$= bitValue \; c'' \times 2^{k+1} + bitValue \; s \times 2^k + wordValue \; ss$$
$$= (bitValue \; c'' \times 2 + bitValue \; s) \times 2^k + wordValue \; ss$$
$$= (bitValue \; x + bitValue \; y + bitValue \; c') \times 2^k + wordValue \; ss$$
$$= (bitValue \; x + bitValue \; y) \times 2^k + (bitValue \; c') \times 2^k + wordValue \; ss)$$
$$= (bitValue \; x + bitValue \; y) \times 2^k + wordValue \; xs + wordValue \; ys$$
$$= (bbitValue \; x \times 2^k + wordValue \; xs) + (bitValue \; y \times 2^k + wordValue \; ys)$$
$$= wordValue \; (x : xs) + wordValue \; (y : ys)$$
$$= \text{rhs (numeric value of inputs to the adder)}$$

□

13.3.4   Binary Comparison

Comparison of binary numbers is just as important as adding them. It is particularly interesting to consider how to implement a comparison circuit, since this problem has some strong similarities and also some strong differences to the ripple carry adder.

First, let's consider the comparison of two bits. This is analogous to starting out with a half adder. The problem is to design a circuit *halfCmp* that compares two one-bit numbers x and y. The result of the comparison should be a triple of bits of the form (lt, eq, gt), where lt is true if $x < y$, eq is true if $x = y$, and gt is true if $x > y$. The problem is pretty simple, as x and y are both just one bit! The type should be

$$halfCmp :: Signal\ a \implies (a, a) \rightarrow (a, a, a),$$

where the input bits x and y are provided as a pair (x, y), and the result triple consists of the result bits $(lt, eq, gt) :: (a, a, a)$. Finding a solution is straightforward:

```
halfCmp :: Signal a => (a,a) -> (a,a,a)
halfCmp (x,y) =
  (and2 (inv x) y,    -- x<y when x=0,y=1
   inv (xor x y),     -- x=y when x=0,y=0 or x=1,y=1
   and2 x (inv y))    -- x>y when x=1,y=0
```

The next problem to consider is that of designing a ripple comparator that takes two words representing binary numbers (the words must have the same size), and returns a triple of three bits (lt, eq, gt) that indicate the result of comparing x and y. The meanings of the output bits are just the same as in the previous problem; the only difference is that now the inputs to the comparator are words rather than bits.

Just as you compare two numbers by looking first at the most significant digits, a binary comparison is performed by moving from left to right through the word. Initially we assume the two words are equal; represent this by $(lt, eq, gt) = (0, 1, 0)$. If the next bit position has $x = 1$ and $y = 0$, then we know that the final result must be $(0, 0, 1)$ regardless of any bits to the right; conversely if $x = 0$ and $y = 1$, then the final result must be $(1, 0, 0)$ regardless of the bit values to the right. However, if x and y have the same value in this bit position, then as far as we know the result is still $(0, 1, 0)$ but that result might be changed later.

The calculation in each bit position requires the two local bits (that is, for position i we need the ith bit of both of the input words). It also requires the result of the comparison for all the bits to the left. The task is performed by a full comparison circuit, which is analogous to the full adder.

```
fullCmp :: Signal a => (a,a,a) -> (a,a) -> (a,a,a)
fullCmp (lt,eq,gt) (x,y) =
  (or2 lt (and3 eq (inv x) y),    -- <
   and2 eq (inv (xor x y)),       -- =
   or2 gt (and3 eq x (inv y)))    -- >
```

Now we can define the ripple comparison circuit, which compares two binary numbers. Its definition is similar to the ripple carry adder, but there are several differences in the circuit pattern required. In the first place, the information flow is left to right for comparison, rather than the right to left order used in addition. Another difference is that for comparison we are interested only in the final horizontally moving value; this would be analogous to wanting the carry output from an addition, but we do not need a result analogous to the sum bits. The standard *foldl* higher order function specifies exactly the circuit pattern needed here. A final difference is that the comparator just takes the two numbers to be compared; it generates the initial horizontal value locally. In contrast, the ripple carry adder takes a carry input. The reason for that is that many applications of adders, such as the ALU of a computer's processor, use carry inputs to provide the ability to add long numbers comprising several words.

```
rippleCmp :: Signal a => [(a,a)] -> (a,a,a)
rippleCmp z = foldl fullCmp (False,True,False) z
```

Exercise 10. Define a full set of test cases for the circuit *halfCmp*, which compares two bits, and execute them using the computer.

Exercise 11. Define three test cases for the *rippleCmp* circuit, with a word size of three bits, demonstrating each of the three possible results. Run your test cases on the computer.

13.4   Suggestions for Further Reading

The application of discrete mathematics to digital circuit design is a large subject. Most of the publications that address this area are aimed more at researchers than students, so some of the references cited here may be difficult to read, but it's interesting to see *real* applications of discrete mathematics.

The ripple carry adder presented in this chapter takes time proportional to the word size, which would make it too slow for practical use on modern machines, where the word size is typically 64 bits. It is possible to design faster adders, but their operation is much more subtle than that of the ripple carry adder. The best way to understand how a fast adder works is to apply mathematical and formal methods to its design [23].

Hydra [22] is a computer hardware description language based on Haskell. The software, as well as a collection of papers on how to use Hydra for circuit design and formal correctness proofs, can be downloaded from the web [22].

13.5   Review Exercises

Exercise 12. Show that the and4 logic gate, which takes four inputs a, b, c, and d and outputs $a \wedge b \wedge c \wedge d$, can be implemented using only and2 gates.

Exercise 13. Work out a complete set of test cases for the full adder, and calculate the expected results. Then simulate the test cases and compare with the predicted results.

Exercise 14. Suppose that a computer has 8 memory locations, with addresses $0, 1, 2, \ldots 7$. Notice that we can represent an address with 3 bits, and the size of the memory is 2^3 locations. We name the address bits $a_0 a_1 a_2$, where a_0 is the most significant bit and a_2 is the least significant. When a memory location is accessed, the hardware needs to send a signal to each location, telling it whether it is the one selected by the address $a_0 a_1 a_2$. Thus there are 8 *select* signals, one for each location, named $s_0, s_1, \ldots, s_7$. Design a circuit that takes as inputs the three address bits a_0, a_1, a_2, and outputs the select signals $s_0, s_1, \ldots, s_7$. *Hint: use demultiplexors, arranged in a tree-like structure.* (Note: modern computers have an address size from 32 to 64 bits, allowing for a large number of locations, but a 3-bit address makes this exercise more tractable!)

Exercise 15. Does the definition of *rippleAdd* allow the word size to be 0? If not, what prevents it? If so, what does it mean?

Exercise 16. Does the definition of *rippleAdd* allow the word size to be negative? If not, what prevents it? If so, what does it mean?

Exercise 17. Note that for the half adder and full adder, we did thorough testing—we checked the output of the circuit for every possible input. Note also that we did not do this for the ripple carry adder, where we just tried out a few particular examples. The task: Explain why it is infeasible to do thorough testing of a ripple carry adder circuit, and estimate how long it would take to test all possible input values for the binary adder in a modern processor where the words are 64 bits wide.

Exercise 18. Computer programs sometimes need to perform arithmetic, including additions and comparisons, on big integers consisting of many words. Most computer processor architectures provide hardware support for this, and part of that hardware support consists of the ability to perform an addition where the carry input is supplied externally, and is not assumed to be 0. Explain why the carry input to the *rippleAdd* circuit helps to implement multiword addition, but we don't need an analogous horizontal input to *rippleCmp* for multiword comparisons.

Appendix A

Software Tools

The Haskell programming language provides excellent support for mathematical computing in general. This book uses programs in Haskell 98, which is standardised, stable, and well-supported. The book also requires a library of definitions that give additional support for the topics of discrete mathematics. The library is called *Software Tools for Discrete Mathematics*, and it consists of a single file called Stdm.hs. You need, therefore, two pieces of software to use a computer along with the book:

- An interactive implementation of Haskell;

- The file Stdm.hs.

Both items are free, and they run on most major computer platforms. All of the software can be downloaded from the web. The web home page for this book contains the stdm.hs file along with full documentation:

www.dcs.gla.ac.uk/ jtod/discrete-mathematics/

The web page also tells you how to download and use various implementations of Haskell.

Many of the programming examples and exercises that appear in this book rely on functions that are defined in Stdm.hs. To run the examples or exercise solutions, you will need to import the Stdm module. You can do this by placing the following line at the beginning of a Haskell sript file:

```
import Stdm
```

You can also import the software tools in most interactive versions of Haskell; for example, with the ghci interpreter, enter the following command:

```
:load Stdm
```

The details of how to use the software tools may depend on which Haskell implementation you are using, so check the web page for up-to-date instructions.

Appendix B

Resources on the Web

Home page for *Discrete Mathematics Using a Computer*. The book's web page contains a variety of useful information, and is an integral part of the book:

- You can download the *Software Tools for Discrete Mathematics*, along with complete documentation;

- There are additional practice problems, solutions, and explanations;

- There are up-to-date pointers to many other relevant web pages.

```
http://www.dcs.gla.ac.uk/~jtod/discrete-mathematics/
```

Instructor's Guide for Discrete Mathematics Using a Computer. The Instructor's Guide is entirely online. A password is required to read it; please contact the authors to obtain access. See the book home page for the current contact address.

```
http://www.dcs.gla.ac.uk/~jtod
   /discrete-mathematics/instructors-guide/
```

Home page for Haskell. This page contains complete and current information on the Haskell language, including free (and open source) compilers and interpreters that you can download, up-to-date pointers to the home pages for all the Haskell compilers and interpreters, the official Haskell language definition, the complete specification for the standard libraries, pointers to books and articles on Haskell and functional programming, and more.

```
http://www.haskell.org/
```

Appendix C

Solutions to Selected Exercises

C.1   Introduction to Haskell

3.

```
isA :: Char -> Bool
isA 'a' = True
isA c   = False
```

4.

```
isHello :: String -> Bool
isHello ('h':'e':'l':'l':'o':[]) = True
isHello str                      = False
```

5.

```
removeSpace :: String -> String
removeSpace []       = []
removeSpace (' ':xs) = xs
removeSpace xs       = xs
```

6.

```
toBool :: Int -> Bool
toBool 1 = True
toBool 0 = False

convert :: [Int] -> [Bool]
convert xs = map toBool xs
```

7.

```
member0 :: String -> Bool
member0 string = or (map (== '0') string)
```

8.

```
foldr max 0 [1,5,3]
  = 1 'max' (5 'max' (3 'max' 0))
  = 1 'max' (5 'max' 3)
  = 1 'max' 5
  = 5
```

9.

```
addJust :: Maybe Int -> Maybe Int -> Maybe Int
addJust (Just a) (Just b) = Just (a + b)
addJust (Just a) Nothing = Just a
addJust Nothing (Just a) = Just a
addJust Nothing Nothing = Nothing

addMaybe :: [Maybe Int] -> [Maybe Int] -> [Maybe Int]
addMaybe lst1 lst2 = zipWith addJust lst1 lst2
```

10.

```
data Metals = Copper | Silver | Gold
            | Tin | Platinum | Bronze
            deriving (Eq, Show)
```

11. The coins can be represented by a list containing one or more elements of the following type:

```
data Coins = OneP Int | TwoP Int | FiveP Int
           | TenP Int | TwentyP Int
           | FiftyP Int | HundredP Int
           deriving (Eq, Show)
```

12.

```
data Universal = BOOL Bool | INT Int | CHAR Char
                 deriving (Eq, Show)
```

13. A separate constructor is needed to indicate the number of elements that are present, so we introduce five constructors Tuple0, etc. It would not be correct to use a list type, for two reasons: the individual elements should be allowed to have different types, and the number of elements must be limited to four.

```
data Tuples a b c d = Tuple0 | Tuple1 a | Tuple2 a b
                    | Tuple3 a b c | Tuple4 a b c d
                    deriving (Eq, Show)
```

15.

```
showMaybe :: Show a => Maybe a -> String
showMaybe Nothing  = []
showMaybe (Just a) = show a
```

16.

```
bitwiseAnd :: [Int] -> [Int] -> [Int]
bitwiseAnd word1 word2 = zipWith bitAnd word1 word2
```

17.

```
[True, False]               "2" ++ "a"
[(3,True), (9,False)]       2 == 3
'a' > 'b'                   [[1],[2],[3]]
```

18. The types of the two paired elements must be the same, according to the type signature given. In addition, the type of the first component of the pair was defined as being of the Num class, while the function f was applied to a pair with a Boolean as its first element.

19. There are two possible constructors in the type, and the function definition handles only one of them. If the other is used, an error occurs.

20.

```
[a | (Just a) <- xs]
```

21.

```
largerThanN :: [Int] -> Int -> [Int]
largerThanN lst n = [e | e <- lst, e > n]
```

22.

```
f :: [Int] -> Int -> [Int]
f lst v = [n | n <- [0..length lst - 1], lst!!n == v]
```

23.

```
[e | e <- [1..20],
    [x | x <- [1..e], x * x == e] == []]
```

24. We define an auxiliary function that compares the current letter with the one we are looking for, and increments the running count if it matches. The main iteration over the string is performed by the foldr.

```haskell
count :: (Eq a, Num b) => a -> a -> b -> b
count letter x acc
 = if letter == x then acc + 1 else acc

countLetters :: Char -> String -> Int
countLetters c lst = foldr (count c) 0 lst
```

25.

```haskell
remove :: Char -> Char -> [Char] -> [Char]
remove ch x acc
 = if x == ch then acc else x:acc

removeEachLetter :: Char -> [Char] -> [Char]
removeEachLetter ch lst
 = foldr (remove ch) [] lst
```

26.

```haskell
rearrange :: [a] -> a -> [a]
rearrange lst x = x:lst

rev :: [a] -> [a]
rev lst = foldl rearrange [] lst
```

27.

```haskell
takeLast :: Maybe a -> a -> Maybe a
takeLast  Nothing x = Just x
takeLast  (Just y) x = Just x

maybeLast :: [a] -> Maybe a
maybeLast lst = foldl takeLast Nothing lst
```

C.3   Recursion

1.

```haskell
copy :: [a] -> [a]
copy []     = []
copy (x:xs) = x : copy xs
```

2.

```haskell
inverse :: [(a,b)] -> [(b,a)]
inverse [] = []
inverse ((a,b):xs) = (b,a) : inverse xs
```

3.

```
merge :: Ord a => [a] -> [a] -> [a]
merge [] bs = bs
merge as [] = as
merge (a:as) (b:bs) =
  if a < b
    then a : merge as (b:bs)
    else b : merge (a:as) bs
```

4.

```
(!!) :: Int -> [a] -> Maybe a
(!!) n [] = Nothing
(!!) 0 (x:xs) = Just x
(!!) n (x:xs) = (!!) (n-1) xs
```

5.

```
lookup :: Eq a => a -> [(a,b)] -> Maybe b
lookup a [] = Nothing
lookup a ((a',b):ps)
 = if a==a' then Just b else lookup a ps
```

6.

```
countElts :: Eq a => a -> [a] -> Int
countElts e []     = 0
countElts e (x:xs) =
  if e == x
    then 1 + countElts e xs
    else countElts e xs
```

7.

```
removeAll :: Eq a => a -> [a] -> [a]
removeAll e [] = []
removeAll e (x:xs) =
  if e == x
    then removeAll e xs
    else x : removeAll e xs
```

8.

```
f :: [a] -> [a]
f [] = []
f (x:xs) = g xs

g :: [a] -> [a]
g [] = []
g (x:xs) = x:f xs
```

9.

```
extract :: [Maybe a] -> [a]
extract []           = []
extract (Nothing:xs) = extract xs
extract (Just x:xs)  = x : extract xs
```

10.

```
loop :: String -> String -> Int -> Maybe Int
loop [] s2 n = Nothing
loop (x:xs) s2 n
  = if length s2 > length (x:xs)
      then Nothing
      else if take (length s2) (x:xs) == s2
              then Just n
              else loop xs s2 (n+1)

f :: String -> String -> Maybe Int
f str1 str2 = loop str1 str2 0
```

11.

```
foldrWith ::
  (a -> b -> c -> c) -> c -> [a] -> [b] -> c
foldrWith f z [] bs = z
foldrWith f z as [] = z
foldrWith f z (a:as) (b:bs) =
  f a b (foldrWith f z as bs)
```

12.

```
mappend :: (a -> [b]) -> [a] -> [b]
mappend f xs = foldr fun [] xs
  where fun x acc = f x ++ acc
```

13.

```
removeDuplicates :: Eq a => [a] -> [a]
removeDuplicates [] = []
removeDuplicates (x:xs)
  = if elem x xs
      then removeDuplicates xs
      else x : removeDuplicates xs
```

14.

```
member :: Eq a => a -> [a] -> Bool
member a []     = False
member a (x:xs) = a == x || member a xs
```

15.

```
intersection :: Eq a => [a] -> [a] -> [a]
intersection [] set2       = []
intersection (x:set1) set2
  = if (elem x set2)
       then x : intersection set1 set2
       else intersection set1 set2
```

16.

```
isSubset :: Eq a => [a] -> [a] -> Bool
isSubset [] set2     = True
isSubset (x:xs) set2 = elem x set2 /\ isSubset xs set2
```

17.

```
isSorted :: Ord a => [a] -> Bool
isSorted []      = True
isSorted (x:[]) = True
isSorted (x:y:xs) = x < y && isSorted (y:xs)
```

C.4 Induction

1. Let m be an arbitrary natural number, and we will prove that the theorem holds for every natural number n by induction over n.

Base case: $a^{0 \times n} = a^0 = 1 = (a^m)^0$.

Induction case: assume the hypothesis $a^{m \times n} = (a^m)^n$. The aim is to establish $a^{m \times (n+1)} = (a^m)^{n+1}$. Transform the left-hand side into the right-hand side using algebra: $a^{m \times (n+1)} = a^{(m \times n)+m} = a^{m \times n} \times a^m = $ (using the induction hypothesis) $(a^m)^n \times a^m = (a^m)^{n+1}$. Therefore $a^{m \times n} = (a^m)^n \to a^{m \times (n+1)} = (a^m)^{(n+1)}$. So by mathematical induction, we have $\forall n \in N.a^{m \times n} = (a^m)^n$.

2. The sum of the first n odd numbers can be written as $\sum_{i=0}^{n-1} (2i+1)$, and we want to prove that this is equal to n^2. It's possible to prove this by induction,

This theorem can also be proved by reusing the theorem we already have, along with some simple algebra, as follows: $\sum_{i=0}^{n-1}(2i + 1) = 2\sum_{i=0}^{n-1} i + \sum_{i=0}^{n-1} 1 = 2\frac{(n-1)n}{2} + n = n^2 - n + n = n^2$.

4. Proof by induction on n.

Base case.

$\sum_{i=1}^{0} \mathit{fib}\ i$
$= 0$
$= 0 + 1 - 1$
$= \mathit{fib}\ 0 + \mathit{fib}\ (0 + 1) - 1$
$= \mathit{fib}\ (0 + 2) - 1$

Induction case. Assume the hypothesis $\sum_{i=1}^{n} fib\ i = fib\ (n+2) - 1$.

$\sum_{i=1}^{n+1} fib\ i$
$= \sum_{i=1}^{n} fib\ i + fib\ (n+1)$
$= fib\ (n+2) - 1 + fib\ (n+1)$
$= fib\ (n+1) + fib\ (n+2) - 1$
$= fib\ (n+3) - 1$
$= fib\ (n+1+2) - 1$

Thus we have proved that the theorem holds for $n = 0$, and that if it holds for arbitrary n it also holds for $n + 1$. Therefore, by induction, the theorem holds for all n in the set of natural numbers.

5.

Proof. Induction over xs. The base case is

$length\ ([]+\!\!+ys)$
$= length\ ys$ $\{\ (+\!\!+).1\ \}$
$= 0 + length\ ys$ $\{\ 0 + x = x\ \}$
$= length\ []\ +\ length\ ys$ $\{\ length.1\ \}$

Assume that $length\ (xs+\!\!+ys) = length\ xs + length\ ys$. The inductive case is

$length\ ((x:xs)+\!\!+ys)$
$= length\ (x:(xs+\!\!+ys))$ $\{\ (+\!\!+).2\ \}$
$= 1 + length\ (xs+\!\!+ys)$ $\{\ length.2\ \}$
$= 1 + (length\ xs\ +\ length\ ys)$ $\{\ hypothesis\ \}$
$= (1 + length\ xs) + length\ ys$ $\{\ (+)\ is\ associative\ \}$
$= length\ (x:xs) + length\ ys$ $\{\ length.2\ \}$

$\square$

6.

Proof. Induction over xs. The base case is

$map\ f\ ([]+\!\!+ys)$
$= map\ f\ ys$ $\{\ (+\!\!+).1\ \}$
$= []+\!\!+ map\ f\ ys$ $\{\ (+\!\!+).1\ \}$
$= map\ f\ []+\!\!+ map\ f\ ys$ $\{\ map.1\ \}$

For the inductive case, assume $map\ f\ (xs+\!\!+ys) = map\ f\ xs +\!\!+ map\ f\ ys$. Then

$map\ f\ ((x:xs)+\!\!+ys)$
$= map\ f\ (x:(xs+\!\!+ys))$ $\{\ (+\!\!+).2\ \}$
$= f\ x\ :\ map\ f\ (xs+\!\!+ys)$ $\{\ map.2\ \}$
$= f\ x\ :\ (map\ f\ xs +\!\!+ map\ f\ ys)$ $\{\ hypothesis\ \}$
$= (f\ x\ :\ map\ f\ xs) +\!\!+ map\ f\ ys$ $\{\ (+\!\!+).2\ \}$
$= map\ f\ (x:xs) +\!\!+ map\ f\ ys$ $\{\ map.2\ \}$

□

7.

Proof. Using the definition of composition, we can rewrite the equation to be proved as:

$$map\ f\ (map\ g\ xs)\ =\ map\ (f.g)\ xs$$

We prove this by induction over xs. The base case is

$$
\begin{aligned}
map\ f\ (map\ g\ []) \\
= map\ f\ [] &\qquad \{\ map.1\ \} \\
= [] &\qquad \{\ map.1\ \} \\
= map\ (f.g)\ [] &\qquad \{\ map.1\ \}
\end{aligned}
$$

Assume that $map\ f\ (map\ g\ xs)\ =\ map\ (f.g)\ xs$. The inductive case is

$$
\begin{aligned}
map\ f\ (map\ g\ (x:xs)) \\
= map\ f\ (g\ x\ :\ map\ g\ xs) &\qquad \{\ map.2\ \} \\
= f\ (g\ x)\ :\ map\ f\ (map\ g\ xs) &\qquad \{\ map.2\ \} \\
= f\ (g\ x)\ :\ map\ (f.g)\ xs &\qquad \{\ \text{hypothesis}\ \} \\
= ((f.g)\ x)\ :\ map\ (f.g)\ xs &\qquad \{\ (.)\ \} \\
= map\ (f.g)\ (x:xs) &\qquad \{\ map.2\ \}
\end{aligned}
$$

□

11.

Theorem 109 (++ is associative). $(xs \mathbin{+\!\!+} ys) \mathbin{+\!\!+} zs = xs \mathbin{+\!\!+} (ys \mathbin{+\!\!+} zs)$

Proof. Induction over xs. The base case is

$$
\begin{aligned}
([] \mathbin{+\!\!+} ys) \mathbin{+\!\!+} zs \\
= ys \mathbin{+\!\!+} zs \\
= [] \mathbin{+\!\!+} (ys \mathbin{+\!\!+} zs)
\end{aligned}
$$

Inductive case.

$$
\begin{aligned}
((x:xs) \mathbin{+\!\!+} ys) \mathbin{+\!\!+} zs \\
= (x:(xs \mathbin{+\!\!+} ys)) \mathbin{+\!\!+} zs \\
= x:((xs \mathbin{+\!\!+} ys) \mathbin{+\!\!+} zs) \\
= x:(xs \mathbin{+\!\!+} (ys \mathbin{+\!\!+} zs)) \\
= (x:xs) \mathbin{+\!\!+} (ys \mathbin{+\!\!+} zs)
\end{aligned}
$$

□

13. The first line of the inductive case says 'Assume $P(n)$, and consider a set containing $n + 1$ horses; call them $h_1, h_2, \ldots, h_{n+1}$.' Consider what happens when $n = 1$. Our set of $n + 1$ horses contains only two of them, h_1 and h_2. Thus the two subsets turn out to be $A = \{h_1\}$ and $B = \{h_2\}$. All the horses in A have the same colour, and all the horses in B have the same colour. So far, so good—but the next sentence says 'Pick one of the horses that is an element of both A and B,' *and it is impossible to do this because when $n = 1$, the sets A and B are disjoint.* The rest of the proof is invalid because it relies on this non-existent horse.

There are two useful lessons to learn from this.

- Whenever a proof says to 'pick an x such that $\ldots$', it is essential to make sure that such an x exists.

- It is helpful to work through the details using a concrete example.

By the way: the flaw explained above is the only error in this proof. *If* the proof worked for the case $n = 1$—if all pairs of horses had the same colour—then *it would indeed be true* that all horses are the same colour.

14.

The length of xss must be finite, otherwise the operation would never terminate and this result would never be reached.

15.

$reverse\ (reverse\ [1, 2, 3])$
$= reverse\ [3, 2, 1]$
$= [1, 2, 3]$

16.

Proof. Induction over xs. The base case:

$reverse([]+\!\!+ys)$
$= reverse\ ys$
$= reverse\ ys+\!\!+[\,]$
$= reverse\ ys+\!\!+reverse\ [\,]$

Inductive case:

$reverse\ ((x : xs)+\!\!+ys)$
$= reverse\ (x : (xs+\!\!+ys))$
$= reverse\ (xs+\!\!+ys)+\!\!+[x]$
$= (reverse\ ys+\!\!+reverse\ xs)+\!\!+[x]$
$= reverse\ ys+\!\!+(reverse\ xs+\!\!+[x])$
$= reverse\ ys+\!\!+(reverse(x : xs))$

So the theorem holds by list induction. □

17.

Proof. Induction over xs. The base case is

$reverse\ (reverse\ [\,])$
$= reverse\ [\,]$
$= [\,]$

For the inductive case, assume for some xs that $reverse\ (reverse\ xs) = xs$. Then

$reverse\ (reverse\ (x : xs))$
$= reverse\ (reverse\ xs\!+\!\!+\![x])$
$= reverse\ [x]\!+\!\!+reverse\ (reverse\ xs)$
$= [x]\!+\!\!+xs$
$= x : xs$

□

18. Consider, for example, the infinite list $nats = [0, 1, 2, 3, \ldots]$. This is easily defined in Haskell, with the expression [0..]. Notice that $nats$ is not an infinite loop; a computation requiring only a finite portion of it will terminate. On the other hand, any computation requiring all of $nats$ will go into an infinite loop; thus $length\ nats$ will never terminate. This is expressed by saying $length\ nats = \bot$, where the symbol $\bot$ (pronounced 'bottom') denotes an undefined value (either an infinite loop or an error).

Now consider the equation $reverse\ (reverse\ nats) = nats$. The left-hand side of the equation is not just a vague, intuitive statement that might be interpreted as leaving a list unchanged. It has a specific meaning that can (and must) be determined by equational reasoning using the definition of $reverse$.

The outermost application of reverse must begin by determining which of the two defining equations is relevant; it does this by performing a case analysis on its argument (which is $reverse\ nats$) to decide whether this is the empty list [] or a non-empty list in the form x:xs.

$reverse\ (reverse\ (nats\ 0))$
$= \textbf{case}\ reverse\ (nats\ 0)\ \textbf{of}$
$\quad [\,] \to [\,]$
$\quad x : xs \to reverse\ xs\!+\!\!+[x]$

Now the computer must evaluate $reverse\ (nats\ 0)$ far enough to decide whether it is the empty list, or a cons expression. This evaluation proceeds as follows:

$reverse\ (nats\ 0)$
$= reverse\ (0 : nats\ 1)$
$= reverse\ (nats\ 1) +\!\!+ [0]$
$= reverse\ (1 : nats\ 2) +\!\!+ [0]$
$= (reverse\ (nats\ 2) +\!\!+ [1]) +\!\!+ [0]$

This evaluation is never going to terminate. The computer will just keep generating larger and larger natural numbers, constructing ever bigger expressions, but it will never actually figure out whether the original value $reverse\ (nats\ 0)$ is empty! Hence no information at all can be obtained from evaluating

$$reverse\ (reverse\ (nats\ 0)).$$

To summarise, we know that:

$$reverse\ (reverse\ (nats\ 0)) \;=\; \bot$$
$$nats\ 0 \;\neq\; \bot$$

A consequence of this, which might be surprising, is that

$$reverse\ .\ reverse \neq id.$$

19.

Base case:

```
length (concat xss)
  = { xss = [] }
length (concat [])
  = { (concat).[] }
length []
  =
0
```

True, as the empty list has no elements, thus its length is 0

Inductive case:

```
length(concat xss)
  = { xss = ys : yss }
length (concat (ys : yss))
  = { (concat).(:) }
length (ys ++ concat yss)
  = { (length).theorem }
length ys + length (concat yss)
```

True, as `length ys` delivers the length of the first element in `xss`, and under the induction assumption $P(n)$, `length (concat yss)` delivers the length of the remaining elements in `xss`.

20.

Assume that the list xs is of type [Bool], that it has $n + 1$ elements, and that True is an element of xs.

We will prove that

$$P(n) \equiv \text{or xs} = \text{True}$$

Base case:

```
or xs
  = { xs = [True] }
or [True]
  = { (:) }
or (True: [])
  = { definition of or }
foldr (||) False (True: [])
  = { (foldr).(:) }
True || (foldr (||) False [])
  = { || commutative }
(foldr (||) False []) || True
  = { || null }
True
```

Inductive case:

```
or xs
  = { definition of or }
foldr (||) False xs
  = { xs = y : ys }
foldr (||) False (y : ys)
  = { (foldr).(:) }
y || (foldr (||) False (ys))
```

There are two cases:

Case 1: y = True

```
True || (foldr (||) False (ys))
  = { || commutative }
(foldr (||) False (ys)) || True
  = { || null }
True
```

Case 2: True *is in* ys

```
y || (foldr () False (ys))
  = { induction assumption }
y || True
  = { || null }
True
```

21.

Assume that the list xs is of type Bool, that it has $n + 1$ elements, and that all elements of xs are True.

We will prove that

$$P(n) \equiv \text{and xs} = \text{True}$$

Base case:

```
and xs
  = { definition of and }
foldr (&&) True xs
  = { xs = [True] }
foldr (&&) True [True]
  = { (:) }
foldr (&&) True (True: [])
  = { (foldr).(:) }
True && (foldr (&&) True [])
  = { (foldr).[] }
True && True
  = { && identity }
True
```

Inductive case:

```
and xs
  = { definition of and }
foldr (&&) True xs
  = { xs = True: ys }
foldr (&&) True (True: ys)
  = { (foldr).(:) }
True (&&) (foldr (&&) True ys)
  = { inductive assumption }
True && True
  = { && identity }
True
```

22.

```
maximum :: [Ord] -> Ord

maximum xs = foldr (max) y ys
           where xs = y:ys
```

23.

Base case:

```
maximum xs
   = { xs = y : [] }
maximum (y : [])
   = { definition of maximum }
foldr (max) y []
   = { (foldr).[] }
y
```

True, as $y >= y$.

Inductive case:

```
maximum xs
   = { xs = y1 : (y2 : ys) }
maximum (y1 : (y2 : ys))
   = { definition of maximum }
foldr (max) y1 (y2 : ys)
   = { (foldr).(:) }
y2 max (foldr (max) y1 ys)
```

Case 1: y2 is the largest element of xs.

```
y2 max (foldr (max) y1 ys)
   = { definition of max }
y2
```

Case 2: y2 is not the largest element of xs.

```
y2 max (foldr (max) y1 ys)
   = { definition of max }
foldr (max) y1 ys
   = { definition of maximum }
maximum (y1 : ys)
```

In case 1, y2 is greater than or equal to any other element in xs by the case 1 assumption.

In case 2, the largest element in xs, maximum (y1 : ys), is greater than or equal to all the other elements in y1 : ys by the induction assumption and greater than or equal to y2 by the case 2 assumption.

24.

```
firstElement :: [a] -> a
firstElement (x:xs) = x

firstElements :: [[a]] -> [a]
```

```
firstElements xs = map (firstElement) xs
```

25.

Assume that the finite list xss is of type [[a]] and that it has length n. We prove that

$$P(n) \equiv \texttt{concat xss} = \texttt{foldr (++) [] xss}$$

Base case:

```
concat xss
  = { xss = [] }
concat []
  = { (concat).[] }
[]
  = { (foldr).[] }
foldr (++) [] []
  = { xss = [] }
foldr (++) [] xss
```

Inductive case:

```
concat xss
  = { xss = ys : yss }
concat ys : yss
  = { (concat).(:) }
ys ++ concat yss
  = { induction assumption }
foldr (++) [] (ys : yss)
  = { xss = ys : yss }
foldr (++) [] xss
```

26.

```
(&&)::Bool -> Bool -> Bool

and :: [Bool] -> Bool
and = foldr (&&) True
```

27.

```
and ([False] ++ xs)
  = { ++.(:) }
and (False : xs)
  = { and.(:) }
foldr (&&) True (False : xs)
  = {foldr.(:)}
False && (foldr (&&) True xs)
  = { && null }
False
```

C.5 Trees

1.

```
data Tree1
  = Tree1Leaf
  | Tree1Node Char Integer Tree1 Tree1 Tree1
```

2.

```
data Tree2
  = Tree2Leaf
  | Tree2Node Integer [Tree2]
```

4.

```
inorderf :: (a->b) -> BinTree a -> [b]
inorderf f BinLeaf = []
inorderf f (BinNode x t1 t2) =
  inorderf f t1 ++ [f x] ++ inorderf f t2
```

8.

```
mapTree :: (a->b) -> BinTree a -> BinTree b
mapTree f BinLeaf = BinLeaf
mapTree f (BinNode a t1 t2) =
  BinNode (f a) (mapTree f t1) (mapTree f t2)
```

9.

```
concatTree :: Tree [a] -> [a]
concatTree Tip = []
concatTree (Node as t1 t2) =
  concatTree t1 ++ as ++ concatTree t2
```

10.

```
zipTree :: Tree a -> Tree b -> Maybe [(a,b)]
zipTree Tip t2                   = Nothing
zipTree t1  Tip                  = Nothing
zipTree (Node a t1 t2) (Node b t3 t4)
  = case (zipTree t1 t3) of
      Nothing -> Nothing
      Just lst1 ->
        case (zipTree t2 t4) of
          Nothing -> Nothing
          Just lst2 -> Just (lst1 ++ [(a,b)] ++ lst2)
```

11. The function traverses the two trees together. If either tree (or both) is empty, then the empty list is returned. If both trees are nonempty, then their top elements a and b are combined by computing f a b, and the function traverses the subtrees recursively. ~

```
zipWithTree ::
  (a -> b -> c) -> Tree a -> Tree b -> [c]
zipWithTree f Tip t2 = []
zipWithTree f t1 Tip = []
zipWithTree f (Node a t1 t2) (Node b t3 t4)
  = zipWithTree f t1 t3
    ++ [f a b]
    ++ zipWithTree f t2 t4
```

12. A straightforward approach just uses existing functions:

```
appendTree1 :: BinTree a -> [a] -> [a]
appendTree1 t ks = inorder t ++ ks
```

However, this solution builds a list of data from the tree and then recopies this data during the concatenation. The solution can be made more efficient using a continuation list:

```
appendTree2 :: BinTree a -> [a] -> [a]
appendTree2 BinLeaf ks = ks
appendTree2 (BinNode x t1 t2) ks =
  appendTree2 t1 (x : appendTree2 t2 ks)
```

C.6   Propositional Logic

1. The answer is a question. If someone were to ask the computer named "Bob" whether it would answer "yes" to the question "Does this bus go to the airport?", would the computer named "Bob" answer "yes"?

These are the cases to consider.

- Case *PfAir* : Suppose the computer named "Bob" is a properly functioning computer and the bus is going to the airport. Then the computer would answer "yes" if it were asked the question posed within the question the computer is asked. So, the correct answer to the question is "yes", and, since the computer functions properly, that is the answer it gives.

- Case *TwAir*: Suppose the computer named "Bob" contains twisted logic and the bus is going to the airport. Then the computer would answer "no" if it were asked the question posed within the question the computer is asked. So, the correct answer to the question is "no", and, since the computer is twisted, the answer it gives is "yes". So far, so good. You get the answer "yes" if the bus is going to the airport, no matter which kind of computer Bob is.

- Case *PfAin*: Suppose the computer named "Bob" is a properly functioning computer and the bus is not going to the airport. Then the computer would answer "no" if it were asked the question posed within the question the computer is asked. So, the correct answer to the question is "no", and, since the computer functions properly, that is the answer it gives.

- Case *TwAin*: Suppose the computer named "Bob" contains twisted logic and the bus is not going to the airport. Then the computer would answer "yes" if it were asked the question posed within the question the computer is asked. So, the correct answer to the question is "yes", and, since the computer is twisted, the answer it gives is "no".

So you get the answer "no" if the bus is not going to the airport and the answer "yes" if it is going to the airport, regardless of whether Bob is a properly functioning computer or a computer with twisted logic.

Another Solution. Would the computer named "Bob" give the answer "yes" to one and only one of the following two questions?

- Q1. Does the computer named Bob have twisted logic?

- Q2. Is this bus going to the airport?

The first solution was based on the logical operation "implication". This solution is based on "exclusive or".

- Case *PfAirXor*: Suppose the computer named "Bob" is a properly functioning computer and the bus is going to the airport. Then the computer would answer "yes" because that is the correct answer to the question about the number of "yes" answers to the Q-questions ("no" for Q1 and "yes" for Q2).

- Case *TwAirXor*: Suppose the computer named "Bob" contains twisted logic and the bus is going to the airport. Then the computer would answer "yes" because the computer named "Bob" would answer "no" to Q1 (because the correct answer is "yes") and "no" to Q2 (because the correct answer is "yes"). Since the answer to neither Q1 nor Q2 is "yes", the correct answer to the question about the number of "yes" answers to the Q-questions is "no", which means the computer, which is twisted, would answer "yes". For the airport bus, the answer is "yes", no matter which type of computer Bob is.

- Case *PfAinXor*: Suppose the computer named "Bob" is a properly functioning computer and the bus is not going to the airport. Then the computer would answer "no" to Q1 and "no" to Q2, and since the answer to neither question is "yes", the properly functioning computer gives the answer "no" to the question about how many of the Q-questions have the answer "yes".

- Case *TwAinXor*: Suppose the computer named "Bob" contains twisted logic and the bus is not going to the airport. Then the computer would answer "no" to Q1 (since the correct answer is "yes") and "yes" to Q2 (since the correct answer is "no"). That means the correct answer to the question about the number of "yes" answers to the Q-questions is "yes". Bob, being twisted, answers "no".

  For the airport bus, the answer is "no', no matter which type of computer Bob is. Therefore, if the computer answers "yes", the first bus is going to the airport. If the computer answers "no", the first bus is not going to the airport, so wait for the second bus.

Both solutions require the computers to know all the bus schedules and their own names. The solution based on exclusive or requires, additionally, the computer named "Bob" to know whether it contains a twisted logic circuit. Bob can get by without this bit of knowledge in the case of the solution based on implication because it can simply feed the questions through its own circuits, observe the answers internally, then run the answer to the larger question through the normal output device.

4.

$$(\text{True} \wedge P) \vee Q$$

- True is a WFF.

- P is a WFF, therefore $\text{True} \wedge P$ is a WFF.

- Q is a WFF. Since $\text{True} \wedge P$ and Q are WFFs, $\text{True} \wedge P \vee Q$ is a WFF.

P	Q	$\text{True} \wedge P$	$(\text{True} \wedge P) \vee Q$
True	True	True	True
True	False	True	True
False	True	False	True
False	False	False	False

The proposition is satisfiable but not a tautology.

5.

- P, Q, and R are WFFs.

- Since P, Q, and R are WFFs, $P \vee Q$, $P \vee R$, and $Q \vee R$ are WFFs.

- Since P and $Q \vee R$ are WFFs, $P \wedge (Q \vee R)$ is a WFF.

- Since $P \vee Q$ and $P \vee R$ are WFFs, $(P \vee Q) \wedge (P \vee R)$ is a WFF.

- Since $(P \vee Q) \wedge (P \vee R)$ and $P \wedge (Q \vee R)$ are WFFs, $(P \vee Q) \wedge (P \vee R) \leftrightarrow P \wedge (Q \vee R)$ is a WFF.

P	Q	R	$P \vee Q$	$P \vee R$	$Q \vee R$	$(P \vee Q) \wedge (P \vee R)$
True	True	True	True	True	True	True
True	True	False	True	True	True	True
True	False	True	True	True	True	True
True	False	False	True	True	False	True
False	True	True	True	True	True	True
False	False	True	False	True	True	False
False	False	False	False	False	False	False

P	Q	R	$P \wedge (Q \vee R)$	$(P \vee Q) \wedge (P \vee R) \leftrightarrow P \wedge (Q \vee R)$
True	True	True	True	True
True	True	False	True	True
True	False	True	True	True
True	False	False	False	False
False	True	True	False	False
False	False	True	False	True
False	False	False	False	True

The proposition is satisfiable but not a tautology.

6.

- P and Q are WFFs.

- Since P and Q are WFFs, $\neg P$ and $\neg Q$ are WFFs.

- Since P, Q, $\neg P$ and $\neg Q$ are WFFs, $P \wedge \neg Q$ and $Q \wedge \neg P$ are WFFs.

- Since P and Q are WFFs, $P \leftrightarrow Q$ is a WFF.

- Since $P \leftrightarrow Q$ is a WFF, $\neg (P \leftrightarrow Q)$ is a WFF.

- Since $P \wedge \neg Q$ and $Q \wedge \neg P$ are WFFs, $(P \wedge \neg Q) \vee (Q \wedge \neg P)$ is a WFF.

- Since $(P \wedge \neg Q) \vee (Q \wedge \neg P)$ and $\neg (P \leftrightarrow Q)$ are WFFs, $((P \wedge \neg Q) \vee (Q \wedge \neg P)) \rightarrow \neg (P \leftrightarrow Q)$ is a WFF.

P	Q	$\neg P$	$\neg Q$	$P \wedge \neg Q$	$Q \wedge \neg P$
True	True	False	False	False	False
True	False	False	True	True	False
False	True	True	False	False	True
False	False	True	True	False	False

P	Q	$(P \wedge \neg Q) \vee (Q \wedge \neg P)$	$P \leftrightarrow Q$	$\neg(P \leftrightarrow Q)$
True	True	False	True	False
True	False	True	False	True
False	True	True	False	True
False	False	False	True	False

P	Q	$((P \wedge \neg Q) \vee (Q \wedge \neg P)) \rightarrow \neg(P \leftrightarrow Q)$
True	True	True
True	False	True
False	True	True
False	False	True

The proposition is a tautology.

7. $(P \rightarrow Q) \wedge (P \rightarrow \neg Q)$

- P and Q are WFFs.
- Since Q is a WFF, $\neg Q$ is a WFF.
- Since P, Q, and $\neg Q$ are WFFs, $P \rightarrow Q$ and $P \rightarrow \neg Q$ are WFFs.
- Since $P \rightarrow Q$ and $P \rightarrow \neg Q$ are WFFs, $(P \rightarrow Q) \wedge (P \rightarrow \neg Q)$ is a WFF.

P	Q	$\neg Q$	$P \rightarrow Q$	$P \rightarrow \neg Q$	$(P \rightarrow Q) \wedge (P \rightarrow \neg Q)$
True	True	False	True	False	False
True	False	True	False	True	False
False	True	False	True	True	True
False	False	True	True	True	True

The proposition is satisfiable but not a tautology.

8. $(P \rightarrow Q) \wedge (\neg P \rightarrow Q)$

- P and Q are WFFs.
- Since P is a WFF, $\neg P$ is a WFF.
- Since P, Q, and $\neg P$ are WFFs, $P \rightarrow Q$ and $\neg P \rightarrow Q$ are WFFs.
- Since $P \rightarrow Q$ and $\neg P \rightarrow Q$ are WFFs, $(P \rightarrow Q) \wedge (\neg P \rightarrow Q)$ is a WFF.

P	Q	$\neg P$	$P \rightarrow Q$	$(\neg P) \rightarrow Q$	$(P \rightarrow Q) \wedge (\neg P \rightarrow Q)$
True	True	False	True	True	True
True	False	False	False	True	False
False	True	True	True	True	True
False	False	True	True	False	False

The proposition is satisfiable but not a tautology.

9. $(P \to Q) \leftrightarrow (\neg Q \to \neg P)$

- P and Q are WFFs.

- Since P and Q are WFFs, $\neg P$ and $\neg Q$ are WFFs.

- Since P, Q, $\neg Q$, and $\neg P$ are WFFs, $P \to Q$ and $\neg Q \to \neg P$ are WFFs.

- Since $P \to Q$ and $\neg Q \to \neg P$ are WFFs, $(P \to Q) \leftrightarrow (\neg Q \to \neg P)$ is a WFF.

P	Q	$\neg P$	$\neg Q$	$P \to Q$	$\neg Q \to \neg P$	$(P \to Q) \leftrightarrow (\neg Q \to \neg P)$
True	True	False	False	True	True	True
True	False	False	True	False	False	True
False	True	True	False	True	True	True
False	False	True	True	True	True	True

The proposition is a tautology.

10.

$$\cfrac{R \quad \cfrac{Q \quad R}{Q \wedge R}{\scriptstyle\{\wedge I\}}}{P \wedge (Q \wedge R)}{\scriptstyle\{\wedge I\}}$$

11. The proof of y will have a symmetrical shape, but the proof of x will appear triangular, with more inference on the right side than on the left. In the general case, the proof of x with 2^n variables will have height 2^n, because every extra variable will require one extra inference above everything else. In contrast, the proof of y with 2^n variables will have height n.

12.

$$\cfrac{\cfrac{\cfrac{(P \wedge Q) \wedge R}{P \wedge Q}{\scriptstyle\{\wedge E_L\}}}{P}{\scriptstyle\{\wedge E_L\}} \quad \cfrac{\cfrac{\cfrac{(P \wedge Q) \wedge R}{P \wedge Q}{\scriptstyle\{\wedge E_L\}}}{Q}{\scriptstyle\{\wedge E_R\}} \quad \cfrac{(P \wedge Q) \wedge R}{R}{\scriptstyle\{\wedge E_R\}}}{Q \wedge R}{\scriptstyle\{\wedge I\}}}{P \wedge (Q \wedge R)}{\scriptstyle\{\wedge I\}}$$

13.

$$\cfrac{\cfrac{\cfrac{P \quad \cfrac{P \quad P \to Q}{Q}{\scriptstyle\{\to E\}}}{P \wedge Q}{\scriptstyle\{\wedge I\}} \quad P \wedge Q \to R \wedge S}{R \wedge S}{\scriptstyle\{\to E\}}}{S}{\scriptstyle\{\wedge E_R\}}$$

15.

$$\cfrac{\boxed{Q}\quad\cfrac{P\quad\boxed{Q}}{P\wedge Q}{\scriptstyle\{\wedge I\}}}{Q\rightarrow P\wedge Q}{\scriptstyle\{\rightarrow I\}}$$

18.

$$\cfrac{\cfrac{P\quad Q}{P\wedge Q}{\scriptstyle\{\wedge I\}}}{(P\wedge Q)\vee(Q\vee R)}{\scriptstyle\{\vee I_L\}}$$

19. We prove that True $\wedge$ True $\rightarrow$ True and then that True $\rightarrow$ True $\wedge$ True, without translating True into False $\rightarrow$ False.

$$\cfrac{\cfrac{\text{True}\wedge\text{True}}{\text{True}}{\scriptstyle\{\wedge E_R\}}}{\text{True}\wedge\text{True}\rightarrow\text{True}}{\scriptstyle\{\rightarrow I\}}$$

$$\cfrac{\cfrac{\text{True}\quad\text{True}}{\text{True}\wedge\text{True}}{\scriptstyle\{\wedge I\}}}{\text{True}\rightarrow\text{True}\wedge\text{True}}{\scriptstyle\{\rightarrow I\}}$$

20. We prove that True $\vee$ False $\rightarrow$ True and then that True $\rightarrow$ False $\vee$ True.

$$\cfrac{\text{True}\vee\text{False}\quad\cfrac{\text{True}}{\text{True}}{\scriptstyle\{ID\}}\quad\cfrac{\text{False}}{\text{True}}{\scriptstyle\{CTR\}}}{\cfrac{\text{True}}{\text{True}\vee\text{False}\rightarrow\text{True}}{\scriptstyle\{\rightarrow I\}}}{\scriptstyle\{\vee E\}}$$

$$\cfrac{\cfrac{\text{True}}{\text{True}\vee\text{False}}{\scriptstyle\{\vee I_L\}}}{\text{True}\rightarrow\text{True}\vee\text{False}}{\scriptstyle\{\rightarrow I\}}$$

23.

- P is represented by P

- $Q\vee FALSE$ is represented by Or Q FALSE

- $Q \to (P \to (P \land Q))$ is represented by `Imp Q (Imp P (And P Q))`

25. Proof by equational reasonong.

$(P \land \mathsf{False}) \lor (Q \land \mathsf{True})$
$\quad = \mathsf{False} \lor (Q \land \mathsf{True})$ $\{\land \text{ null}\}$
$\quad = \mathsf{False} \lor Q$ $\{\land \text{ identity}\}$
$\quad = Q \lor \mathsf{False}$ $\{\text{commutativity}\}$
$\quad = Q$ $\{\lor \text{ identity}\}$

27. Proof by equational reasoning.

$(P \land ((Q \lor R) \lor Q)) \land S$
$\quad = S \land (P \land ((Q \lor R) \lor Q))$ $\{\land \text{ commutative}\}$
$\quad = S \land (((Q \lor R) \lor Q) \land P)$ $\{\land \text{ commutative}\}$
$\quad = S \land ((Q \lor (R \lor Q)) \land P)$ $\{\lor \text{ associative}\}$
$\quad = S \land ((Q \lor (Q \lor R)) \land P)$ $\{\lor \text{ commutative}\}$
$\quad = S \land (((Q \lor Q) \lor R) \land P)$ $\{\lor \text{ associative}\}$
$\quad = S \land ((Q \lor R) \land P)$ $\{\lor \text{ idempotent}\}$
$\quad = S \land ((R \lor Q) \land P)$ $\{\lor \text{ commutative}\}$

30. Proof by equational reasoning.

$(A \lor B) \land B$
$\quad = \{\lor \text{ } identity\}$
$(A \lor B) \land (B \lor \mathsf{False})$
$\quad = \{\lor \text{ } comm\}$
$(B \lor A) \land (B \lor \mathsf{False})$
$\quad = \{\lor \text{ } over \land\}$
$B \lor (A \land \mathsf{False})$
$\quad = \{\land \text{ } null\}$
$B \lor \mathsf{False}$
$\quad = \{\lor \text{ } identity\}$
B

31. Solution by equational reasoning.

$(\neg A \wedge B) \vee (A \wedge \neg B)$
  $= \{\vee \ over \wedge\}$
$((\neg A \wedge B) \vee A) \wedge ((\neg A \wedge B) \vee \neg B)$
  $= \{\vee \ comm\}$
$(A \vee (\neg A \wedge B)) \wedge (\neg B \vee (\neg A \wedge B))$
  $= \{\vee \ over \wedge\}$
$((A \vee \neg A) \wedge (A \vee B)) \wedge ((\neg B \vee \neg A) \wedge (\neg B \vee B))$
  $= \{\vee \ comm\}$
$((A \vee \neg A) \wedge (A \vee B)) \wedge ((\neg B \vee \neg A) \wedge (B \vee \neg B))$
  $= \{\vee \ compl\}$
$(\text{True} \wedge (A \vee B)) \wedge ((\neg B \vee \neg A) \wedge \text{True})$
  $= \{\wedge \ comm\}$
$((A \vee B) \wedge \text{True}) \wedge ((\neg B \vee \neg A) \wedge \text{True})$
  $= \{\wedge \ identity applied twice\}$
$(A \vee B) \wedge (\neg B \vee \neg A)$
  $= \{DM\}$
$(A \vee B) \wedge \neg(B \wedge A)$
  $= \{\wedge \ comm\}$
$(A \vee B) \wedge \neg(A \wedge B)$

32. The problem is solved by equational reasoning:
$\neg(A \wedge B)$
  $= \{double \ negation applied twice\}$
$\neg(\neg\neg A \wedge \neg\neg B)$
  $= \{DM\}$
$\neg(\neg(\neg A \vee \neg B))$
  $= \{double \ negation\}$
$\neg A \vee \neg B$

33. Solution by equational reasoning.

$(A \vee B) \wedge (\neg A \vee C) \wedge (B \vee C)$
  $= \{\vee \ identity\}$
$(A \vee B) \wedge (\neg A \vee C) \wedge ((B \vee C) \vee \text{False})$
  $= \{\wedge \ compl\}$
$(A \vee B) \wedge (\neg A \vee C) \wedge ((B \vee C) \vee (A \wedge \neg A))$

  $= \{\vee \ over \wedge\}$
$(A \vee B) \wedge (\neg A \vee C) \wedge ((B \vee C \vee A) \wedge (B \vee C \vee \neg A))$
  $= \{\vee \ comm\}$
$(A \vee B) \wedge (\neg A \vee C) \wedge ((A \vee B \vee C) \wedge (\neg A \vee B \vee C))$

  $= \{\wedge \ comm\}$
$(\neg A \vee C) \wedge (A \vee B) \wedge ((A \vee B \vee C) \wedge (\neg A \vee B \vee C))$
  $= \{\wedge \ assoc\}$
$(\neg A \vee C) \wedge ((A \vee B) \wedge (A \vee B \vee C)) \wedge (\neg A \vee B \vee C)$

$$= \{\wedge \; comm\}$$
$$((A \vee B) \wedge (A \vee B \; \vee \; C)) \; \wedge \; (\neg A \vee C) \wedge (\neg A \vee B \; \vee \; C)$$
$$= \{\wedge \; assoc\}$$
$$((A \vee B) \wedge (A \vee B \; \vee \; C)) \; \wedge \; ((\neg A \vee C) \wedge (\neg A \vee B \; \vee \; C))$$
$$= \{\vee \; over \; \wedge\}$$
$$((A \vee (B \wedge (B \; \vee \; C))) \; \wedge \; (\neg A \vee (C \wedge (B \; \vee \; C)))$$
$$= \{\wedge \; comm\}$$
$$(A \vee ((B \; \vee \; C) \wedge B)) \; \wedge \; (\neg A \vee ((B \; \vee \; C) \wedge C))$$
$$= \{\vee \; comm\}$$
$$(A \vee ((C \vee B) \wedge B)) \; \wedge \; (\neg A \vee ((B \; \vee \; C) \wedge C))$$
$$= \{absorbtion, appliedtwice\}$$
$$(A \vee B) \; \wedge \; (\neg A \vee C)$$

34. A solution in natural deduction style:

$$\cfrac{\cfrac{A \wedge \neg A}{A}\{\wedge E_L\} \quad \cfrac{A \wedge \neg A}{\neg A}\{\wedge E_R\}}{\text{False}}\{\rightarrow E\}$$

A solution in the proof-checker notation:

```
hwThm1 = Theorem [A 'And' (Not A)] (FALSE)

proof_hwThm1 =
  (Assume(A 'And' (Not A))
  {----------------------------} 'AndEL'
             A,
                  Assume(A 'And' (Not A))
                  {--------------------------------} 'AndER'
                  (Not A))
  {----------------------------------------------------------}'ImpE'
                  (FALSE)
```

35.

```
hwThm2 = Theorem [A] ((A 'Imp' FALSE) 'Imp' FALSE)

proof_hwThm2 =
  (Assume A,
                  Assume(A 'Imp' FALSE))
  {------------------------------------------------------} 'ImpE'
                  FALSE
  {-------------------------------------------------------} 'ImpI'
        ((A 'Imp' FALSE) 'Imp' FALSE)
```

$$\cfrac{\cfrac{A \quad A \rightarrow \text{False}}{\text{False}}\{\rightarrow E\}}{(A \rightarrow \text{False}) \rightarrow \text{False}}\{\rightarrow I\}$$

36.

```
hwThm3 = Theorem [A, A 'Imp' B, B 'Imp' C, C 'Imp' D] (D)

proof_hwThm3 =
  (((Assume A, Assume(A 'Imp' B))
     {-------------------------} 'ImpE'
     B,        Assume(B 'Imp' C))
    {------------------------------------} 'ImpE'
         C,               Assume(C 'Imp' D))
     {-------------------------------------------} 'ImpE'
                  D
```

$$\cfrac{\cfrac{\cfrac{A \quad A \to B}{B}{\{\to E\}} \quad B \to C}{C}{\{\to E\}} \quad C \to D}{D}{\{\to E\}}$$

37.

```
hwThm6 = Theorem [A 'Imp' B] ((B 'Imp' FALSE) 'Imp' (A 'Imp' FALSE))

proofThm6 =

     (((Assume A, Assume(A 'Imp' B))
     {-----------------------------} 'ImpE'
            B),      Assume(B 'Imp' FALSE))
     {------------------------------------------------} 'ImpE'
                FALSE
     {-------------------------------------} 'ImpI'
              (A 'Imp' FALSE)
     {-------------------------------------} 'ImpI'
     ((B 'Imp' FALSE) 'Imp' (A 'Imp' FALSE))
```

$$\cfrac{\cfrac{\cfrac{A \quad A \to B}{B}{\{\to E\}} \quad B \to \text{False}}{\text{False}}{\{\to E\}}}{\cfrac{A \to \text{False}}{(B \to \text{False}) \to (A \to \text{False})}{\{\to I\}}}{\{\to I\}}$$

38.

$$\cfrac{\cfrac{\cfrac{A \quad A \to B}{B}{\{\to E\}} \quad \cfrac{A \quad A \to \neg B}{\neg B}{\{\to E\}}}{\text{False}}{\{\to E\}}}{A \to \text{False}}{\{\to I\}}$$

39.

$$
\cfrac{\cfrac{\cfrac{A \quad A \to \text{False}}{\text{False}}\{\to E\}}{\cfrac{B}{A \to B}\{CTR\}}\{\to I\} \quad \cfrac{\cfrac{\cfrac{A \quad A \to \text{False}}{\text{False}}\{\to E\}}{\cfrac{\neg B}{A \to \neg B}\{CTR\}}\{\to I\}}{(A \to B) \land (A \to \neg B)}\{\land I\}
$$

41.

$$
\cfrac{\cfrac{P \quad Q}{P \land Q}\{\land I\} \quad \cfrac{R \quad S}{R \land S}\{\land I\}}{(P \land Q) \land (R \land S)}\{\land I\}
$$

42.

$$
\cfrac{\cfrac{\cfrac{\boxed{P \land Q}}{P}\{\land E_L\} \quad P \to R}{R}\{\to E\}}{P \land Q \to R}\{\to I\}
$$

43. We prove that True ∨ True → True and then that True → True ∨ True.

$$
\cfrac{\cfrac{\text{True} \lor \text{True} \quad \cfrac{\text{True}}{\text{True}}\{ID\} \quad \cfrac{\text{True}}{\text{True}}\{ID\}}{\text{True}}\{\lor E\}}{\text{True} \lor \text{True} \to \text{True}}\{\to I\}
$$

$$
\cfrac{\cfrac{\text{True}}{\text{True} \lor \text{True}}\{\lor I_L\}}{\text{True} \to \text{True} \lor \text{True}}\{\to I\}
$$

45. A list comprehension can generate a truth table for you.

```
logicExprValue1 = [((a,b),logicExpr1 a b) |
                    a <- [False,True],
                    b <- [False,True]
                   ]

logicExprValue2 = [((a,b),logicExpr2 a b) |
                    a <- [False,True],
                    b <- [False,True]
                   ]
```

46.

```
logicExprValue3 = [((a,b,c),logicExpr3 a b c) |
                    a <- [False,True],
                    b <- [False,True],
                    c <- [False,True]
                   ]

logicExprValue4 = [((a,b,c),logicExpr4 a b c) |
                    a <- [False,True],
                    b <- [False,True],
                    c <- [False,True]
                   ]
```

47.

```
distribute :: Logic -> Logic
distribute (And a (Or b c)) = Or (And a b) (And a c)
distribute (Or a (And b c)) = And (Or a b) (Or a c)

deMorgan :: Logic -> Logic
deMorgan (Not (Or a b)) = And (Not a) (Not b)
deMorgan (Not (And a b)) = Or (Not a) (Not b)
deMorgan (And (Not a) (Not b)) = Not (Or a b)
deMorgan (Or (Not a) (Not b)) = Not (And a b)
```

48.

$$
\begin{aligned}
(C \wedge A \wedge B) &\vee C \\
&= C \vee (C \wedge A \wedge B) && \{\vee \text{ commutative}\} \\
&= (C \vee C) \wedge (C \vee A) \wedge (C \vee B) && \{\vee \text{ distributes over } \wedge\} \\
&= C \wedge (C \vee A) \wedge (C \vee B) && \{\vee \text{ idempotent}\} \\
&= C \wedge ((C \vee A) \wedge (C \vee B)) && \{\wedge \text{ associative}\} \\
&= C \wedge (C \vee (A \wedge B)) && \{\vee \text{ distributes over } \wedge\}
\end{aligned}
$$

49.

$$
\begin{aligned}
C \vee (A \wedge (B &\vee C)) \\
&= C \vee ((A \wedge B) \vee (A \wedge C)) && \{\wedge \text{ distributes over } \vee\} \\
&= C \vee ((A \wedge C) \vee (A \wedge B)) && \{\vee \text{ commutative}\} \\
&= (C \vee (A \wedge C)) \vee (A \wedge B) && \{\vee \text{ associative}\} \\
&= ((C \vee A) \wedge (C \vee C)) \vee (A \wedge B)) && \{\vee \text{ distributes over } \wedge\} \\
&= ((C \vee A) \wedge C) \vee (A \wedge B) && \{\vee \text{ idempotent}\}
\end{aligned}
$$

C.7 Predicate Logic

1.

$$\begin{aligned}
(a) \quad & F(1) \wedge F(2) \wedge F(3) \\
(b) \quad & F(1) \vee F(2) \vee F(3) \\
(c) \quad & \big(G(1,1) \wedge G(1,2) \wedge G(1,3)\big) \\
& \vee \big(G(2,1) \wedge G(2,2) \wedge G(2,3)\big) \\
& \vee \big(G(3,1) \wedge G(3,2) \wedge G(3,3)\big)
\end{aligned}$$

2.

$$\begin{aligned}
& \big(F(1,5) \vee F(1,6)\big) \\
\wedge \; & \big(F(2,5) \vee F(2,6)\big) \\
\wedge \; & \big(F(3,5) \vee F(3,6)\big) \\
\wedge \; & \big(F(4,5) \vee F(4,6)\big)
\end{aligned}$$

3.

- There is an even number.
  $\exists x.\ E(x)$

- Every number is either even or odd.
  $\forall x.\ (E(x) \vee O(x))$

- No number is both even and odd.
  $\forall x.\ \neg(E(x) \wedge O(x))$

- The sum of two odd numbers is even.
  $\forall x.\ \forall y.\ \big(O(x) \wedge O(y) \rightarrow E(x+y)\big)$

- The sum of an odd number and an even number is odd.
  $\forall x.\ \forall y.\ \big(E(x) \wedge O(y) \rightarrow O(x+y)\big)$

4.

- Chickens are birds.
  $\forall x.\ C(x) \rightarrow B(x)$

- Some doves can fly.
  $\exists x.\ D(x) \wedge F(x)$

- Pigs are not birds.
  $\forall x.\ P(x) \rightarrow \neg B(x)$

- Some birds can fly, and some can't.
  $(\exists x.\ B(x) \wedge F(x)) \wedge (\exists x.\ B(x) \wedge \neg F(x))$

- An animal needs wings in order to fly.
  $\forall x.\ (\neg W(x) \rightarrow \neg F(x))$

- If a chicken can fly, then pigs have wings.
  $$(\exists x.\ C(x) \land F(x))\ \rightarrow\ (\forall x.\ P(x) \rightarrow W(x))$$

- Chickens have more feathers than pigs do.
  $$\forall x.\ \forall y.\ (C(x) \land P(y)) \rightarrow M(x,y)$$

- An animal with more feathers than any chicken can fly.
  $$\forall x.\ \Big((A(x) \land (\forall y.\ (C(y) \land M(x,y))))\ \rightarrow F(x)\Big)$$

5.

- $\forall x.\ (\exists y.\ \text{wantsToDanceWith}\ (x,y))$
  Everybody has someone they want to dance with.

- $\exists x.\ (\forall y.\ \text{wantsToPhone}\ (y,x))$
  There is someone whom everybody wants to call.

- $\exists x.\ (\text{tired}\ (x) \land \forall y.\ \text{helpsMoveHouse}\ (x,y))$
  There is a person who is tired, and who helps everyone to move house.

6.

```
forall [1,2,3] (==2)

= 1==2 /\ 2==2 /\ 3==2

= False /\ True /\ False

= False

forall [1,2,3] (< 4)

= (1 < 4) /\ (2 < 4) /\ (3 < 4)

= True /\ True /\ True

= True
```

7.

```
exists [0,1,2] (==2)

= 0==2 \/ 1==2 \/ 2==2

= False \/ False \/ True

= True
```

```
exists [1,2,3] (> 5)

= (1 > 5) \/ (2 > 5) \/ (3 > 5)

= False \/ False \/ False

= False
```

8.

1. $\forall x.(\exists y.p(x,y))$
   $= \forall x.\ x = 1 + 1 \vee x = 2 + 1$
   $= (1 = 1 + 1 \vee 1 = 2 + 1) \wedge (2 = 1 + 1 \vee 2 = 2 + 1)$
   $= (\text{False} \vee \text{False}) \wedge (\text{True} \vee \text{False})$
   $= \text{False} \wedge \text{True}$
   $= \text{False}$

2. $\exists x.(\exists y.p(x,y))$
   $= \exists x.(x = 1 + 1) \vee (x = 2 + 1)$
   $= (1 = 1 + 1 \vee 1 = 2 + 1) \vee (2 = 1 + 1 \vee 2 = 2 + 1)$
   $= (\text{False} \vee \text{False}) \vee (\text{True} \vee \text{False})$
   $= \text{False} \vee \text{True}$
   $= \text{True}$

3. $\exists x.(\forall y.p(x,y))$
   $= \exists x.(x = 1 + 1 \wedge x = 2 + 1)$
   $= (1 = 1 + 1 \wedge 1 = 2 + 1) \vee (2 = 1 + 1 \wedge 2 = 2 + 1)$
   $= (\text{False} \wedge \text{False}) \vee (\text{True} \wedge \text{False})$
   $= \text{False} \vee \text{False}$
   $= \text{False}$

4. $\forall x, y.p(x,y)$
   $= \forall x.(x = 1 + 1 \wedge x = 2 + 1)$
   $(1 = 1 + 1 \wedge 1 = 2 + 1) \wedge (2 = 1 + 1 \wedge 2 = 2 + 1)$
   $= (\text{False} \wedge \text{False}) \wedge (\text{True} \wedge \text{False})$
   $= \text{False} \wedge \text{False}$
   $= \text{False}$

9. The solution is based on showing that for *any* arbitrary value of x, say p, we can infer $G(p)$; hence we can infer $\forall x.G(x)$.

Theorem 110. $\forall x.F(x), \forall x.F(x) \to G(x) \vdash \forall x.G(x)$

Proof.

$$
\cfrac{\cfrac{\cfrac{\forall x.F(x)}{F(p)}\{\forall E\} \quad \cfrac{\forall x.F(x) \to G(x)}{F(p) \to G(p)}\{\forall E\}}{G(p)}\{\to E\}}{\forall x.G(x)}\{\forall I\}
$$

$\square$

10.

$$\exists x.\exists y.F(x,y) \vdash \exists y.\exists x.F(x,y)$$

$$
\cfrac{\exists x.\exists y.F(x,y) \quad \cfrac{\cfrac{\exists y.F(p,y) \quad \cfrac{F(p,q)}{\exists x.F(x,q)}\{\exists I\}}{\exists x.F(x,q)}\{\exists E\}}{\exists y.\exists x.F(x,y)}\{\exists I\}}{\exists y.\exists x.F(x,y)}\{\exists E\}
$$

11.

$$\forall y.\exists x.F(x,y) \vdash \exists x.\forall y.F(x,y)$$

WRONG!

Counterexample:

Let $F(x,y) = x == y$ with $U = \{1,2\}$

$\forall y.\exists x.F(x,y)$
$= \forall y.(1 = y \lor 2 = y)$
$= (1 = 1 \lor 2 = 1) \land (1 = 2 \lor 2 = 2)$
$= (\text{True} \lor \text{False}) \land (\text{False} \lor \text{True})$
$= (\text{True} \land \text{True})$
$= \text{True}$

but on the other side:

$\exists x.\forall y.F(x,y)$
$= \exists x.(x = 1 \land x = 2)$
$= (1 = 1 \land 1 = 2) \lor (2 = 1 \land 2 = 2)$
$= (\text{True} \land \text{False}) \lor (\text{False} \land \text{True})$

$$= \text{False} \lor \text{False}$$
$$= \text{False}$$

The equations do not return the same result, therefore the sequent is not correct.

12.

Theorem 111. $\forall x.(F(x) \land G(x)) \vdash (\forall x.F(x)) \land (\forall x.G(x))$

Proof.

$$\cfrac{\cfrac{\cfrac{\forall x.F(x) \land G(x)}{F(p) \land G(p)}\{\forall E\}}{\cfrac{F(p)}{\forall x.F(x)}\{\forall I\}}\{\land E_L\} \qquad \cfrac{\cfrac{\cfrac{\forall x.F(x) \land G(x)}{F(q) \land G(q)}\{\forall E\}}{\cfrac{G(q)}{\forall x.G(x)}\{\forall I\}}\{\land E_R\}}{\forall x.F(x) \land \forall x.G(x)}\{\land I\}$$

$\square$

13. There will be 1000 terms containing F. In general, if the quantifiers are nested k deep, and the universe contains n elements, then the innermost term will occur n^k times.

16.

$$P \to \forall x.f(x) \vdash \forall x.P \to f(x)$$

$$\cfrac{\cfrac{\cfrac{\cfrac{P \quad P \to \forall x.f(x)}{\forall x.f(x)}\{\to E\}}{f(c)}\{\forall E\}}{P \to f(c)\{c \text{ arbitrary}\}}\{\to I\}}{\forall x.P \to f(x)}\{\forall I\}$$

17.

- $\forall x.f(x) \lor \forall x.g(x) \vdash \forall x.f(x) \lor g(x)$

Counterexample:

Let $f(x) = x == 1$ and $g(x) = x == 2$ with $U = \{1, 2\}$

$$\forall x.f(x) \lor \forall x.g(x)$$
$$= (1 = 1 \land 2 = 1) \lor (1 = 2 \land 2 = 2)$$
$$= (\text{True} \land \text{False}) \lor (\text{False} \land \text{True})$$
$$= \text{False} \lor \text{False}$$

$=$ False

However, on the other side:

$\forall x.f(x) \vee g(x)$
$(1 = 1 \vee 1 = 2) \wedge (2 = 1 \vee 2 = 2)$
$=$ (True $\vee$ False) $\wedge$ (False $\vee$ True)
$=$ True $\wedge$ True
$=$ True

Thus the law cannot be used as an equation.

- $\exists x.f(x) \wedge g(x) \vdash \exists x.f(x) \wedge \exists x.g(x)$

Counterexample:

Let $f(x) = x == 1$ and $g(x) = x == 2$ with $U = \{1, 2\}$

$\exists x.f(x) \wedge g(x)$
$= (1 = 1 \wedge 1 = 2) \vee (2 = 1 \wedge 2 = 2)$
$=$ (True $\wedge$ False) $\vee$ (False $\wedge$ True)
$=$ False $\vee$ False
$=$ False

On the other side:

$\exists x.f(x) \wedge \exists x.g(x)$
$= (1 = 1 \vee 1 = 2) \wedge (1 = 2 \vee 2 = 2)$
$=$ (True $\vee$ False) $\wedge$ (False $\vee$ True)
$=$ True $\wedge$ True
$=$ True

Thus the law cannot be used as an equation.

19.

Assume that lists xs and ys are finite lists of type [a] and that xs is of length n.

$P(n) \equiv (\text{length xs } = n) \rightarrow$
$\qquad (\forall \text{ z.z} \in \text{xs} \vee z \in \text{ys} \rightarrow z \in (\text{xs ++ ys}))$

C.8 Set Theory

13.

$(A' \cup B)' \cap C'$
$$= A'' \cap B' \cap C'$$
$$= A \cap B' \cap C'$$
$$= A \cap (B \cup C)'$$

$A - (B \cup C)'$
$$= A \cap (B \cup C)''$$
$$= A \cap (B \cup C)$$

$(A \cap B) \cup (A \cap B')$
$$= A \cap (B \cup B')$$
$$= A \cap U$$
$$= A$$

$A \cup (B - A)$
$$= A \cup (B \cap A')$$
$$= (A \cup B) \cap (A \cup A')$$
$$= (A \cup B) \cap U$$
$$= A \cup B$$

$A - B$
$$= A \cap B'$$
$$= B' \cap A$$
$$= B' \cap A''$$
$$= B' - A'$$

$(A \cap B) - (A \cap C)$
$$= (A \cap B) \cap (A \cap C)'$$
$$= (A \cap B) \cap (A' \cup C')$$
$$= ((A \cap B) \cap A') \cup ((A \cap B) \cap C')$$
$$= (A \cap B \cap A') \cup (A \cap B \cap C')$$
$$= \varnothing \cup (A \cap B \cap C')$$
$$= A \cap B \cap C'$$
$$= A \cap (B - C)$$

$A - (B \cup C)$
$$= A \cap (B \cup C)'$$
$$= A \cap (B' \cap C')$$
$$= A \cap A \cap B' \cap C'$$
$$= (A \cap B') \cap (A \cap C')$$
$$= (A - B) \cap (A - C)$$

$$A \cap (A' \cup B)$$
$$= (A \cap A') \cup (A \cap B)$$
$$= \varnothing \cup (A \cap B)$$
$$= A \cap B$$

$$(A - B') \cup (A - C')$$
$$= (A \cap B'') \cup (A \cap C'')$$
$$= (A \cap B) \cup (A \cap C)$$
$$= A \cap (B \cup C)$$

14.

```
smaller :: Ord a => a -> [a] -> Bool
smaller x [] = True
smaller x (y:xs) = x < y

powerSet :: (Ord a, Eq a) => [a] -> [[a]]
powerSet set = normalizeSet (foldr g [[]] set)
  where g x acc =
          [x:epset | epset <- acc,
              not (elem x epset) && smaller x epset]
          ++ acc
```

15. $(A \cup B)' = A' \cap B'$
$$= (U - A) \cap (U - B)$$
$$= ((A \cup A') - A) \cap ((B \cup B') - B)$$
$$= ((A \cup A') \cap A') \cap ((B \cup B') \cap B')$$

16.

```
isSubset :: Eq a => [a] -> [a] -> Bool
isSubset set1 set2 = null [e | e <- set1, not (elem e set2)]
```

17. The arguments are in the wrong order.

18. The second definition of e shadows the first, so the result is the second set. One way to define it correctly is

```
intersection :: [a] -> [a] -> [a]
intersection set1 set2 = [e | e <- set1, e `elem` set2]
```

19.

```
union :: Eq a => [a] -> [a] -> [a]
union set1 set2
  = set1 ++ [e | e <- set2, not (elem e set1)]
```

20. Yes, when $A \subseteq B$ and $C \subseteq A$.

21. Both unions contain C, so C must be empty, and A and B must be disjoint. For example, $A = \{1, 2, 3\}, B = \{4, 5, 6\}, C = \varnothing$.

22.

$A \cap B$
$= \{ \text{ defn } \cap \}$
$\{x | x \in A \wedge x \in B\}$
$= \{ \wedge \text{ commutative } \}$
$\{x | x \in B \wedge x \in A\}$
$= \{ \text{ defn } \cap \}$
$B \cap A$

23.

A *** B = B *** A

24.

$(A \cup B) \cup C$
$= \{ \text{ defn } \cup \}$
$\{x \,|(x \in A \vee x \in B) \vee x \in C\}$
$= \{ \vee \text{ associative } \}$
$\{x | x \in A \vee (x \in B \vee x \in C)\}$
$= \{ \text{ defn } \cup \}$
$A \cup (B \cup C)$

25.

$A - B''$
$= \{ \text{ double complement } \}$
$\{x | x \in A \wedge x \notin B''\}$
$= \{ \text{ defn. } - \}$
$\{x | x \in A \wedge \neg(x \in U \wedge x \notin B')\}$
$= \{ \text{ defn. complement } \}$
$\{x | x \in A \wedge \neg(x \in U \wedge \neg(x \in B'))\}$
$= \{ \text{ defn. } \notin \}$
$\{x | x \in A \wedge \neg(x \in U \wedge \neg(x \in U \wedge x \notin B))\}$
$= \{ \text{ defn. complement } \}$
$\{x | x \in A \wedge \neg(x \in U \wedge \neg(x \in U \wedge \neg(x \in B)))\}$
$= \{ \text{ defn. } \notin \}$
$\{x | x \in A \wedge \neg(x \in U \wedge (\neg(x \in U) \vee (x \in B)))\}$
$= \{ DM, \text{ double negation } \}$
$\{x | x \in A \wedge (\neg(x \in U) \vee \neg(\neg(x \in U) \vee (x \in B)))\}$
$= \{ DM \}$
$\{x | x \in A \wedge (x \notin U \vee (\neg\neg(x \in U) \wedge \neg(x \in B)))\}$
$= \{ DM \}$
$\{x | x \in A \wedge (x \notin U \vee (x \in U \wedge x \notin B))\}$
$= \{ \text{ double negation, defn. } \notin \}$
$\{x | x \in A \wedge (x \in U \wedge x \notin B)\}$

$$= \{ \lor \; null, \; defn. \; \notin \}$$
$$\{x | x \in A \land x \in U - B\}$$
$$= \{ \; defn.- \}$$
$$A \cap B'$$
$$= \{ \; defn. \; complement, \cap \}$$

26.

$$A \cup (B \cap C)$$
$$= \{ \; defn \; \cap, \cup \}$$
$$\{x | x \in A \lor (x \in B \land x \in C)\}$$
$$= \{ \lor \; over \; \land \}$$
$$\{x | (x \in A \lor x \in B) \land (x \in A \lor x \in C)\}$$
$$= \{ \; def. \cap, \; \cup \}$$
$$(A \cup B) \cap (A \cup C)$$

27.

$$(A \cap B)'$$
$$= \{ \; defn. \; complement \}$$
$$\{x | x \in U \land x \notin (A \cap B)\}$$
$$= \{ \; defn. \; \notin, \cap \}$$
$$\{x | x \in U \land \neg (x \in A \land x \in B)\}$$
$$= \{ \; DM \}$$
$$\{x | x \in U \land (\neg (x \in A) \lor \neg (x \in B))\}$$
$$= \{ \; defn. \; \notin \}$$
$$\{x | x \in U \land (x \notin A \lor x \notin B)\}$$
$$= \{ \; \land \; over \; \lor \}$$
$$\{x | (x \in U \land x \notin A) \lor (x \in U \land x \notin B)\}$$
$$= \{ \; defn. \; - \}$$
$$\{x | (x \in U - A) \lor (x \in U - B)\}$$
$$= \{ \; defn. \; complement, \cup \}$$
$$A' \cup B'$$

C.9   Inductively Defined Sets

19. The base case appears at the end of the stream, so it cannot be used to start the inductive process of calculating successive elements. So, the stream does not have a printable value.

20. It will never terminate and return the accumulator in which the stream of naturals is constructed.

21. No. The problem is that the stream may present an infinite number of naturals before the one we are interested in can be reached. For example, in [1,3..] ++ [2,4..] all the odd naturals appear before any even number.

22. Given a number n, the set R of roots of n is defined as follows:

- $n^1 \in R$

- $n^{1/m} \in R \to n^{1/m+1} \in R$

- Nothing is in R unless it can be shown to be in R by a finite number of uses of the base and induction rules.

23.

1. $n^0 \in P$ by the base case

2. By the induction rule, $n^0 \in P \to n^1 \in P$ and so by *Modus Ponens*, $n^1 \in P$.

3. By the induction rule, $n^1 \in P \to n^2 \in P$ and so by *Modus Ponens*, $n^2 \in P$.

4. By the induction rule, $n^2 \in P \to n^3 \in P$ and so by *Modus Ponens*, $n^3 \in P$.

24. If n is a positive multiple of 2 then yes, otherwise no.

25. The odd positive integers.

26. By rule (1), $0 \in S$
By rule (2) $0 \in S \to 2 \in S$ and so by *Modus Ponens* $2 \in S$.
By rule (2) $2 \in S \to 4 \in S$ so $4 \in S$.

27.

1. "" $\in YYS$

2. "" $\in YYS \to$ "yy" $\in YYS$ and so by *Modus Ponens*, "yy" $\in YYS$.

3. "yy" $\in YYS \to$ "yyyy" $\in YYS$ and so by *Modus Ponens*, "yyyy" $\in YYS$.

28.

```
zs = "" : map ('z':) zs
```

29. The set SS of strings of spaces of length less than or equal to n is defined as follows:

1. "" $\in SS$

2. $ss \in SS \land length\, ss < n \to$ ' ': $ss \in SS$

3. Nothing is in SS unless it can be shown to be in SS by a finite number of uses of rules (1) and (2).

30.

```
ss :: Integer -> [String]
ss 0 = []
ss n = take n (repeat ' ') : ss (n-1)
```

31. The set of sets of naturals SSN, each of which is missing a distinct natural number, is defined inductively as follows:

1. $N - \{0\} \in SSN$

2. $(N - \{n\}) \in SSN \rightarrow (N - \{n+1\} \in SSN$

3. Nothing is in SSN unless it can be shown to be in SSN by a finite number of uses of rules (1) and (2).

32.

1. $I - \{-1\} \in SSI-$

2. $I - \{-1\} \rightarrow I - \{-2\} \in SSI-$

3. $I - \{-2\} \rightarrow I - \{-3\} \in SSI-$

4. By *Modus Ponens*, $I - \{-3\} \in SSI-$.

33.

1. $-1 \in ONI$

2. $-1 \in ONI \rightarrow -3 \in ONI$ so $-3 \in ONI$.

3. $-3 \in ONI \rightarrow -5 \in ONI$ so $-5 \in ONI$.

4. $-5 \in ONI \rightarrow -7 \in ONI$ so $-7 \in ONI$.

34.

```
decrement :: Integer -> Integer
decrement x = x - 1

ni = -1 : map decrement ni
```

35. No, because 0 will be paired first with every element in [0..], which is infinitely long.

36. The set is $\{0, 1\}$.

C.10   Relations

45.

- (a) Yes.

- (b) This depends upon whether the relationship is defined by DNA, in which case all of us are somewhat related to each other, or last name, in which case some people named Smith are not actually closely related physically. 'No' seems a reasonable answer.

- (c) No, because equivalence relations are reflexive, and you cannot be bigger than yourself.

- (d) Yes.

46. Yes. Here is an example:

```
[(1,2),(2,3),(3,1)]
```

47. The empty relation.

48.

```
checkPowers :: (Eq a, Show a) => Digraph a -> Bool
checkPowers (set,relation)
 = any (setEq relation)
      [relationalPower (set,relation) n
          | n <- [2..length (domain relation)]]
```

49. No. If it did, then the end of each loop would have to have a reflexive loop, because the relation is transitive. Thus it would not be irreflexive.

50. Yes. For example

```
[(1,2),(2,3),(3,1)]
```

has a transitive closure that is reflexive and symmetric.

51.

```
fewerArcs :: (Eq a, Show a) => Digraph a -> Bool
fewerArcs (set,relation)
  = all
     (< (length relation))
       [length (relationalPower (set,relation) n)
           | n <- [2..length (domain relation)]]
```

52.

```
isSmaller :: (Ord a, Show a) => Digraph a -> Bool
isSmaller (set,relation)
  = let (symset,symrelation) =
          symmetricClosure (set,relation)
     in length relation < length symrelation
```

53. The last is not a topological sort.

54. Yes, for example this is reflexive, symmetric, and antisymmetric:

```
[(1,1),(2,2),(3,3)]
```

55. The number of arcs the transitive closure will add is $1+2+...+n-1$. The symmetric closure will double each of these, so the total is $2(1+2+...+n-1)$.

56. Power n.

57. No, because the function could not examine every arc in the relation.

58. Given a partial order, assume that it has a cycle of length n. Because it is transitive, it also has cycles of lengths 1 to $n - 1$. But that means that it has a cycle of length 2, which cannot be because a partial order is antisymmetric. So it cannot both be a partial order and have cycles greater than 1.

59. No. Some have powers that repeat, such as the relation

```
[(1,2),(2,1)]
```

60. A linear order.

61. Yes. For example, the composition of these two relations

```
[(Red,Blue),(Blue,Green),(Green,Yellow),(Yellow,Red)]
```

```
[(Blue,Red),(Green,Blue),(Yellow,Green),(Red,Yellow)]
```

yields

```
[(Red,Red),(Blue,Blue),(Green,Green),(Yellow,Yellow)]
```

62. The empty relation and the equality relation.

C.11   Functions

26. The functions `surjExp`, `injExp`, and `bijExp` each take two domains and images and test the corresponding hypothesis.

```
surjectiveExperiment :: (Eq a, Show a) =>
  Set a -> Set a -> Set (a, FunVals a) ->
  Set a -> Set a -> Set (a, FunVals a) -> Bool
surjectiveExperiment dom_f co_f f dom_g co_g g
  = isSurjective dom_f co_f f /\
    isSurjective dom_g co_g g
    ==>
    isSurjective dom_g co_f
    (functionalComposition f g)

surjExp :: Set Int -> Set Int ->
           Set Int -> Set Int -> Bool
surjExp domain_f image_f domain_g image_g
  = let f = [(x,Value y) | x <- domain_f, y <- image_f]
        g = [(x,Value y) | x <- domain_g, y <- image_g]
    in
    surjectiveExperiment [1..10] [1..10] f
                         [1..10] [1..10] g
```

```
injectiveExperiment :: (Eq a, Show a) =>
  Set a -> Set a -> Set (a, FunVals a) ->
  Set a -> Set a -> Set (a, FunVals a) -> Bool
injectiveExperiment dom_f co_f f dom_g co_g g
  = isInjective dom_f co_f f /\
    isInjective dom_g co_g g
    ==>
    isInjective dom_g co_f
    (functionalComposition f g)

injExp :: Set Int -> Set Int ->
          Set Int -> Set Int -> Bool
injExp domain_f image_f domain_g image_g
  = let f = [(x,Value y) | x <- domain_f, y <- image_f]
        g = [(x,Value y) | x <- domain_g, y <- image_g]
    in injectiveExperiment [1..10] [1..10] f
                           [1..10] [1..10] g

bijectiveExperiment :: (Eq a, Show a) =>
  Set a -> Set a -> Set (a, FunVals a) ->
  Set a -> Set a -> Set (a, FunVals a) -> Bool
bijectiveExperiment dom_f co_f f dom_g co_g g
  = isBijective dom_g co_f
    (functionalComposition f g)
    ==>
    isSurjective dom_f co_f f /\
    isInjective dom_g co_g g

bijExp :: Set Int -> Set Int ->
          Set Int -> Set Int -> Bool
bijExp domain_f image_f domain_g image_g
  = let f = [(x,Value y) | x <- domain_f, y <- image_f]
        g = [(x,Value y) | x <- domain_g, y <- image_g]
    in bijectiveExperiment [1..10] [1..10] f
                           [1..10] [1..10] g
```

27. 1,1,4,4

28. The function f is a permutation, g is an identity function, and h is the inverse of f.

29. constant function, injection, constant function.

30. Only g is a function. f maps 3 to two values, and g does not map 5 to any value.

31. g is a total function, and the other two are partial functions.

32. This can be solved by making a simple table for each function showing the results it returns for all inputs, and then referring to the definition of surjection.

```
h o h    yes
f o g    no
g o f    yes
h o f    domains don't match
g o h    domains don't match
```

33.

```
f o g    no
g o f    yes
h o f    no
```

34.

```
g o g    yes
h o f    no
f o h    yes
```

35. The first function is partial, because it does not terminate when given
False, and so does not produce a result given that domain value. The second
function is total.

36.

```
isFun :: (Eq a, Show a) => Set (a,b) -> Bool
isFun ps = normalForm (map fst ps)
```

37.

```
isInjection :: (Eq a, Eq b, Show a, Show b)
          => Set (a,b) -> Bool
isInjection ps
  = normalForm (map fst ps) /\ normalForm (map snd ps)
```

39. You only need to know its type. If the domain type is the same as that
of the codomain, then it might be the identity function, otherwise it certainly
isn't. There are many properties of functions that can be deduced solely from
the function type.

40.

```
compare f g h = ((g.f) a) == (h a)
```

41. Yes, for natural numbers. If it receives a negative integer, it will loop
indefinitely.

42. Yes, for all integers.

43.

For problems 2 and 3, we will use the bullet/target metaphor of functions. In this metaphor, the domain of the function is regarded as the set of bullets. The codomain (also known as the range) is regarded as the set of targets. When x comes from the domain and y comes from the codomain and $y = f(x)$, the target y is said to be hit by the bullet x.

Following this metaphor, we can define some of the terminology as follows:

- Function. A function is a subset of the Cartesian product of the set of bullets with the set of targets that does not use the same bullet for more than one target.

- Surjective. A function is surjective if it hits all of the targets.

- Injective. A function is injective if it hits no targets more than once.

1. Yes, f is a function because it is a subset of $A \times B$ with the property that whenever both (a, b) and (a, c) are elements of f, it is the case that $b = c$. That's what it means to be a function of type $A \to B$. Or, using the metaphor, f is a function because no single bullet hits two or more targets.

2. No, f is not injective because $f(1) = 7$ and $f(4) = 7$. So, $f^{-1}(\{7\})$ contains two elements (1 and 4), and inverse images of singleton sets never have more than one element for injective functions. Or, using the metaphor, f is not injective because it hits the target 7 with bullet 1 and bullet 4.

3. No, f is not surjective because $f(A)$ is $\{6, 7, 9, 10\}$, which is lacking an element of $B(8)$. The image of the domain is the entire range when a function is surjective. Or, using the metaphor, f is not surjective because it fails to hit the target 8.

4. No, g is not a function because it contains the pairs $(6,b)$ and $(6,d)$, and functions are not allowed to contain two pairs with the same first element and different second elements. Or, using the metaphor, g is not a function because the bullet 6 hits two targets (b and d).

44.

1. The codomain of f is $P(A)$, which contains 2^n elements (where 2^n stands for 2 to the power n). That is, the function f has 2^n targets. The domain of f contains n elements. That is, the function f has n bullets. Since a function must use at least one bullet per target, and since 2^n always exceeds n, f doesn't have enough bullets to hit all the targets, so it can't be surjective.

2. The composition $g.f$ is defined when g is defined on the image $f(A)$. So, X is a subset of $P(A)$. Furthermore, since $(g.f)(A)$ is a subset of A, and since $(g.f)(A) = g(f(A))$, Y, the codomain of g, must contain at least those elements of A that are in $(g.f)(A)$. That's one way to look at it. Here's another way: The composition $(g.f)$ is defined when the codomain of f is the domain of g. Taking this position, $X = P(A)$ and Y, the codomain of g contains at least $g(f(A))$, which, by hypothesis, is a subset of A. Either position is ok. It depends on how the term "composition" is defined. Because we were not precise about when to consider the composition defined, either answer is ok.

3. Again, it depends on the details of the definition of composition. Taking the first view, the domain of g must contain, at least, the set $f(A)$. It may contain only this set, in which case it can be injective because the number of targets f can hit cannot exceed the number of bullets in A, so g will not have more bullets than targets and can be defined to hit each potential target at most once. On the other hand, if $P(A)$ is considered to be the domain of g, then g will have too many bullets for A. So, if g is to be injective, its codomain must contain elements other than those of A. The intersection of A and the codomain of g will contain, at least, $(g.f)(A)$.

4. Define $g : f(A) \to A$ as follows. For each s in $f(A)$, let $g(s)$ be one of the elements in $f^{-1}(s)$. Any element will do. Just pick one, and let that be $g(s)$. There must be at least one such element because s is in $f(A)$. Then $(f.g) : f(A) \to f(A)$ will be bijective.

C.13   Discrete Mathematics in Circuit Design

1.

$$
\begin{aligned}
f\ a\ b\ c\ = \\
(\neg a \wedge \neg b \wedge \neg c) \\
\vee\ (\neg a \wedge \neg b \wedge c) \\
\vee\ (a \wedge \neg b \wedge c) \\
\vee\ (a \wedge b \wedge \neg c) \\
\vee\ (a \wedge b \wedge c)
\end{aligned}
$$

4. The inputs are a and b. The outputs are the carry output, c, which comes from the truth table for *and2*, and the sum output, s, which comes from the truth table for *xor2*. The calculations show that for all inputs, the left-hand and right-hand sides of the equation in the theorem have the same value.

a	b	c	s	lhs	rhs
0	0	0	0	0	0
0	1	0	1	1	1
1	0	0	1	1	1
1	1	1	0	2	2

9.

```
Example: addition of 13+41=54 using 6-bit words
  13 = 001101
  41 = 101001
 sum = 110110 = 54, with carry out = 0
 rippleAdd False [(False,True),(False,False),(True,True),
                  (True,False),(False,False),(True,True)]
The expected result is
  (False, [True,True,False,True,True,False])
```

10.

```
Test cases for halfCmp
halfCmp (False,False) -- (False,True,False)  since x=y
halfCmp (False,True) -- (True,False,False)  since x<y
halfCmp (True,False) -- (False,False,True)  since x>y
halfCmp (True,True) -- (False,True,False)  since x=y
```

11.

```
Test cases for rippleCmp
rippleCmp [(False,False),(False,True),(True,False)]
    -- (True,False,False) since x<y
rippleCmp [(False,False),(True,True),(False,False)]
    -- (False,True,False) since x=y
rippleCmp [(False,False),(True,False),(False,True)]
    -- (False,False,True) since x>y
```

12. Define

```
andfour a b c d = and2 (and2 a b) (and2 c d)
```

The output should be 1 if all of the inputs are 1; note that in this case, and2 a b gives 1, and so does and2 c d, so the circuit outputs 1 as it should. In all other cases, either and2 a b will be 0, or and2 c d will be 0, or both, so the andfour circuit produces 0, as it should.

13.

```
Test cases for the full adder, with expected results
  fullAdd (False,False) False   -- 0 0
  fullAdd (False,True) False   -- 0 1
```

```
fullAdd (True,False) False    -- 0 1
fullAdd (True,True) False     -- 1 0
fullAdd (False,False) True    -- 0 1
fullAdd (False,True) True     -- 1 0
fullAdd (True,False) True     -- 1 0
fullAdd (True,True) True      -- 1 1
```

15. Yes. The *rippleAdd* circuit is defined using the higher order function *mscanr*, which is defined even when the word input is empty. In this case, the sum word is just [], and the carry output is the same as the carry input. Notice that you can connect an m-bit adder with an n-bit adder and obtain an $(m + n)$-bit adder, and this works for all natural numbers m and n.

16. No, because *mscanr* (and therefore *rippleAdd*) requires a list input, and lists cannot have a negative length.

17. You can't test an adder thoroughly because there is an exponential growth in the size of the truth table as the word size grows. For an n-bit word, there are 2×2^{2n} lines in the truth table. The exponent is $2n$ because there are two bits for each position, and the initial factor of 2 accounts for the possibility that the carry input is either 0 or 1. For a current generation processor, where the word size is 64, the truth table would have 2^{129} lines, which is larger than the size of the known universe (it hardly matters what unit of measurement you choose!)

Bibliography

[1] *Computer-Aided Reasoning: An Approach.* Kluwer Academic Publishers, 2000.

[2] Harold Abelson, Gerald Jay Sussman, and Julie Sussman. *Structure and Interpretation of Computer Programs.* The MIT Press, 1985.

[3] Martin Aigner and Günter M. Ziegler. *Proofs from THE BOOK.* Springer, 1998.

[4] Richard Bird. *Introduction to Functional Programming using Haskell.* Prentice Hall, second edition, 1998.

[5] Richard Bird and Oege de Moor. *Algebra of Programming.* Prentice Hall, 1997.

[6] Stanley N. Burris. *Logic for Mathematics and Computer Science.* Prentice Hall, 1998.

[7] C. J. Date. *An Introduction to Database Systems.* Addison Wesley Longman, 1999.

[8] Antony Davie. *An Introduction to Functional Programming Systems Using Haskell.* Cambridge University Press, 1992.

[9] Michael Downward. *Logic and Declarative Language.* Taylor & Francis, 1998.

[10] Herbert B. Enderton. *A Mathematical Introduction to Logic.* Academic Press, 1972.

[11] Herbert B. Enderton. *Elements of Set Theory.* Academic Press, 1977.

[12] Arthur Engel. *Problem-Solving Strategies.* Springer, 1998.

[13] Jean-Yves Girard, Yves Lafont, and Paul Taylor. *Proofs and Types*, volume 7 of *Cambridge Tracts in Theoretical Computer Science.* Cambridge University Press, 1989.

[14] Derek Goldrei. *Classic Set Theory*. Chapman & Hall, 1996.

[15] Ronald L. Graham, Donald E. Knuth, and Oren Patashnik. *Concrete Mathematics*. Addison-Wesley, 1989.

[16] Jan Gullberg. *Mathematics from the Birth of Numbers*. Norton, 1997.

[17] Douglas R. Hofstadter. *Gödel, Escher, Bach: An Eternal Golden Braid*. Penguin, 1979.

[18] Paul Hudak. *The Haskell School of Expression: Learning Functional Programming through Multimedia*. Cambridge University Press, 2000.

[19] Geraint Jones and Mary Sheeran. Relations and refinement in circuit design. In *3rd Refinement Workshop*, Workshops in Computing. Springer, 1991.

[20] Donald E. Knuth, Tracy Larrabee, and Paul M. Roberts. *Mathematical Writing*. Number 14 in MAA Notes. The Mathematical Association of America, 1989.

[21] E. J. Lemmon. *Beginning Logic*. Chapman & Hall, 1965.

[22] John O'Donnell. The hydra computer hardware description language. www.dcs.gla.ac.uk/ jtod/hydra/.

[23] John T. O'Donnell and Gudula Rünger. Derivation of a logarithmic time carry lookahead addition circuit. *Journal of Functional Programming*, 14(6):697–713, November 2004. Cambridge University Press.

[24] Chris Okasaki. *Purely Functional Data Structures*. Cambridge University Press, 1998.

[25] John O'Leary, Xudong Zhao, Rob Gerth, and Carl-Johan Seger. Formally verifying IEEE compliance of floating-point hardware. *Intel Technology Journal*, 1999.

[26] Kenneth A. Ross and Chalres R. B. Wright. *Discrete Mathematics*. Prentice Hall, 1999.

[27] Simon Singh. *The Code Book: The Science of Secrecy from Ancient Egypt to Quantum Cryptography*. Fourth Estate, 1999.

[28] Raymond Smullyan. *Forever Undecided: A Puzzle Guide to Gödel*. Oxford University Press, 1987.

[29] Donald F. Stanat and David F. McAllister. *Discrete Mathematics in Computer Science*. Prentice Hall, 1977.

[30] Patrick Suppes. *Axiomatic Set Theory*. Dover, 1972.

[31] Simon Thompson. *Type Theory and Functional Programming*. Addison-Wesley Publishing Company, 1991.

[32] Simon Thompson. *Haskell: The Craft of Functional Programming*. Addison-Wesley, second edition, 1999.

[33] Daniel J. Velleman. *How To Prove It: A Structured Approach*. Cambridge University Press, 1994.

Index